Lecture Notes in Mathematics

1912

Editors:
J.-M. Morel, Cachan
F. Takens, Groningen
B. Teissier, Paris

T0202984

Vasile Berinde

Iterative Approximation
of Fixed Points

 Springer

Author

Vasile Berinde

Department of Mathematics
  and Computer Science
Faculty of Sciences
North University of Baia Mare
Victoriei Nr. 76
430122 Baia Mare, Romania
e-mail: vberinde@ubm.ro
    vasile_berinde@yahoo.com

Library of Congress Control Number: 2007925692

Mathematics Subject Classification (2000): 47H10, 47J25, 47H09, 65J15, 54H25

2nd rev. and enlarged edition
Originally (1st edition) published by Editura Efemeride, Baia Mare, Romania, 2002

ISSN print edition: 0075-8434
ISSN electronic edition: 1617-9692
ISBN-10 3-540-72233-5 Springer Berlin Heidelberg New York
ISBN-13 978-3-540-72233-5 Springer Berlin Heidelberg New York

DOI 10.1007/978-3-540-72234-2

Springer is a part of Springer Science+Business Media
springer.com
© Springer-Verlag Berlin Heidelberg 2007

Typesetting by the author and SPi using a Springer LaTeX package
Cover design: WMXDesign GmbH, Heidelberg

Printed on acid-free paper   SPIN: 12057177   41/SPi   5 4 3 2 1 0

To my parents

Preface to the Second Edition

This is a revised and enlarged version of the second printing (with up-dated bibliography, 2004) of the first edition, published by *Efemeride* in 2002.

All chapters of the book were practically revised in a certain extent and some new Sections were also added with the aim to improve the coverage of the topic and to attain its main aim: to summarize in a gradual and natural way the most significant contributions to the approximation of fixed points of nonlinear contractive type mappings, by presenting, for each important iterative method, some of the most relevant, interesting and actual results.

Only *constructive* fixed point theorems are mainly the subject of the book. A constructive fixed point theorem establishes not only the existence (and possibly, uniqueness) of the fixed points, but also provides a method for approximating these fixed points and, moreover, offers information on the data dependence of the fixed points (or, alternatively, on the stability of the fixed point iterative methods).

Main Changes in the Second Edition

1. Since the first edition had no exercises explicitly formulated, we selected and included in the new edition a number of 111 Exercises, Applications and Miscellaneous Results, distributed to all chapters, which completes the topic treated in each chapter or indicate other related directions of research.

2. A number of 7 new sections were added (3.5, 5.5, 6.4, 6.5, 9.3, 9.4, 9.5) or enlarged; section 4.4 merged section 4.2 to form a new section 4.2. Practically, all sections were significantly revised. Section 6.3 changed the title from "Ergodic fixed point iteration procedures" to "Ergodic and other fixed point iteration procedures"; section 7.3 changed the name from "Continuous dependence of the fixed points" to "Data dependence of fixed points"; Chapter 8 changed the title from "Applications of some fixed point iteration procedures" to "Iterative solution of nonlinear operator equations", to indicate more clearer the area of applications.

3. We also simplified several proofs and corrected many of the typos.

4. We enlarged and improved Chapter 9 with some very recent new results related to the numerical comparison of fixed point iteration procedures.

5. We added other numerical examples in Chapter 9, obtained by means of the software package FIXPOINT.

6. We inserted new information in the Bibliographical Comments sections in Chapters 3-9.

7. We up-dated significantly the bibliography: more than 500 new entries were added; at the same time, some of the old entries in the first edition, now considered to be not directly related to the main topic, were eliminated. In comparison to the first edition, which had about 1050 references at the bibliography, in the present edition it considerably increased: it contains now more than 1575 titles. The bibliography itself could show how dynamic this field of research is: 1481 titles, representing 94% of the whole bibliography, were published in the last 35 years (1970-2005); 1294 of the latter, representing 82% of the whole bibliography, were published in the last 25 years (1980-2005); 1059 of them, representing 67% of the whole bibliography, were published in the last 15 years (1990-2005), while 876 titles, that is, almost 50% of the total bibliography, were published in the last 10 years (1995-2005).

The decade 1990-1999 has doubled the bibliography of the previous one (1980-1989), while the last half decade 2000-2004 produced much more than the whole decade 1990-1999.

Note that, the very recent publications (on 2005, 2006 and 2007) are partially covered in the present list of references, with only 54 titles.

Main Merits of the Present Edition

The main merits of the current edition consist not only in a better presentation of the material, but especially in the fact that we tried to introduce and systematically apply some firm criteria of evaluating, judging and presenting the vast material existing in literature.

This enabled us, in Sections 5.5, 6.4, 6.5, 9.3, 9.4 and 9.5, to indicate some new directions of investigation of real and significant interest in the subject, and also to mention those topics which, in our opinion, are less important for theoretical and numerical purposes.

Chapter 9, devoted to error analysis of iterative methods, as well as sections 3.5, 5.5, 6.4, 6.5, include very recent, new and important results that could put into a new light the future research in the area.

In order to give an overview of the huge research work, see the data above, emphasis is put mainly on the generic results regarding the main topic, but the author's intention was to produce an as in-depth and up-to-date coverage as possible of the most significant 400 recent articles in that area.

From a huge amount of bibliography - more than 1575 entries are included in the present edition, as mentioned before - in principle only innovative research was selected and presented in the book.

More Acknowledgments

I want to thank again Professor Ioan A. Rus, this time for carefully reading the first edition and making numerous and valuable remarks and suggestions for improving the book. I thank also Dr. Sorin Iuliu Pop for the help given at the completion of the bibliography.

Thanks are due to my PhD students Ioana Banc, Marina Bic, Natalia Jurja and Monica Lauran for reading the manuscript carefully and providing a list of typos which have now been corrected.

Baia Mare   *Vasile BERINDE*
                                                                    December 22, 2006

Preface to the First Edition

The literature of the last four decades abounds with papers which establish fixed point theorems for selfmaps or nonselfmaps satisfying a variety of contractive type conditions on several ambient spaces.

Having in view that many of the most important nonlinear problems of applied mathematics reduce to solving a given equation which in turn may be reduced to finding the fixed points of a certain operator, on the one hand, and the fact that contractive (Lipschitzian) type conditions naturally arise for many of these problems, on the other hand, the metrical fixed point theory has developed significantly in the second part of the XXth century.

A plethora of *metrical* fixed point theorems have been obtained, more or less important from a theoretical point of view, which establish usually the existence, or the existence and uniqueness of fixed points for a certain contractive operator. Among these fixed point theorems, only a small number are important from a practical point of view, that is, they offer a *constructive* method for finding the fixed points. Among the last ones only a few give information on the error estimate (the rate of convergence) of the method.

However, from a practical point of view it is important not only to know the fixed point exists (and, possibly, is unique), but also to be able to construct that fixed point(s). As the constructive methods used in metrical fixed point theory are prevailingly *iterative* procedures, that is, *approximate* methods, it is also of crucial importance to have a priori or / and a posteriori error estimates (or, alternatively, rate of convergence) for such a method.

Starting from these numerical commands, the book aims to survey some of the most used fixed point iteration procedures: the Picard iteration, the Krasnoselskij iteration, the Mann iteration, the Ishikawa iteration etc.

The present version of the book arose out of a rather long personal research experience as well as of a Master degree course "Methods for approximating fixed points" and of a graduate course entitled "Fixed point theory".

The last one was taught by the author to students in the Mathematics programmes at the North University Baia Mare, since 1996.

In author's opinion, the monograph is undoubtedly a *provisional* introductory approach to iterative approximation of fixed points.

With a view to its next improved and revised version(s), we shall welcome any comments, remarks, suggestions and additional bibliographical references coming with criticism from the readers.

Acknowledgments

I am deeply indebted to Professor Ioan A. Rus from "Babes-Bolyai" University in Cluj-Napoca, who guided me patiently in the field of fixed point theory from the very beginning of my MSc Dissertation, continuing with the research included in my PhD Thesis, and extended even today. I take this opportunity to thank him heartedfully.

It is impossible to acknowledge individually colleagues and friends to whom I am indebted for support in writing this monograph. I must, however, express my appreciation and thanks to Acad. Petar Kenderov from the Institute of Mathematics, Bulgarian Academy of Sciences, Sofia, to Dr. Jaime Zavala Carvajal, Pontificia Universidad Catolica de Valparaiso, Chile, to Dr. Peter Kortesi from the University of Miskolc, Hungary and to Dr. Goetz Pfeiffer, from National University of Ireland in Galway, for the excellent conditions they offered me during my visits at their institutions, when some parts of this book have been written and various bibliographical references were provided to me.

I am also indebted to many scientists whose research work formed a basis for this monograph. I wish to express my thanks to all of them, and to each in a measure proportional to my indebtedness. Amongst them, I particularly want to thank Professor B.E. Rhoades from Indiana University, U.S.A., who has sent me the reprints of his considerable and long term work in the field of approximating fixed points.

Last but most of all, I would like to express my deepest gratitude to Zoiţa, my wife, for her patient support and insistent pushing me toward my desk in order to finish the book, as well as to Mădălina and Ruxandra, my daughters, who contributed directly and in different manners to the accomplishment of this book.

Baia Mare

Vasile BERINDE
January 18, 2002

Contents

Introduction

A possible starting point in judging the merits of the book would be the idea that it is a drop in an ocean of intensive and extensive research work. Consequently, our aim was to present, as clearly and completely as possible, a survey of the basic results in iterative approximation of fixed points.

In order to meet the taste of the majority of scientists interested in this area, our intention was to produce an in-depth and up-to-date coverage of about 400 recent publications out of more than 1575 entries in the reference list. However, it would have been impossible to cover consistently the diversity of research work that has been done in the field of iterative approximation of fixed points and related areas.

The diversity of results on this topic comes mainly from three directions:

1. The variety of the underlying spaces where the operators are defined;

2. The variety of contractiveness assumptions and/or topological properties associated with these operators;

3. The variety of assumptions on the parameters that define a certain fixed point iteration procedure. Sometimes these parameters depend also on the geometry of the ambient space and/or on the properties of the considered operator.

Therefore, the author is perfectly aware of the risks he has taken when designing the book. It is doubtful that the structure, contents and organization of the material in each Chapter or Section will meet all the needs and horizons of the specialists working in this area.

As a general rule, emphasis is put only on some generic results regarding the main topic, since it would be impossible to aim for complete coverage. Usually, for each iterative fixed point procedure, some of the most interesting, representative and significant results are completely presented, while some others are formulated as exercises or are only briefly mentioned.

Moreover, in some chapters and sections we did not always include the most general result related to a certain topic, but the most accessible one. In these circumstances, we tried to stress on the most clear result when possible and to mention the other more general results. Simultaneously we tried to illustrate the diversity of the results, and so to avoid presenting the convergence results of different iterative processes in the same or in a similar setting.

No matter how narrow its topic, a book cannot be written in a self-contained manner when space limits are imposed. This is the reason why we preferred to include some (auxiliary) results without (detailed) proofs, and to insert other much more diversified results instead. The readers interested in knowing the details should consult the appropriate references, as the bibliography, with its more than 1575 references, provides additional sources of results and approaches on the approximation of fixed points.

In order to make reading as fluent as possible, we generally tried to avoid bibliography citations in the text of the sections. Instead we have supplemented each chapter with a special section containing a set of "Bibliographical Comments", where many literature citations are given and other related results are sometimes mentioned. Including a result in a certain section does not mean it is the most general in that area: in several circumstances the taste of the author was simply the dominant reason, when we tried to mention the similar more general or most important results.

Despite the considerable amount of overlapping research work on the Ishikawa and Mann iteration procedures, we however decided to have a distinct chapter for each one, where specific results were also included. Apart from the sections "Exercises ans Miscellaneous results", in some sections at least one proof or parts of the proof are left for the reader to be completed.

Throughout the book we adopted the following numbering system: in each Chapter the Definitions, Lemmas and Theorems are numbered using two digits, while the equations are numbered using one digit only. For example, Theorem 3.6 or Definition 4.5 or Lemma 7.2 denote the sixth theorem included in Chapter 3, the fifth definition in Chapter 4 and the second lemma in Chapter 7, respectively. When references to them are needed, examples are also numbered, in the same described manner. On the contrary, when referring to a certain equation we shall say, for example, equation (3) in Chapter 4 instead of equation (4.3).

In writing non-English author names, we ignored the specific diacritical signs. So, Hadzić and Păvăloiu will be written simply as Hadzic and Pavaloiu, respectively. For Krasnosel'skij we preferred the form Krasnoselskij, even though in some sources other variants (e.g., Krasnoselskii) can be found.

Concluding the introduction, we want to stress on the main merit of this book: the very fact that it was written down. However, we hope that, by gathering and systematizing various significant results in the dynamic field of fixed point iteration procedures, we provide a useful tool for many postgraduate and PhD students as well as for any interested researchers.

1

Pre-Requisites of Fixed Points

It is the purpose of this chapter to provide the terminology, basic concepts and notations from fixed point theory used throughout the book. They are presented without proofs (for their extensive treatment we refer the readers to any monograph in the list of references). We shall also illustrate how a fixed point restatement of certain functional equations could be concretely done.

1.1 The Background of Metrical Fixed Point Theory

Let X be a nonempty set and $T : X \to X$ a selfmap. We say that $x \in X$ is a fixed point of T if

$$T(x) = x$$

and denote by F_T or $Fix\,(T)$ the set of all fixed points of T.

Example 1.1. 1) If $X = \mathbb{R}$ and $T(x) = x^2 + 5x + 4$, then $F_T = \{-2\}$;
2) If $X = \mathbb{R}$ and $T(x) = x^2 - x$, then $F_T = \{0, 2\}$;
3) If $X = \mathbb{R}$ and $T(x) = x + 2$, then $F_T = \emptyset$;
4) If $X = \mathbb{R}$ and $T(x) = x$, then $F_T = \mathbb{R}$.

Let X be any set and $T : X \to X$ a selfmap. For any given $x \in X$, we define $T^n(x)$ inductively by $T^0(x) = x$ and $T^{n+1}(x) = T\,(T^n(x))$; we call $T^n(x)$ the nth *iterate of x under T*. In order to simplify the notations we will often use Tx instead of $T(x)$.

The mapping $T^n(n \geq 1)$ is called the nth *iterate* of T. For any $x_0 \in X$, the sequence $\{x_n\}_{n \geq 0} \subset X$ given by

$$x_n = Tx_{n-1} = T^n x_0, \qquad n = 1, 2, ... \tag{1}$$

is called *the sequence of successive approximations with the initial value x_0*. It is also known as the *Picard iteration* starting at x_0.

For a given selfmap the following properties obviously hold:
1) $F_T \subset F_{T^n}$, for each $n \in \mathbb{N}^*$;
2) $F_{T^n} = \{x\}$, for some $n \in \mathbb{N}^* \Rightarrow F_T = \{x\}$;
The reverse of 2) is not true, in general, as shown by the next example.

Example 1.2. Let $T : \{1, 2, 3\} \rightarrow \{1, 2, 3\}$, $T(1) = 3$, $T(2) = 2$ and $T(3) = 1$. Then $F_{T^2} = \{1, 2, 3\}$ but $F_T = \{2\}$.

The fixed point theory is concerned with finding conditions on the structure that the set X must be endowed as well as on the properties of the operator $T : X \rightarrow X$, in order to obtain results on:
a) the existence (and uniqueness) of fixed points;
b) the data dependence of fixed points;
c) the construction of fixed points.

The ambient spaces X involved in fixed point theorems cover a variety of spaces: lattice, metric space, normed linear space, generalized metric space, uniform space, linear topological space etc., while the conditions imposed on the operator T are generally metrical or compactness type conditions. In order to introduce the most important ones, we need some minimal functional analysis background.

Metric spaces

Definition 1.1. Let X be a non-empty set. A mapping $d : X \times X \rightarrow \mathbb{R}_+$ is called a *metric* or a *distance* on X provided that
(d_1) $d(x, y) = 0 \Leftrightarrow x = y$; ("separation axiom")
(d_2) $d(y, x) = d(x, y)$, for all $x, y \in X$; ("symmetry")
(d_3) $d(x, z) \leq d(x, y) + d(y, z)$, for all $x, y, z \in X$ (*"the triangle inequality"*).

A set X endowed with a metric d is called *metric space* and is denoted by (X, d).

Example 1.3.
1) $X = \mathbb{R}$; $d(x, y) = |x - y|$, $\forall x, y \in \mathbb{R}$, where $|\cdot|$ denotes the absolute value, is a metric (a distance) on $\mathbb{R}$;

2) $X = \mathbb{R}^n$; $d(x, y) = \left[\sum_{i=1}^{n} (x_i - y_i)^2\right]^{1/2}$, for all

$$x = (x_1, x_2, ..., x_n), y = (y_1, y_2, ..., y_n) \in \mathbb{R}^n,$$

is a metric on $\mathbb{R}^n$, called the *euclidean metric*. The next two mappings:

$$\delta(x, y) = \sum_{i=1}^{n} |x_i - y_i|, \quad x, y \in \mathbb{R}^n,$$

$$\rho(x, y) = \max_{1 \leq i \leq n} |x_i - y_i|, \quad x, y \in \mathbb{R}^n,$$

are also metrics on $\mathbb{R}^n$;

3) Let $X = \{f : [a,b] \to \mathbb{R} \mid f \text{ is continuous}\}$. We define $d : X \times X \to \mathbb{R}_+$ by

$$d(f,g) = \max_{x \in [a,b]} |f(x) - g(x)|, \quad \text{for all} \quad f, g \in X.$$

Then d is a metric on X (called the *Chebyshev metric*); the metric space (X, d) is usually denoted by $C[a, b]$;

4) Let X be as at 3) and $\delta : X \times X \to \mathbb{R}_+$ be given by

$$\delta(f,g) = \max_{x \in [a,b]} \left(|f(x) - g(x)| \, e^{-\tau|x-x_0|} \right),$$

for all $f, g \in X$ where $\tau > 0$ is a constant and $x_0 \in [a, b]$ is fixed.

Then δ is a metric on X, called the *Bielecki metric*, and the metric space (X, δ) is usually denoted by $B[a, b]$.

Definition 1.2. Let (X, d) be a metric space. The topology having as basis the family of all open balls, $B(x; r)$, $x \in X$, $r > 0$, is called the *topology induced* by the metric d.

Definition 1.3. Two metrics d_1 and d_2 defined on the set X are called *equivalent metrics* if they induce the same topology on X.

Remarks.

1) Two metrics d_1 and d_2 are metrically equivalent if there exist two constants $m > 0$, $M > 0$ such that

$$md_1(x, y) \leq d_2(x, y) \leq Md_1(x, y), \quad \text{for all} \quad x, y \in X;$$

2) In Example 1.3, the metrics d, δ and ρ from 2) are equivalent; the metrics d from 3) and ρ from 4) are equivalent as well.

Definition 1.4. Let $\{x_n\}_{n=0}^{\infty}$ be a sequence in a metric space (X, d). We say that the sequence $\{x_n\}_{n=0}^{\infty}$

a) is *convergent to* $a \in X$ if, for any $\varepsilon > 0$, there exists $n_0 = n_0(\varepsilon)$ such that

$$d(x_n, a) < \varepsilon, \quad \text{for any } n \in \mathbb{N}, \, n \geq n_0.$$

b) is *fundamental* or *Cauchy sequence* if, for any $\varepsilon > 0$, there exists $n_0 = n_0(\varepsilon)$ such that

$$d(x_n, x_{n+p}) < \varepsilon, \quad \text{for all } n \in \mathbb{N}, \, n \geq n_0, \text{and any } p \in \mathbb{N}^*.$$

Remark. In a metric space, any convergent sequence is a Cauchy sequence, too, but the reverse is not generally true.

Definition 1.5. A metric space (X, d) is called *complete* if any Cauchy sequence in X is convergent.

1) Using the metrics given in Example 1.3, the following are complete metric spaces: $(\mathbb{R}, |\cdot|)$; $(\mathbb{R}^n, d)$; $(\mathbb{R}^n, \delta)$; $(\mathbb{R}^n, \rho)$; $C[a,b]$; $B[a,b]$;

2) If $\mathbb{Q}$ denotes the rationals in $\mathbb{R}$, then $(\mathbb{Q}, |\cdot|)$ is not a complete metric space.

Definition 1.6. Let (X, d) be a metric space. A mapping $T : X \to X$ is called:

(C_1) *Lipschitzian* (or *L-Lipschitzian*) if there exists $L > 0$ such that

$$d(Tx, Ty) \leq L \cdot d(x, y), \text{ for all } x, y \in X;$$

(C_2) *(strict) contraction* (or *a-contraction*) if T is a-Lipschitzian, with $a \in [0, 1)$;

(C_3) *nonexpansive* if T is 1-Lipschitzian;

(C_4) *contractive* if $d(Tx, Ty) < d(x, y)$, for all $x, y \in X$, $x \neq y$;

(C_5) *isometry* if $d(Tx, Ty) = d(x, y)$, for all $x, y \in X$.

Example 1.5.

1) $T : \mathbb{R} \to \mathbb{R}$ $T(x) = x/2 + 3$, $x \in \mathbb{R}$, is a strict contraction and $F_T = \{6\}$;

2) The function $T : [1/2, 2] \to [1/2, 2]$, $T(x) = 1/x$, is 4-Lipschitzian with $F_T = \{1\}$, while the functions T in Example 1.1, 3)-4) are all isometries;

3) $T : [1, +\infty] \to [1, +\infty]$, $T(x) = x + \dfrac{1}{x}$, is contractive and $F_T = \emptyset$.

The following theorem is of fundamental importance in the metrical fixed point theory and will be considered in an extended form in Chapter 2.

Theorem 1.1. (Contraction mapping principle)

Let (X, d) be a complete metric space and $T : X \to X$ be a given contraction. Then T has a unique fixed point p, and

$$T^n(x) \to p \ (as \ n \to \infty), \ for \ each \ x \in X.$$

There are various generalizations of the contraction mapping principle, roughly obtained in two ways:

1) by weakening the contractive properties of the map and, possibly, by simultaneously giving to the space a sufficiently rich structure, in order to compensate the relaxation of the contractiveness assumptions;

2) by extending the structure of the ambient space.

Several fixed point theorems have been also obtained by combining the two ways previously described or by adding supplementary conditions.

Remarks.

1) The conclusion of Theorem 1.1 is not valid if we consider "T contractive" instead of "T strict contraction", as shown by Example 1.5, part 3), but if we ask that (X, d) is a compact metric space, then the conclusion still holds (see Theorem 2.2, in Chapter 2);

2) One of the most important way in extending Theorem 1.1 consists of replacing the strict contractive condition (C_2) by a similar but weaker condition:

$$(C_6) \qquad d(Tx, Ty) \leq \varphi(d(x, y)), \qquad x, y \in X,$$

where $\varphi : \mathbb{R}_+ \to \mathbb{R}_+$ is a certain comparison function preserving some essential properties of the function appearing in (C_2), $\varphi(t) = at, 0 \leq a < 1$, see Chapter 2, Definition 2.3;

An alternative is to extend (C_2) to the following more general condition:

$$(C_7) \quad d(Tx, Ty) \leq \varphi(d(x, y), d(x, Tx), d(x, Ty), d(y, Tx), d(y, Ty)), \ x, y \in X,$$

where $\varphi : \mathbb{R}_+^5 \to \mathbb{R}_+$ stands for a 5-dimensional comparison function (see Section 2.6).

Normed spaces

Definition 1.7. Let E be a real (complex) vector space. A *norm* on E is a mapping $\|\cdot\| : E \times E \to \mathbb{R}_+$ having the following properties
(n_1) $\|x\| = 0 \Leftrightarrow x = 0$, the null element of E;
(n_2) $\|\lambda\, x\| = |\lambda| \cdot \|x\|$, for any $x \in E$ and any scalar λ;
(n_3) $\|x + y\| \leq \|x\| + \|y\|$, for all $x, y \in E$ ("*the triangle inequality*").
The pair $(E, \|\cdot\|)$ is called *normed (linear) space.*

Remarks.
1) If $\|\cdot\|$ is a norm on the (linear) vector space E, then $d : E \times E \to \mathbb{R}_+$, given by

$$d(x, y) = \|x - y\|, \qquad x, y \in E, \tag{2}$$

is a distance on E. This shows that any normed space can be always regarded as a metric space with respect to the distance induced by the norm;
2) A *Banach space* is a normed space which is complete (as a metric space).

Example 1.6.
1) The examples given in the previous paragraph, *Metric spaces,* are in fact all normed spaces, and the distances introduced in those examples are obtained from the corresponding norms by the process (2). The normed linear spaces obtained in this way are complete and hence are Banach spaces;
2) Let $I = [a, b]$ be a closed bounded interval in $\mathbb{R}$ and $E = C_\mathbb{R}(I)$ the vector space of all real-valued continuous functions on I. Then $\|\cdot\|_1 : E \times E \to \mathbb{R}_+$,

$$\| f \|_1 = \int_a^b |f(x)| \, dx, \quad f \in E,$$

is a norm on E. The normed space $(E, \|\cdot\|_1)$ is not complete (i.e., E is not a Banach space).

From the previous remark 1) we deduce that all concepts related to the norm in a normed space could be adapted from the metric space setting, including the contraction mapping principle (as it was originally formulated by Banach) and all contractive type conditions. One of these conditions, namely, the nonexpansiveness condition, is of particular interest in Banach spaces: if T is assumed to be only nonexpansive, that is

$$\|Tx - Ty\| \leq \|x - y\|, \text{ for all } x, y \in E,$$

then T need not have a fixed point.

By endowing the space with a sufficiently rich geometric structure, it is however possible to guarantee the existence of fixed points for nonexpansive operators.

Definition 1.8. A Banach space $(E, \|\cdot\|)$ is called *uniformly convex* if, given any $\varepsilon > 0$, there exists $\delta > 0$ such that for all $x, y \in E$ satisfying $\| x \| \leq 1, \| y \| \leq 1$, and $\| x - y \| \geq \varepsilon$, we have

$$\frac{1}{2} \| x + y \| < 1 - \delta.$$

Example 1.7. $E = \mathbb{R}^n$ endowed with the euclidean norm $\| x \| = \left(\sum x_i^2\right)^{1/2}$, $x = (x_1, x_2, ..., x_n) \in \mathbb{R}^n$, is uniformly convex, while, endowed with the norm $\|x\| = \sum_{i=1}^{n} |x_i|$, it is not.

Definition 1.9. A subset C of a real vector space E is called *convex* if, for any pair of points x, y in C, the closed segment with the extremities x, y, that is, the set $\{\lambda x + (1 - \lambda) y : \lambda \in [0,1]\}$ is contained in C. A subset C of a real normed space is called *bounded* if there exists $M > 0$ such that $\|x\| \leq M$, for all $x \in C$.

Theorem 1.2. *Let C be a closed, bounded, and convex subset of a uniformly convex Banach space and $T : C \to C$ a nonexpansive map. Then T has a fixed point.*

But, even though T is nonexpansive and has a fixed point, it is possible that the Picard iteration (1) no longer converge to the fixed point, as shown by the next example.

Example 1.8. Let $C = [0,1]$ and $T : [0,1] \to [0,1]$, $Tx = 1 - x$, for all $x \in [0,1]$. Then T is nonexpansive, T has a unique fixed point, $F_T = \left\{\frac{1}{2}\right\}$, but, for any $x_0 = a \neq \frac{1}{2}$, the Picard iteration (1) yields an oscillatory sequence $a, 1 - a, a, 1 - a, ...$.

Therefore, in order to compute the desired fixed point, it is necessary to consider other iteration procedures, as it will be shown in the next Section.

Definition 1.10. A linear normed space E is called *strictly convex* if $x, y \in E$ with $\|x\| = \|y\| = 1$ and $\|(1 - \lambda) x + \lambda y\| = 1$ for a $\lambda \in (0, 1)$ holds if and only if $x = y$.

This is equivalent to the condition that the unit sphere (or any sphere) contains no line segments. In such a space, any three points x, y, z satisfying $\|x - z\| + \|z - y\| = \|x - y\|$ must lie on a line, i.e., if $\|x - z\| = r_1, \|z - y\| = r_2$, and $\|x - y\| = r = r_1 + r_2$, then $z = \frac{r_1}{r} x + \frac{r_1}{r} y$.

Definition 1.11. Let E be a real Banach space. The space E^* of all linear continuous functionals on E is called the *dual* space of E. For $f \in E^*$ and $x \in E$ the value of f at x is denoted by $\langle f, x \rangle$ and is called the *duality pairing*.

1) The dual E^* is a Banach space with respect to the norm

$$\|f\|_* = \sup \{ \langle f, x \rangle : \|x\| \leq 1 \},$$

usually denoted by $\|.\|$;

2) The dual space of E^* is E^{**}, the *bidual* space of E. Since, in general, $E \subseteq E^{**}$, we say that E is *reflexive* if $E = E^{**}$;

3) A uniformly convex Banach space is strictly convex and reflexive. The concepts of uniformly convex and strictly convex Banach spaces are equivalent in finite dimensional spaces, since balls in such spaces are compact.

Definition 1.12. Let E^* be the dual space of a real Banach space. The multivalued mapping $J : E \to \mathcal{P}(E^*)$ defined by

$$Jx = \{ f \in E^* : \langle f, x \rangle = \|x\| \cdot \|f\|, \|x\| = \|f\| \}$$

is called the *normalized duality mapping* of E.

Remarks.

1) It is well known that if E^* is strictly convex, then J is single-valued. It will be consequently denoted by j in the sequel;

2) For reflexive Banach spaces, the assumption on strict convexity is not an essential restriction, since E and E^* can be equivalently re-normed as strictly convex spaces such that the duality mapping is preserved.

Example 1.9. 1) The space

$$l_p(\mathbb{R}) = \left\{ x = (x_n)_{n \geq 1} \subset \mathbb{R} \mid \sum_{n=1}^{\infty} |x_n|^p < \infty \right\}$$

endowed with the norm

$$\|x\| = \left(\sum_{n=1}^{\infty} |x_n|^p \right)^{1/p}, \quad x \in l_p,$$

is a Banach space for all $p \geq 1$;

2) Similarly, the space $L_p(\mathbb{R})$ of all p-integrable functions is a Banach space, for all $p \geq 1$, with respect to the corresponding norm ($\sum$ is replaced by the integral);

It is well known that, for any reflexive Banach space E, $L^p(E)$ with $1 < p < \infty$ is uniformly convex and hence reflexive, but L_1, L_∞, as well as $C[0,1]$, are not reflexive spaces.

In a Banach space E, beside the *strong convergence* defined by the norm, i.e., $\{x_n\}_{n=0}^\infty \subset E$ *converges strongly* to a if and only if $\|x_n - a\| \to 0$, as $n \to \infty$ (which is denoted by $x_n \to a$), we shall often consider the weak convergence, corresponding to the weak topology in E. We say that $\{x_n\} \subset E$ *converges weakly* to a if for any $f \in E^*$

$$\langle f, x_n \rangle \to \langle f, a \rangle, \text{ as } n \to \infty.$$

We denote this by $x_n \rightharpoonup x$ $(n \to \infty)$.

Remarks.
1) In L_p spaces the weak convergence of a sequence $\{x_n\}_{n=0}^\infty$ to a, together with the convergence of the norms ($\|x_n\| \to \|a\|$), implies the strong convergence of $\{x_n\}_{n=0}^\infty$ to a;
2) Any weakly convergent sequence $\{x_n\}_{n=0}^\infty$ in a Banach space is bounded. Further, if $x_n \rightharpoonup a$, then $\|a\| \leq \liminf \|x_n\|$.

When the contrary is not explicitly specified, throughout the book we shall simply consider that the strong convergence is involved.

Since conditions of pseudo-contractive type are very useful additional assumptions in approximating fixed points of Lipschitzian mappings, we summarize in the sequel the most important concepts of this kind.

Definition 1.13. Let E be an arbitrary real Banach space. A mapping T with domain $D(T)$ and range $R(T)$ in E is called
(a) *strong pseudocontraction* if there exists $k > 0$ such that for all $x, y \in D(T)$ there exists $j(x, y) \in J(x - y)$ such that

$$\langle (I - T)x - (I - T)y, \ j(x - y) \rangle \geq k \cdot \|x - y\|^2 ;$$

(b) *pseudocontractive* if for each $x, y \in D(T)$ there exists $j(x - y) \in J(x - y)$ such that
$$\langle (I - T)x - (I - T)y, \ j(x - y) \rangle \geq 0,$$

where J is the normalized duality mapping.

Pseudo-contractive mappings are firmly connected with another important class of operators, i.e., the class of accretive operators.

Definition 1.14. A mapping U with domain and range in E is called
a) *strongly accretive* if there exists a positive number k such that for each
$x, y \in D(U)$ there is a $j(x - y) \in J(x - y)$ such that

$$\langle Ux - Uy, j(x - y) \rangle \geq k \left\| x - y \right\|^2 ;$$

b) *accretive* if for each $x, y \in D(U)$ we have

$$\langle Ux - Uy, j(x - y) \rangle \geq 0.$$

Remarks.
1) By comparing Definitions 1.13 and 1.14, we remark that an operator T
is (strongly) pseudo-contractive if and only if $(I - T)$ is (strongly) accretive;
2) The concepts of pseudo-contractive and accretive operators can be
equivalently defined as follows:
(i) T is *strongly pseudocontractive* if there exists $t > 1$ such that, for all
$x, y \in D(T)$ and $r > 0$, the following inequality holds

$$\| x - y \| \leq \| (1 + r)(x + y) - rt(Tx - Ty) \| ;$$

(ii) T is *pseudocontractive* if $t = 1$ in the previous inequality;
(iii) T is *strongly accretive* if there exists $k > 0$ such that the inequality

$$\| x - y \| \leq \| x - y + r \left[(T - kI)x - (T - kI)y \right] \|$$

holds for all $x, y \in D(U)$ and $r > 0$;
(iv) T is *accretive* if $k = 0$ in the previous inequality.

Definition 1.15. A Banach space E is called *smooth* if, for every $x \in E$
with $\| x \| = 1$, there exists a unique $f \in E^*$ such that $\| f \| = \langle f, x \rangle = 1$. The
modulus of smoothness of E is the function $\rho_E : [0, \infty) \to [0, \infty)$, defined by

$$\rho_E(\tau) = \sup \left\{ \frac{1}{2} \left(\| x + y \| + \| x - y \| \right) - 1 : \ x, y \in E, \| x \| = 1, \| y \| = \tau \right\}.$$

The Banach space E is called *uniformly smooth* if

$$\lim_{\tau \to 0} \frac{\rho_E(\tau)}{\tau} = 0$$

and, for $q > 1$, E is said to be *q-uniformly smooth* if there exists a constant
$c > 0$ such that

$$\rho_E(\tau) \leq c\tau^2, \quad \tau \in [0, \infty).$$

Example 1.10. The L_p and l_p spaces have smoothness properties as
follows:

$$L_p \ (\text{or } l_p) \ \text{is} \ \begin{cases} p - \text{uniformly smooth, if } \ 1 < p \leq 2 \\ 2 - \text{uniformly smooth, if } \ p \leq 2. \end{cases}$$

In proving some convergence theorems for various iteration procedures, the following lemma will be used.

Lemma 1.1. *Let E be a uniformly smooth Banach space. Then there exists a nondecreasing continuous function $b : [0, \infty) \to [0, \infty)$ satisfying the following conditions:*

(i) $b(ct) \leq cb(t)$, for all $c \geq 1$;

(ii) $\lim_{t \to 0^+} b(t) = 0$;

(iii) $\|x + y\|^2 \leq \|x\|^2 + 2Re \langle y, j(x) \rangle + \max\{\|x\|, 1\} \cdot \|y\| \, b(\|y\|)$, for all $x, y \in E$.

For other results concerning the geometry of Banach spaces, see the monographs on the subject in the reference list.

Hilbert spaces

Hilbert spaces are the most important examples of uniformly convex Banach spaces that serve as very natural ambient spaces for various fixed point iteration procedures.

Definition 1.16. Let H be a real vector space. An *inner product* is a functional $\langle \cdot, \cdot \rangle : H \times H \to \mathbb{R}$ satisfying:

(p_1) $\langle x, x \rangle \geq 0$, for all $x \in H$ and $\langle x, x \rangle = 0$ if and only if $x = 0$, the null vector in H;

(p_2) $\langle x, y \rangle = \langle y, x \rangle$, for all $x, y \in H$;

(p_3) $\langle ax + by, z \rangle = a \langle x, z \rangle + b \langle y, z \rangle$, for each $x, y, z \in H$ and all $a, b \in \mathbb{R}$.

If $\langle \cdot, \cdot \rangle$ is an inner product on H, then the function $x \to \langle x, x \rangle^{1/2}$ defines a norm on H, called the *norm induced by the inner product*. The pair $(H, \langle \cdot, \cdot \rangle)$ is called a *prehilbertian space*.

A prehilbertian space that is complete (with respect to the metric corresponding to the norm induced by the scalar product) is called *Hilbert space*.

Remarks

1) Any Hilbert space is a uniformly convex Banach space;

2) It is then clear that all notions introduced in Banach spaces can be reformulated by replacing the duality pairing by the inner product. The Hilbert space setting will be preferred for most convergence theorems, even though these results are valid in a more general setting, i.e., in Banach spaces with certain geometric properties;

3) For example, in a Hilbert space, a *pseudocontraction* T is a map satisfying

$$\|Tx - Ty\|^2 \leq \|x - y\|^2 + \|Tx - Ty - (x - y)\|^2,$$

which is equivalent to

$$\langle Tx - Ty, x - y \rangle \leq \|x - y\|^2 \Leftrightarrow \langle (I - T)x - (I - T)y, x - y \rangle \geq 0;$$

4) T is *strictly (strongly) pseudocontractive* on C if there exists a constant $k < 1$ such that

$$\|Tx - Ty\|^2 \leq \|x - y\|^2 + k \|(I - T) x - (I - T) y\|^2, \ \forall x, y \in C.$$

Difference inequalities

In proving several convergence theorems we shall use various elementary results concerning recurrent inequalities. We collect in the following most of them as lemmas, without proofs.

Lemma 1.2. *Let $\{x_n\}_{n=0}^\infty$ be a sequence of nonnegative real numbers and let $\{a_n\}_{n=0}^\infty$ be a real sequence in $[0, 1]$ such that*

$$\sum_{n=0}^\infty a_n = \infty.$$

(i) If for a given $\epsilon > 0$ there exists a positive integer n_0 such that

$$x_{n+1} \leq (1 - a_n)x_n + \epsilon \cdot a_n, \ \text{for all } n \geq n_0,$$

then we have $0 \leq \limsup\limits_{n \to \infty} x_n \leq \epsilon$.

(ii) If there exists a positive integer n_1 such that

$$x_{n+1} \leq (1 - a_n)x_n + a_n b_n, \ \text{for all } n \geq n_0,$$

where $b_n \geq 0$ for all $n = 0, 1, 2, \ldots$ and $b_n \to 0$ as $n \to \infty$, then we have

$$\lim_{n \to \infty} x_n = 0.$$

Lemma 1.3. *Let $\{a_n\}_{n=0}^\infty, \{b_n\}_{n=0}^\infty, \{c_n\}_{n=0}^\infty$ be sequences of nonnegative numbers satisfying*

$$a_{n+1} \leq (1 - \omega_n)a_n + b_n + c_n, \ \text{for all } n \geq 0,$$

where $\{\omega_n\}_{n=0}^\infty \subset [0, 1]$. If

$$\sum_{n=0}^\infty \omega_n = \infty, \ b_n = o(\omega_n) \text{ and } \sum_{n=0}^\infty c_n < \infty,$$

then

$$\lim_{n \to \infty} a_n = 0.$$

Lemma 1.4. *Let $\{a_n\}_{n=0}^\infty$ be a sequence of nonnegative numbers satisfying*

$$a_{n+1} \leq (1 + \delta_n)a_n - \lambda_n \frac{\Phi(a_{n+1})}{1 + \Phi(a_{n+1}) + a_{n+1}} \cdot a_n, \ \text{for all } n \geq 0,$$

where $\Phi : [0, \infty) \to [0, \infty)$ is a strictly increasing function with $\Phi(0) = 0$, and $\{\lambda_n\}_{n=0}^{\infty}, \{\delta_n\}_{n=0}^{\infty}$ are sequences of nonnegative numbers satisfying

$$(i) \sum_{n=0}^{\infty} \lambda_n = \infty; \quad (ii) \sum_{n=0}^{\infty} \delta_n < \infty.$$

Then

$$\lim_{n \to \infty} a_n = 0.$$

Lemma 1.5. Let $\{a_n\}_{n=0}^{\infty}$ be a sequence of nonnegative numbers satisfying

$$a_{n+1} \leq (1 + \delta_n) a_n - \lambda_n \frac{\Phi(a_{n+1})}{1 + \Phi(a_{n+1}) + a_{n+1}} \cdot a_n + \theta_n, \text{ for all } n \geq 0,$$

where $\Phi : [0, \infty) \to [0, \infty)$ is a strictly increasing function with $\Phi(0) = 0$, and $\{\lambda_n\}_{n=0}^{\infty}, \{\delta_n\}_{n=0}^{\infty}, \{\theta_n\}_{n=0}^{\infty}$ are sequences of nonnegative numbers satisfying

$$(i) \sum_{n=0}^{\infty} \lambda_n = \infty; \quad (ii) \sum_{n=0}^{\infty} \delta_n < \infty; \quad (iii) \sum_{n=0}^{\infty} \theta_n < \infty.$$

Then

$$\lim_{n \to \infty} a_n = 0.$$

Remarks.

1) It is easy to see that Lemma 1.5 follows by Lemma 1.3 for

$$\omega_n = -\delta_n + \lambda_n \frac{\Phi(a_{n+1})}{1 + \Phi(a_{n+1}) + a_{n+1}}, \ n \geq 0,$$

while Lemma 1.4 is obtained from Lemma 1.3 for

$$\omega_n = -\delta_n + \lambda_n \frac{\Phi(a_{n+1})}{1 + \Phi(a_{n+1}) + a_{n+1}} \text{ and } c_n = 0, \ n \geq 0;$$

2) In the case $\omega_n = 1 - q$, for all $n \geq 0$, with $0 \leq q < 1$ and $c_n = 0, \ n \geq 0$, we can obtain from Lemma 1.3 a stronger result.

Lemma 1.6. Let $\{a_n\}_{n=0}^{\infty}, \{b_n\}_{n=0}^{\infty}$ be sequences of nonnegative numbers and $0 \leq q < 1$, so that

$$a_{n+1} \leq q a_n + b_n, \text{ for all } n \geq 0.$$

(i) If $\lim_{n \to \infty} b_n = 0$, then $\lim_{n \to \infty} a_n = 0$.

(ii) If $\sum_{n=0}^{\infty} b_n < \infty$, then

$$\sum_{n=0}^{\infty} a_n < \infty.$$

Remark. If $q = 1$, then the above result holds in a weaker form, as shown by the next Lemma.

Lemma 1.7. *Let $\{a_n\}_{n=0}^{\infty}, \{b_n\}_{n=0}^{\infty}$ be sequences of nonnegative numbers satisfying*

$$a_{n+1} \leq a_n + b_n, \text{ for all } n \geq 0.$$

(i) *If $\sum\limits_{n=0}^{\infty} b_n < \infty$, then $\lim\limits_{n \to \infty} a_n$ exists.*

(ii) *If $\sum\limits_{n=0}^{\infty} b_n < \infty$ and $\{a_n\}_{n=0}^{\infty}$ has a subsequence converging to zero,* then

$$\lim_{n \to \infty} a_n = 0.$$

We end this section by stating a property that holds in any Hilbert space.

Lemma 1.8. *Let x, y, z be points in a Hilbert space and $\lambda \in [0, 1]$. Then*

$$\| \lambda x + (1 - \lambda) y - z \|^2 = \lambda \| x - z \|^2 + (1-\lambda) \| y - z \|^2 - \lambda (1-\lambda) \| x - y \|^2.$$

1.2 Fixed Point Iteration Procedures

Picard iteration

Let (X, d) be a metric space, $D \subset X$ a closed subset of X (we often have $D = X$) and $T : D \to D$ a selfmap possessing at least one fixed point $p \in F_T$. For a given $x_0 \in X$ we consider the sequence of iterates $\{x_n\}_{n=0}^{\infty}$ determined by the successive iteration method

$$x_n = T(x_{n-1}) = T^n(x_0), \qquad n = 1, 2, ... \tag{3}$$

We are interested in obtaining (additional) conditions on T, D, and X, as general as possible, and which should guarantee the (strong) convergence of the iterates $\{x_n\}_{n=0}^{\infty}$ to a fixed point of T in D.

As we already mentioned, the sequence defined by (3) is known as the *sequence of successive approximations* or, simply, *Picard iteration*.

Moreover, if the Picard iteration converges to a fixed point of T, we will be interested in evaluating the error estimate (or, alternatively, the rate of convergence) of the method, that is, in obtaining a *stopping criterion* for the sequence of successive approximations.

When the contractive conditions are slightly weaker, then the Picard iterations need not converge to a fixed point of the operator T, and some other iteration procedures must be considered.

All the next fixed point iteration schemes are introduced in a real normed space $(E, \|\cdot\|)$. Let $T : E \to E$ be a selfmap, $x_0 \in E$ and $\lambda \in [0,1]$.

The sequence $\{x_n\}_{n=0}^{\infty}$ given by

$$x_{n+1} = (1 - \lambda)x_n + \lambda T x_n, \quad n = 0, 1, 2, \ldots \qquad (4)$$

will be called the *Krasnoselskij iteration procedure* or, simply, *Krasnoselskij iteration*.

It is easy to see that the Krasnoselskij iteration $\{x_n\}_{n=0}^{\infty}$ given by (4) is exactly the Picard iteration corresponding to the averaged operator

$$T_\lambda = (1 - \lambda)I + \lambda \cdot T, \qquad I = \text{ the identity operator} \qquad (5)$$

and that for $\lambda = 1$ the Krasnoselskij iteration reduces to Picard iteration. Moreover, we have $Fix\,(T) = Fix\,(T_\lambda)$, for all $\lambda \in (0,1]$.

In Chapter 2, the Picard iteration will be studied in connection with conditions of strict contractiveness type, while in Chapter 3 the Krasnoselskij iteration will be mainly associated with Lipschitzian and pseudocontractive type conditions.

Mann and Ishikawa iterations

The normal *Mann iteration procedure* or *Mann iteration*, starting from $x_0 \in E$, is the sequence $\{x_n\}_{n=0}^{\infty}$ defined by

$$x_{n+1} = (1 - a_n)x_n + a_n T x_n, \quad n = 0, 1, 2, \ldots, \qquad (6)$$

where $\{a_n\}_{n=0}^{\infty} \subset [0,1]$ satisfies certain appropriate conditions.

If we consider

$$T_n = (1 - a_n)I + a_n \cdot T,$$

then we have $Fix\,(T) = Fix\,(T_n)$, for all $a_n \in (0,1]$.

If the sequence $a_n = \lambda(const)$, then the Mann iterative process obviously reduces to the Krasnoselskij iteration.

Originally, the Mann iteration was defined in a matrix formulation, see Chapter 4 in this book, for more details.

The *Ishikawa iteration scheme* or, simply, *Ishikawa iteration* was first used to establish the strong convergence to a fixed point for a Lipschitzian and pseudo-contractive selfmap of a convex compact subset of a Hilbert space.

It is defined by $x_0 \in X$ and

$$x_{n+1} = (1 - a_n)x_n + a_n T\left[(1 - b_n)x_n + b_n T x_n\right], \quad n = 0, 1, 2, \ldots, \qquad (7)$$

where $\{a_n\}_{n=0}^{\infty}, \{b_n\}_{n=0}^{\infty} \subset [0,1]$ satisfy certain appropriate conditions.

In the last three decades both Mann and Ishikawa schemes have been successfully used by various authors to approximate fixed points of various classes of operators in Banach spaces.

If we rewrite (7) in a system form

$$\begin{cases} y_n = (1 - b_n)x_n + b_n T x_n, \\ x_{n+1} = (1 - a_n)x_n + a_n T y_n, \quad n = 0, 1, 2, ..., \end{cases} \tag{8}$$

then we can regard the Ishikawa iteration as a sort of two-step Mann iteration, with two different parameter sequences.

Despite this apparent similarity and the fact that, for $b_n = 0$, Ishikawa iteration reduces to the Mann iteration, there is not a general dependence between convergence results for Mann iteration and Ishikawa iteration.

Recently, some authors considered the so called *modified Mann iteration,* respectively *modified Ishikawa iteration,* by replacing the operator T by its n-th iterate T^n.

For example, the modified Ishikawa iteration is defined by

$$\begin{cases} y_n = (1 - b_n)x_n + b_n T^n x_n \\ x_{n+1} = (1 - a_n)x_n + a_n T^n y_n, \quad n = 0, 1, 2, ... \end{cases} \tag{9}$$

Very recently, the so called *Ishikawa and Mann iteration procedures with errors,* for nonlinear mappings were introduced as follows:

(a) Let K be a nonempty subset of a Banach space E and $T : K \to E$ be an operator. The sequence $\{x_n\}_{n=0}^{\infty}$ defined by $x_0 \in K$ and

$$\begin{cases} x_{n+1} = (1 - a_n)x_n + a_n T y_n + u_n, \\ y_n = (1 - b_n)x_n + b_n T x_n + v_n, \quad n = 0, 1, 2, ... \end{cases} \tag{10}$$

where (i) $\{a_n\}_{n=0}^{\infty}$ and $\{b_n\}_{n=0}^{\infty}$ are some sequences in $(0, 1)$, satisfying appropriate conditions and (ii) $\{u_n\}_{n=0}^{\infty}, \{v_n\}_{n=0}^{\infty}$ are sequences in K such that

$$\sum \|u_n\| < \infty, \ \sum \|v_n\| < \infty, \tag{11}$$

is called *Ishikawa iteration process with errors.*

The *Mann iteration with errors* is similarly defined and could be obtained from (10) by taking $b_n = 0$.

In spite of the fact that the fixed point iteration procedures are designed for numerical purposes, and hence the consideration of errors is of both theoretical and practical importance, however it seems that the iteration process with errors introduced by (10) is not quite satisfactory from a practical point of view.

Indeed, the conditions (11) imply, in particular, that the errors tend to zero, which is not suitable for the randomness of the occurrence of errors in practical computations.

As a correction to the previous definition, the same concept was introduced in a different way.

(b) Let K be a nonempty convex subset of E and let $T : K \to E$ be a mapping. For any given $x_0 \in K$, the sequence $\{x_n\}_{n=0}^{\infty}$ defined iteratively by

$$\begin{cases} x_{n+1} = a_n x_n + b_n T y_n + c_n u_n \\ y_n = a_n' x_n + b_n' T x_n + c_n' v_n, n = 0, 1, 2, \dots \end{cases} \qquad (12)$$

where $\{a_n\}_{n=0}^{\infty}, \{b_n\}_{n=0}^{\infty}, \{c_n\}_{n=0}^{\infty}, \{a_n'\}_{n=0}^{\infty}, \{b_n'\}_{n=0}^{\infty}, \{c_n'\}_{n=0}^{\infty}$ are sequences in the interval $(0,1)$ such that $a_n + b_n + c_n = 1 = a_n' + b_n' + c_n'$, and $\{u_n\}_{n=0}^{\infty}, \{v_n\}_{n=0}^{\infty}$ are bounded sequences in K, for all $n = 0, 1, 2, \dots$, is called the *Ishikawa iteration with errors*.

The *Mann iteration with errors* could be obtained from (12) by taking formally $b_n = b_n' = 0$, for all integers $n \geq 0$.

Other important fixed point iteration procedures

Let E be a Banach space, and suppose T is a mapping of E into E. The *Kirk's iteration procedure* is defined by $x_0 \in E$ and

$$x_{n+1} = \alpha_0 x_n + \alpha_1 T x_n + \alpha_2 T^2 x_n + \dots + \alpha_k T^k x_n,$$

where k is a fixed integer, $k \geq 1$, $\alpha_i \geq 0$, for $i = 0, 1, \dots, k$, $\alpha_1 > 0$ and

$$\alpha_0 + \alpha_1 + \dots + \alpha_k = 1.$$

This scheme reduces to Picard iteration, for $k = 0$, and to Krasnoselskij iteration, for $k = 1$.

The Kirk, Krasnoselskij, Mann and Ishikawa iteration procedures are mainly used to generate successive approximations for fixed points of various classes of mappings in normed linear spaces, for which the Picard iteration does not converge.

Let H be a Hilbert space and C be a closed, bounded, and convex subset of H containing 0. The sequence $\{x_n\}_{n=0}^{\infty}$ defined by $x_0 \in C$, and

$$x_n = T_n^{n^2} x_{n-1}, \qquad n = 1, 2, \dots,$$

where $T_n x = \dfrac{n}{n+1} T x$, $n \geq 1$, will be called the *Figueiredo iteration procedure*.

It is known that the Figueiredo iteration converges strongly to a fixed point of nonexpansive operators $T : C \to C$.

There are also several other fixed point iteration schemes, constructed as Cesaro means (ergodic type iterations), as well as both linear and nonlinear generalizations of them.

Let T be a selfmap of a Hilbert space H, and $\alpha = \{\alpha_n\}_{n=0}^{\infty}$ be a sequence in $[0,1]$. The sequence $\{A_n^{\alpha}\}_{n=0}^{\infty}$ defined inductively by $A_0^{\alpha} x = x$ and

$$A_{n+1}^{\alpha} x = \alpha_{n+1} x + (1 - \alpha_{n+1}) T \, A_n^{\alpha} x, \qquad n = 0, 1, 2, \dots$$

will be called the *Halpern iteration scheme*.

If T is positively homogeneous (i.e., $T(tx) = t\,Tx$, for any $t \geq 0$ and $x \in H$) and $\alpha_n = \dfrac{1}{n+1}$, $n \geq 0$, then we have

$$A_n^\alpha = \frac{1}{n+1} S_n x,$$

where $S_0 x = x$, $S_{n+1} = x + T(S_n\,x)$, which shows that for this special choice of $\alpha = \{\alpha_n\}_{n=0}^\infty$, $\{A_n^\alpha\}_{n=0}^\infty$ is a nonlinear generalization of the Cesaro averages.

One can also consider another iteration scheme, $\{A_n^\alpha\}_{n=0}^\infty$, suggested by Wittmann, given by

$$A_0^\alpha = x, \quad A_{n+1}^\alpha x = \alpha_{n+1} x + T\left((1 - \alpha_{n+1})\, A_n^\alpha x\right),$$

which reduces to the Halpern one if T is positively homogeneous.

The main aim of the next chapters of the book is to survey the most important convergence theorems for some of the aforementioned fixed point iteration procedures, in different contexts and under several metrical assumptions.

1.3 Fixed Point Formulation of Typical Functional Equations

Many important nonlinear problems of applied mathematics can be described in a unitary manner by the following scheme.

For a given object f, find another object x satisfying two conditions:
(i) The object x belongs to a given class X of objects;
(ii) The object x is in a certain relation R to the object f.

An object x satisfying these conditions will be called the *solution* of the given problem. This problem can be described by

$$\{x \in X : x\ R\ f\}. \tag{13}$$

Examples.
1) Find a real solution of the equation $x^5 - x - 1 = 0$. Here $f \equiv f(x) = x^5 - x - 1$, $X = \mathbb{R}$ and the relation R expresses the fact that x and f are related by the given equation.
2) The initial value problem for a first order ordinary differential equation

$$\begin{cases} y' = \varphi(t, y) \\ y(t_0) = y_0 \end{cases}$$

fit the scheme (13). Indeed, here we have $f = (\varphi, t_0, y_0)$, $X = C(I)$, where $t_0 \in I \subset \mathbb{R}$, x is the function $y : I \to \mathbb{R}$ and R is given by the previous system of conditions.

In turn, any problem of the form (13) can be written equivalently as a fixed point problem

$$x = Tx, \tag{14}$$

where $T : E \to E$ is a corresponding operator, that allows us to use constructive fixed point tools in obtaining the desired solution.

Consequently, the main aim of the present Section is to illustrate, on some important typical functional equations from applied mathematics, how we can convert them into equivalent fixed point problems. This will, in part, motivate our interest in the study of fixed point iteration procedures.

Single nonlinear equations

Efficiently finding roots of nonlinear equations is of major importance and has significant applications in numerical mathematics. In contrast to the case of linear systems of equations, direct methods for solving nonlinear equations are usually available only for a few special cases. Consequently, we need to resort to iterative methods. According to the mathematical importance of this problem, there exists a vast and dense literature related to iterative methods. Basically, for the equation

$$F(x) = 0, \tag{15}$$

where $F : D \subset \mathbb{R}^n \to \mathbb{R}^n$ is a given operator, we can consider several iterative methods for computing approximate solutions of it.

One of the most used method is to write (15) equivalently in the form (14), where T is a certain operator associated to F, in such a way that, by considering a certain fixed point iteration scheme (usually the Picard iteration), we obtain a sequence that converges to a solution of (15).

The operator T is usually called *iteration function*. There are several methods for constructing iteration functions. If we restrict to real functions of a real single-variable, then one of the most used algorithms for obtaining T is the well-known Newton's method, which is based on the iteration function

$$Tx = x - \frac{F(x)}{F'(x)}.$$

Example 1.11. Consider the polynomial equation

$$x^5 - x - 1 = 0 \tag{16}$$

that can be written in the form (14) in many different ways. Here there are three of them:

$$(i)\ x = x^5 - 1;\ (ii)\ x = \sqrt[5]{x+1};\ (iii)\ x = \frac{4x^5 + 1}{5x^4 - 1}.$$

It is easy to see that (16) has a unique solution in the interval $[1, \infty)$.

Denote:

$$T_1(x) = x^5 - 1, \; T_2(x) = \sqrt[5]{x+1} \text{ and } T_3(x) = \frac{4x^5 - 1}{5x^4 - 1}, \; x \in [1, \infty).$$

Then the Picard iteration associated to T_1 does not converge, whatever the initial approximation $x_0 \in [1, \infty)$, while in the case of T_2 or T_3, it does. In fact, it is easy to show that T_2 is a $\frac{1}{5}$-contraction.

As it could be verified, the iteration function T_3 has been obtained by the Newton's algorithm. The next table shows the first iterations for the three iterative processes defined by the iteration functions T_1, T_2 an T_3, respectively, and for certain initial guesses x_0.

$x_{n+1} = T_1 \, x_n$	$x_{n+1} = T_2 \, x_n$	$x_{n+1} = T_3 x_n$
$x_0 = 1$	$x_0 = 1$	$x_0 = 1$
....	....	....
$x_1 = 0$	$x_1 = 1.149$	$x_1 = 1.25$
$x_2 = -1$	$x_2 = 1.165$	$x_2 = 1.178$
$x_3 = -2$	$x_3 = 1.167$	$x_3 = 1.168$
$x_4 = -33$	$x_4 = 1.167$	$x_4 = 1.167$
$x_5 = -39135394$	$x_5 = 1.167$	$x_5 = 1.167$

$x_{n+1} = T_1 \, x_n$	$x_{n+1} = T_2 \, x_n$	$x_{n+1} = T_3 x_n$
		$x_0 = 10$
$x_0 = 1.167$	$x_0 = 10$	$x_1 = 8$
$x_1 = 1.164$	$x_1 = 1.615$	$x_2 = 6.401$
$x_2 = 1.141$	$x_2 = 1.212$	$x_3 = 5.121$
$x_3 = 0.936$	$x_3 = 1.172$	$x_4 = 4.098$
$x_4 = -0.282$	$x_4 = 1.168$	$x_5 = 3.282$
$x_5 = -1.002$	$x_5 = 1.167$	$x_6 = 2.632$
		...
		$x_{12} = 1.168$
		$x_{13} = 1.167$

The next Theorem gives a recipe for constructing high-order methods of Newton type for approximating roots of F.

Theorem 1.3. Set $F_1(x) = F(x)$, and for each $m \geq 2$ recursively define

$$F_m(x) = \frac{F_{m-1}(x)}{\left[F'_{m-1}(x)\right]^{1/m}}.$$

Then the function

$$G_m(x) = x - \frac{F_{m-1}(x)}{F'_{m-1}(x)}$$

defines an iteration function whose order of convergence for simple roots is m.

Remarks.

1) For $m = 2$, from Theorem 1.3 we obtain the iteration function in the classical Newton or Newton-Raphson method;

2) For $m = 3$, we obtain

$$G_3(x) = x - FF'/(F'^2 - FF''/2),$$

which is the iteration function involved in Halley's method etc.

The following problem arises: for a given F, how to construct an operator T, such that the equation (15) is equivalent to the fixed point problem (14) and T satisfies a certain contractive condition ? (Note that the Newton iteration function is not a strict contraction but a quasi-contraction).

Integral equations

In the class of operator equations that can be naturally reformulated in terms of a fixed point problem, the integral and integro-differential equations play an important role. For f and K given functions, we shall consider here only a simple integral equation of the form

$$y(x) = f(x) + \int_0^1 K(x, s, y(x), y(s)) ds, \quad x \in [0, 1]. \tag{17}$$

Similar considerations will apply to more general equations involving, for example, derivatives of the unknown function y, or to higher-dimensional problems involving unknown functions depending on two or more variables.

Equations of the form (17) arise in a variety of contexts. For example, in connection with a problem of radiation transfer, we are led to the equation

$$y(x) = 1 + \int_0^1 \frac{s\, y(s)\, y(x)}{s + x} \varphi(s)\, ds,$$

where φ is given. A special but important case of (17) is the Urysohn equation

$$y(x) = 1 + \int_0^1 K(x, s, y(s))\, ds,$$

or the nonlinear Fredholm integral equation

$$y(x) = f(x) + \lambda \int_0^1 K(x, s, y(s))\, ds,$$

where $\lambda \in \mathbb{R}$ is a given number.

If we search a continuous solution for one of the aforementioned equations, say for (18), then we can reformulate it as a fixed point problem, under appropriate assumptions.

Let us assume:

(a) $K : [0, 1] \times [0, 1] \times I \to \mathbb{R}$ $(I \subset \mathbb{R})$ is a continuous mapping, bounded on this domain; $K(x, s, z)$ is called the *kernel* of the integral equation;

(b) K is L-Lipschitzian with respect to the third variable, that is, there exists $L > 0$ such that

$$|K(x, s, z_1) - K(x, s, z_2)| \leq L\,|z_1 - z_2|, \text{ for each } x, s \in [0, 1] \text{ and } z_1, z_2 \in I;$$

(c) $f : [0, 1] \to \mathbb{R}$ is continuous;
(d) $\lambda \in \mathbb{R}$ is a given number;
(e) $\varphi : [0, 1] \to I$ is the *unknown* function, supposed to be continuous.

Let X be the space of all functions $\varphi : [0, 1] \to \mathbb{R}$ which satisfy:
(i) φ is continuous; (ii) $\varphi(x) \in I \subset \mathbb{R}$, for each $x \in [0, 1]$.
We consider X endowed with the (Chebyshev) metric

$$d(\varphi_1, \varphi_2) = \max_{x \in [0,1]} |\varphi_1(x) - \varphi_2(x)|, \quad \varphi_1, \varphi_2 \in X.$$

By Example 1.3, 3) in Section 1.1, we know that $X = C[0, 1]$ is a complete metric space. We define on X the operator T given by

$$(T\varphi)(x) = \lambda \int_0^1 K(x, s, \varphi(s))\, ds + f(x), \quad \forall\, x \in [0, 1]. \tag{19}$$

It is obvious that T maps X into itself (K and f continuous implies $T\varphi$ is continuous, too) and hence $T(X) \subset X$.

So, the integral equation (18) is equivalent to the fixed point problem

$$\varphi = T\,\varphi,$$

where T is defined by (19). Moreover, T is Lipschitzian and, under appropriate assumptions on λ, T is even a strict contraction. Indeed,

$$|(T\,\varphi_1)(x) - (T\,\varphi_2)(x)| = \left| \lambda \left[\int_0^1 K(x, s, \varphi_1(s))\, ds - \int_0^1 K(x, s, \varphi_2(s))\, ds \right] \right| \leq$$

$$\leq |\lambda| \cdot \int_0^1 |K(x, s, \varphi_1(s)) - K(x, s, \varphi_2(s))|\, ds \leq |\lambda| \cdot L \int_0^1 |\varphi_1(s) - \varphi_2(s)|\, ds.$$

But

$$|\varphi_1(s) - \varphi_2(s)| \leq \max_{x \in [0,1]} |\varphi_1(x) - \varphi_2(x)| = d(\varphi_1, \varphi_2), \text{ for each } s \in [0, 1]$$

and hence, for each $x \in [0, 1]$ and all $\varphi_1, \varphi_2 \in X$, we have

$$|(T\,\varphi_1)(x) - (T\,\varphi_2)(x)| \le L \cdot |\lambda| \cdot d(\varphi_1, \varphi_2),$$

which leads to

$$\max_{x \in [0,1]} |(T\,\varphi_1)(x) - (T\,\varphi_2)(x)| \le L \cdot |\lambda| \cdot d(\varphi_1, \varphi_2),$$

that holds for all $\varphi_1, \varphi_2 \in X$. Therefore, we have

$$d(T\,\varphi_1, T\,\varphi_2) \le L \cdot |\lambda| \cdot d(\varphi_1, \varphi_2), \quad \varphi_1, \varphi_2 \in X,$$

which shows that T is $L \cdot |\lambda|$-Lipschitzian.

Remark. If we choose λ such that $|\lambda| < \dfrac{1}{L}$, then T is in fact a strict contraction, and then, by the mapping contraction theorem, T has a unique fixed point, which is the unique solution of the integral equation (18), and this solution can be obtained by the Picard iteration.

Similar considerations could be done for *Volterra integral equations*. We shall illustrate this for the following Volterra integral equation of the second kind

$$y(x) = f(x) + \lambda \int_a^x K(x, s, y(s))\, ds, \quad x \in [0, T], \tag{20}$$

where K, f, λ and y are defined similarly to the previous integral equation.

There is a classical way to prove that, if K is Lipschitzian with respect to the third variable, then (20) has a unique solution in the set of continuous functions. By denoting

$$(T\,\varphi)(x) = \lambda \int_a^x K(x, s, \varphi(s))\, ds + f(x), \quad \text{for all } x \in [a, b], \tag{21}$$

we can write (20) equivalently into the fixed point form

$$\varphi = T\,\varphi.$$

Let us consider $B[a, b] = \{ f : [a, b] \to R \mid f \text{ continuous} \}$, the space of all continuous functions on $[a, b]$, endowed with the Bielecki metric

$$\delta(\,f, g) = \max_{x \in [a,b]} \left(|\,f(x) - g(x)| \cdot e^{-\tau(x-a)} \right), \quad f, g \in B[a, b], \ \tau > 0.$$

Then $T : B[a, b] \to B[a, b]$, given by (21), is a strict contraction. Indeed,

$$|(T\,\varphi_1)(x) - (T\,\varphi_2)(x)| \le |\lambda| \cdot \int_a^x |K(x, s, \varphi_1(s)) - K(x, s, \varphi_2(s))|\, ds$$

$$\leq |\lambda|\, L \int_a^x |\varphi_1(s) - \varphi_2(s)|\, ds = |\lambda|\, L \int_a^x |\varphi_1(s) - \varphi_2(s)|\, e^{-\tau(s-a)} e^{\tau(s-a)}\, ds \leq$$

$$\leq |\lambda|\, L\delta(\varphi_1, \varphi_2) \int_a^x e^{\tau(s-a)}\, ds \leq |\lambda|\, L \frac{e^{\tau(x-a)} - 1}{\tau} \delta(\varphi_1, \varphi_2) <$$

$$< |\lambda|\, L\delta(\varphi_1, \varphi_2) \frac{e^{\tau(x-a)}}{\tau},$$

for all $\varphi_1, \varphi_2 \in B[a,b]$, $x \in [a,b]$ and $\tau > 0$, which leads to

$$|(T\,\varphi_1)(x) - (T\,\varphi_2)(x)|\, e^{-\tau(x-a)} \leq \frac{|\lambda| \cdot L}{\tau} \cdot \delta(\varphi_1, \varphi_2), \forall \varphi_1, \varphi_2 \in B[a,b], x \in [a,b].$$

Taking the maximum in the left-hand side, it results

$$\delta(T\,\varphi_1, T\,\varphi_2) \leq \frac{|\lambda|\, L}{\tau} \cdot \delta(\varphi_1, \varphi_2), \ \forall \varphi_1, \varphi_2 \in B[a,b], \tau > 0.$$

We now choose a number τ such that $\tau > |\lambda| \cdot L$, i.e., such that

$$\frac{|\lambda| \cdot L}{\tau} < 1,$$

and then $T : B[a,b] \to B[a,b]$ will be a strict contraction.

By applying the contraction mapping principle, we deduce that equation (20) has a unique solution $y^* \in B[a,b]$. Moreover, defining a sequence of functions $\{y_n\}$ inductively by choosing any $y_0 \in B[a,b]$ and setting

$$y_{n+1}(x) = f(x) + \int_a^x K(x, s, y_n(s))\, ds,$$

the sequence $\{y_n\}$, which is actually the associated Picard iteration, converges uniformly on $[a,b]$ to the unique solution y^* of the equation.

Ordinary Differential Equations

The *initial value problem* for a first order O(rdinary) D(ifferential) E(quation)

$$\begin{cases} y' = f(x,y) \\ y(x_0) = y_0 \end{cases} \tag{22}$$

may be written equivalently as a Volterra integral equation

$$y(x) = y_0 + \int_{x_0}^x f(s, y(s))\, ds.$$

The *initial value problem* for the following second order ODE

$$\begin{cases} y'' = f(x) \\ y(x_0) = 0, \quad y'(x_0) = 0 \end{cases} \tag{23}$$

can be written equivalently in a ready fixed point form as

$$y(x) = \int_{x_0}^{x} (x - s)\, f(s)\, ds,$$

again a Volterra integral equation.

A two-point boundary value problem

$$\begin{cases} y'' = f(x, y) \\ y(a) = A, \quad y(b) = B \end{cases} \tag{24}$$

may be put into the equivalent integral form

$$y(x) = \frac{x-a}{b-a} B + \frac{b-x}{b-a} A - \int_a^b G(x, s)\, f(s, y(s))\, ds,$$

where $G : [a, b] \times [a, b] \to \mathbb{R}$

$$G(x, s) = \begin{cases} \dfrac{(s-a)(b-x)}{b-a}, & \text{if } a \leq s \leq x \leq b \\[2mm] \dfrac{(x-a)(b-s)}{b-a}, & \text{if } a \leq x \leq s \leq b \end{cases} \tag{25}$$

is the Green function associated to the homogeneous problem

$$y'' = 0, \quad y(a) = 0, \quad y(b) = 0.$$

Under appropriate assumptions on f (continuous and Lipschitzian with respect to the last variable), it is an easy task to show that, for all the problems (22), (23) and (24) considered here, the corresponding integral operators fulfill a certain contractive condition and hence we can study these equations under the fixed point formulation, by using an appropriate fixed point technique.

1.4 Bibliographical Comments

§1.1.

Theorem 1.1 is due to Banach [Ban22]. It is an abstraction of the classical method of successive approximations, see also the Comments in Chapter 2. In the metric space setting, Theorem 1.1 is called *contraction mapping theorem* or *Banach's theorem* or *theorem of Picard-Banach* or *theorem of Picard-Banach-Caccioppoli*. For the complete formulation of Banach's fixed point theorem, including both *a priori* and *a posteriori* estimates as well as the rate of convergence estimate, see Theorem 2.1 in Chapter 2.

For the general concepts, examples and remarks presented in this Section we used several monographs and articles in the reference list. For those concepts strictly connected to fixed point theory, see the monographs Berinde [Be97a], Dugundji and Granas [DuG82], Hadzic [Had77], Istratescu [Ist73], [Ist81], Rus [Ru79c], [Rus01] and Taskovic [Tas86], where one can also find various generalizations of the contraction mapping principle.

Important examples of these kind of theorems are associated to names as Boyd and Wong, Browder, Krasnoselskij and Stechenko, Rhoades, Rus and many others (see Berinde [Be97a], Rus [Ru79c]). The most important fixed point theorems of these kind have been obtained by Kannan, Zamfirescu, Ciric, Reich, Rus and many others (see Berinde [Be97a], Rus [Ru79c], [Rus01]).

The fact that a nonexpansive operator in a Banach space need not have a fixed point was pointed out in Petryshyn and Williamson [PWi73], p. 460.

Theorem 1.2 was obtained independently by Browder [Br65a], Kirk [Kir65] and Gohde [Goh65] in 1965. A proof of this result in the Hilbert space setting is given in Chapter 3, Theorem 3.1.

Example 1.8 is taken from Rhoades [Rho91], while Lemma 1.1 is due to Reich [Re78a].

The general concepts in metric, Banach and Hilbert spaces are collected from the monographs and articles in the reference list.

We mention the source of the lemmas presented at the end of the section: Lemma 1.2 appears in many papers. In the form given here, it corresponds to Lemma 2 in Sharma, S. and Deshpande [SD02a]; Lemma 1.3 is given in Liu, L.S. [LL95b]; Lemma 1.4 is taken from Osilike [Os99a]; Lemma 1.5 is taken from Yin, Liu, Z. and Lee, B.S. [YLL00]; Lemma 1.6 is adapted after Theorem 1.2.1 in Berinde [Be97a]; Lemma 1.7, part (i) is given in Tan and Xu, H.K. [TX93a] while part (ii) appears in Chidume and Moore [ChM99]; Lemma 1.8 is taken from Ishikawa [Ish74].

§1.2.

The method of successive approximations appears to have been introduced by Liouville [Lio37] and used by Cauchy. It was developed systematically for the first time by Picard [Pic90] in his classical and well-known proof of the existence and uniqueness of the solution of initial value problems for ordinary differential equations, dating back in 1890.

Krasnoselskij iteration, in the particular case $\lambda = \dfrac{1}{2}$, was first introduced by Krasnoselskij [Kra55] in 1955, and in the general form by Schaefer [Sch57] in 1957.

The original Mann iteration was defined in a matrix formulation by Mann [Man53] in 1953.

Ishikawa [Ish74] introduced his iteration process in a paper published in 1974. The Ishikawa iterations with errors were considered very recently by Liu, L.S. [LL95a], [LL95b] in the form (10) and by Xu, Y.G. [XuY98] in the form (11). For more details on Mann and Ishikawa iterations, see the Bibliographical Comments in Chapters 4 and 5.

The Halpern fixed point iteration procedure was introduced by Wittmann [Wit92].

§1.3.

The material in this Section is classical. Some special concepts and results are taken from Mikhlin [Mik91], Dugundji and Granas [DuG82], Kalantari and Gerlach [KaG00] (Theorem 1.3), as well as from some author's unpublished lectures notes.

Exercises and Miscellaneous Results

1.1. Show that the functions $d : X \times X \to \mathbb{R}_+$ defined in Example 1.3 are metrics on $X = \mathbb{R}^n$.

1.2. Show that the metrics d, δ, ρ defined in Example 1.3, 2), are (metrically) equivalent.

1.3. Show that the metrics d in Example 1.3, 3), and ρ in Example 1.3, 4), are metrically equivalent. Show that a sequence $\{f_n\}$ converges to f in $C[a, b]$ if and only if $\{f_n\}$ converges uniformly to f.

1.4. Show that the following functions are metrics in the space $X = \mathbb{R}$:
(a) $d(x, y) = 2 \cdot |x - y|$;   (b) $d(x, y) = |x^3 - y^3|$.

1.5. Show that $d(x, y) = |xy|$ does not define a metric in $\mathbb{R}$.

1.6. Let $\mathbb{R}^2 \setminus \{O\}$ denote the punctured plane. Define $d(x, y)$ as follows:

$$d(x, y) = |r_1 - r_2| + |\theta|,$$

where $r_1 =$ the Euclidean distance from x to O, $r_2 =$ the Euclidean distance from y to O, where O is the origin, and θ is the smallest angle subtended by the two straight lines connecting x and y to the origin. Show that d is a metric.

1.7. Two metric spaces (X_1, d_1) and (X_2, d_2) are *equivalent* if there is a function $h : X_1 \to X_2$ which is one-to-one and onto (i.e., it is invertible), such that the metric $\widetilde{d_1}$ on X_1 defined by

$$\widetilde{d_1} = d_2(h_1(x), h_2(y)), \text{ for all } x, y \in X_1$$

is equivalent to d_1.

Let $X_1 = [1, 2]$ and $X_2 = [0, 1]$ and let d_1 denote the Euclidean metric in X_1 and let $d_2(x, y) = 2 \cdot |x - y|$ in X_2. Show that (X_1, d_1) and (X_2, d_2) are equivalent metric spaces.

1.8. On the set $X = (0, 1] = \{x \in \mathbb{R} : 0 < x \leq 1\}$ define two metrics by

$$d_1(x, y) = |x - y| \text{ and } d_2(x, y) = \left| \frac{1}{x} - \frac{1}{y} \right|.$$

Show that (X, d_1) and (X, d_2) are not equivalent metric spaces.

1.9. Let $S \subset X$ be a subset of a metric space (X, d). A point $x \in X$ is called a *limit point* of S if there is a sequence $\{x_n\}_{n=1}^{\infty}$ of points $x_n \in S \setminus \{x\}$ such that $\lim_{n \to \infty} x_n = x$. The *closure* of S, denoted by $\overline{S}$, is defined by $\overline{S} = S \cup \{\text{limit points of } S\}$. S is closed if $S = \overline{S}$. Show that if $h : X_1 \to X_2$ makes the metric spaces (X_1, d_1) and (X_2, d_2) equivalent, then the statements:
(a) $x \in X_1$ is a limit point of $S \subset X$, and (b) $h(x) \in X_2$ is a limit point of $h(S) \subset X_2$, are equivalent.

1.10. Let A be the "filled" square in $\mathbb{R}^2$,

$$A = \{x = (x_1, x_2) \in \mathbb{R}^2 : 0 \leq x_1 \leq 1, 0 \leq x_2 \leq 1\}.$$

Find all of the limit points of the set

$$\{x_n = (1/n + (-1)^n, 1/n + (-1)^{2n}) : n = 1, 2, 3 \dots\}$$

in the metric space (A, d), where d is the Euclidean metric.

1.11. Let S be a subset of a complete metric space (X, d). Then (S, d) is a metric space and (S, d) is complete if and only if S is closed in X.

1.12. A subset S of a metric space (X, d) is *compact* if every infinite sequence $\{x_n\}_{n=1}^{\infty}$ in S contains a subsequence having a limit in S.
(a) Let S be a subset of a compact metric space. Show that ∂S (i.e., the boundary of S) is compact;
(b) Show that any compact metric space is complete.

1.13. Let d and ρ be as in Example 1.3, 2), and consider $T : \mathbb{R}^2 \to \mathbb{R}^2$, given by

$$T(x,y) = \left(\frac{4}{5}x + \frac{4}{5}y, \frac{1}{10}x + \frac{1}{10}y\right), (x,y) \in \mathbb{R}^2.$$

(a) Show that T is not a contraction with respect to the metric d;

(b) Show that T is a $\dfrac{9}{10}$-contraction with respect to the metric δ.

1.14. Show that $C[a,b]$ and $B[a,b]$ defined in Example 1.3 are complete metric spaces. Are they equivalent metric spaces ?

1.15. Let $X = C[-1,1]$ and $T : X \to X$ be given by

$$Tx(t) = \min\{1, \max\{-1, x(t) + 2t\}\}, t \in [-1,1].$$

Show that T is nonexpansive but, due to the fact that T maps unit ball into its boundary and since either $Tx(t) > x(t)$ for some $t > 0$ or $Tx(t) < x(t)$ for some $t < 0$, T cannot have a fixed point.

1.16. Let C_0 be the space of real sequences convergent to 0.
(a) Show that $\|x\| = \sup_i |x_i|$, $x = (x_1, x_2, \ldots, x_n, \ldots)$, is a norm on C_0;
(b) Let $K = \{x \in C_0 : \|x\| \leq 1\}$ and define $T : K \to K$ by $Tx = (1, x_1, x_2, \ldots, x_n, \ldots)$. Show that T is nonexpansive and has no fixed points.

1.17. Show that $\mathbb{R}^2$ endowed with the Euclidean norm, i.e., that induced by the metric d from Example 1.3, 2), is uniformly convex and endowed with the norm induced by the metric δ in the same example, is not.

1.18. Prove individually each of the Lemmas 1.1-1.8.

1.19. For T given in Example 1.8, show that the Krasnoselskij iteration converges to the unique fixed point of T, for any $x_0 \in [0,1]$ and any $\lambda \in (0,1]$, though Picard iteration does not converges for any $x_0 \neq \dfrac{1}{2}$.

1.20. Show that if $G(x,s)$ is the Green function defined by equation (25), then:
(a) $0 \leq G(x,s) \leq \dfrac{b-a}{4}$, for all $x, s \in [a,b]$;
(b) $\int_b^a G(x,s)ds \leq \dfrac{(b-a)^2}{8}$, for all $x \in [a,b]$.

1.21. Show that the mapping $T : \left[\frac{1}{2}, 2\right] \to \left[\frac{1}{2}, 2\right]$, $Tx = \dfrac{1}{x}$, $x \in \left[\frac{1}{2}, 2\right]$, with the usual norm is not a strict contraction, but is pseudocontractive and Lipschitzian. Is T strongly pseudocontractive ?

2

The Picard Iteration

The main aim of this chapter is to present some basic convergence theorems regarding the Picard iteration for various contractive type mappings.

2.1 Banach's Fixed Point Theorem

The contraction mapping principle, whose short statement was given in Section 1.1 (Theorem 1.1) and usually called *theorem of Banach* or *theorem of Picard-Banach-Caccioppoli*, will be reformulated here in its complete form.

Theorem 2.1. *Let (X, d) be a complete metric space and $T : X \to X$ be an $a-$contraction, that is an operator satisfying*

$$d(Tx, Ty) \leq a\, d(x, y), \quad \text{for any} \ \ x, y \in X \tag{1}$$

with $a \in [0, 1)$ fixed. Then
(i) T has a unique fixed point, that is, $F_T = \{x^*\}$;
(ii)   The Picard iteration associated to T, i.e., the sequence $\{x_n\}_{n=0}^{\infty}$, defined by

$$x_n = T(x_{n-1}) = T^n(x_0), \quad n = 1, 2, \ldots, \tag{2}$$

converges to x^*, for any initial guess $x_0 \in X$;
(iii)   The following a priori and a posteriori error estimates hold:

$$d(x_n, x^*) \leq \frac{a^n}{1-a} \cdot d(x_0, x_1), \quad n = 0, 1, 2, \ldots \tag{3}$$

$$d(x_n, x^*) \leq \frac{a}{1-a} \cdot d(x_{n-1}, x_n), \quad n = 0, 1, 2, \ldots \tag{4}$$

(iv)   The rate of convergence is given by

$$d(x_n, x^*) \leq a \cdot d(x_{n-1}, x^*) \leq a^n \cdot d(x_0, x^*), \quad n = 1, 2, \ldots \tag{5}$$

Proof. There is at most one fixed point, i.e., $card\ F_T \leq 1$. Indeed, assuming $x^*, y^* \in F_T$, $x^* \neq y^*$, since $0 \leq a < 1$, we get the contradiction

$$d(x^*, y^*) = d(Tx^*, Ty^*) \leq a \cdot d(x^*, y^*) < d(x^*, y^*).$$

To prove the existence of the fixed point, we will show that, for any given $x_0 \in X$, the Picard iteration $\{x_n\}_{n=0}^{\infty}$ is a Cauchy sequence. Notice that, by (1), we have

$$d(x_2, x_1) = d(Tx_1, Tx_0) \leq a\, d(x_1, x_0),$$

and by induction,

$$d(x_{n+1}, x_n) \leq a^n\, d(x_1, x_0), \quad n = 0, 1, 2, \ldots \tag{6}$$

Thus, for any numbers $n, p \in \mathbb{N}$, $p > 0$, we have

$$d(x_{n+p}, x_n) \leq \sum_{k=n}^{n+p-1} d(x_{k+1}, x_k) \leq \sum_{k=n}^{n+p-1} a^k\, d(x_1, x_0) \leq \frac{a^n}{1-a} \cdot d(x_1, x_0). \tag{7}$$

Since $0 \leq a < 1$, it results that $a^n \to 0$ (as $n \to \infty$), which together with (7) shows that $\{x_n\}_{n=0}^{\infty}$ is a Cauchy sequence. But (X, d) is a complete metric space, therefore $\{x_n\}_{n=0}^{\infty}$ converges to some $x^* \in X$.

On the other hand, any Lipschitzian mapping is continuous. So denoting

$$\lim_{n \to \infty} x_n = x^*,$$

we find

$$x^* = \lim_{n \to \infty} x_{n+1} = \lim_{n \to \infty} T(x_n) = T\left(\lim_{n \to \infty} x_n\right) = Tx^*,$$

which gives $x^* = Tx^*$, i.e., x^* is a fixed point of T.

This shows that for any $x_0 \in X$, the Picard iteration converges in X and its limit is a fixed point of T. Since T has at most one fixed point, we deduce that, for every choice of $x_0 \in X$, the Picard iteration converges to the same value x^*, that is, the unique fixed point of T. So we proved (i) and (ii).

To prove (iii) we use (7),

$$d(x_{n+p}, x_n) \leq \frac{a^n}{1-a} \cdot d(x_0, x_1), \quad \text{for all } p \in \mathbb{N}^*,$$

and the continuity of the metric and so, by letting $p \to \infty$, we find

$$d(x_n, x^*) = d(x^*, x_n) = \lim_{p \to \infty} d(x_{n+p}, x_n) \leq \frac{a^n}{1-a} \cdot d(x_0, x_1), \quad n \geq 0$$

and so (3) is proved.

To obtain the *a posteriori* estimation (4), let us notice that by (1) we have

$$d(x_{n+1}, x_n) \leq a \, d(x_n, x_{n-1})$$

and, by induction,

$$d(x_{n+k}, x_{n+k-1}) \leq a^k \, d(x_n, x_{n-1}), \quad k \in \mathbb{N}^*,$$

so

$$d(x_{n+p}, x_n) \leq (a + a^2 + \ldots + a^p) \, d(x_n, x_{n-1}) \leq \frac{a}{1-a} \, d(x_n, x_{n-1}).$$

By letting $p \to \infty$ in the last inequality we get exactly (4).   □

Remarks.

1) The *a priori* estimate (3) shows that, when starting from an initial guess $x_0 \in X$, the approximation error of the nth iterate is completely determined by the contraction coefficient a and the initial displacement $d(x_1, x_0)$;

2) Similarly, the *a posteriori* estimate shows that, in order to obtain the desired error approximation of the fixed point by means of Picard iteration, that is, to have $d(x_n, x^*) < \epsilon$, we need to stop the iterative process at the first step n for which the displacement between two consecutive iterates is at most $(1 - a)\varepsilon/a$;

So, the *a posteriori* estimation offers a direct stopping criterion for the iterative approximation of fixed points by Picard iteration, while the *a priori* estimation indirectly gives a stopping criterion;

3) It is easy to see that the *a posteriori* estimation is better than the *a priori* one, in the sense that from (4) we can obtain (3), by means of (6);

4) Each of the three estimations given in Theorem 2.1 shows that the convergence of the Picard iteration is at least as quick as that of the geometric series $\sum a^n$. This explains why in Example 1.11 the iterative process defined by means of the iteration function T_2 (that is, the Picard iteration) is so quick (quicker than Newton's iteration). However, as shown by (5), the convergence rate of Picard iteration for any contraction is *linear*;

5) In most of the cases, the contraction condition (1) is not satisfied in the whole space X, but only locally. In this context, a local version of the contraction mapping principle is very useful for certain practical purposes.

Corollary 2.1. *Let (X, d) be a complete metric space and*

$$B(y_0, R) = \{x \in X \, | d(x, y_0) < R\}$$

be the open ball. Let $T : B(y_0, R) \to X$ be an a-contraction, such that

$$d(Ty_0, y_0) < (1 - a)R.$$

Then T has a fixed point that can be obtained using the Picard iterative scheme, starting from any $x_0 \in B(y_0, r)$.

Proof. We show that any closed ball $\overline{B} = \overline{B}(y_0, r)$, $r < R$, is an invariant set with respect to T, that is $T(\overline{B}) \subset \overline{B}$. To prove this, let us consider $x \in \overline{B}$. Then $d(x, y_0) \leq R$, and from

$$d(Tx, y_0) \leq d(Tx, Ty_0) + d(Ty_0, y_0) \leq a \cdot d(x, y_0) + (1 - a) \cdot R$$

we obtain

$$d(Tx, y_0) \leq a \cdot R + (1 - a) \cdot R = R,$$

which shows that $Tx \in \overline{B}$. Since $\overline{B}$ is complete, we can apply now Theorem 2.1 to get the conclusion.   □

Definition 2.1. Let (X, d) be a complete metric space. A mapping $T : X \rightarrow X$ is called (*strict*) *Picard mapping* if there exists $x^* \in X$ such that $F_T = \{x^*\}$ and

$$T^n(x_0) \rightarrow x^* \quad \text{(uniformly) for all } x_0 \in X.$$

Example 2.1. If (X, d) is a complete metric space, then any contraction $T : X \rightarrow X$ is a Picard mapping.

The next sections of this chapter will show some other important examples of Picard mappings.

2.2 Theorem of Nemytzki-Edelstein

By weakening the contraction condition to a contractive one, the conclusions of Theorem 2.1 are no longer valid, as the next example shows.

Example 2.2. If $X = [1, \infty)$ and $T : X \rightarrow X$, $T(x) = x + \dfrac{1}{x}$, then:

1) T is not a contraction;
2) T is contractive;
3) $F_T = \emptyset$;
4) The Picard iteration associated to T does not converge, for any $x_0 \in [1, \infty)$.

Indeed, if the Picard iteration $\{x_n\}_{n=0}^{\infty}$, $x_{n+1} = x_n + \dfrac{1}{x_n}$, $n \geq 0$ would be convergent, then its limit l would satisfy $\dfrac{1}{l} = 0$, which is impossible.

However, it is possible to impose some additional conditions on the ambient space, in order to ensure that a contractive mapping is a Picard operator, as the following theorem shows.

Theorem 2.2. *Let (X, d) be a compact metric space and $T : X \rightarrow X$ be a contractive operator. Then T is a strict Picard operator.*

Proof. Recall that a metric space is compact if and only if every family of closed subsets of X with finite intersection property (i.e., any finite number of sets in the family has a nonempty intersection) has a nonempty intersection. From the contractiveness of T we have that $card\ F_T \leq 1$.

Let $x_0 \in X$ and $\{x_n\}_{n=0}^{\infty}$, $x_n = T^n x_0$, $n \geq 0$, be the Picard iteration associated to T.

Since (X, d) is compact, it results that there exists a subsequence $\{x_{n_k}\}_{k=0}^{\infty}$ of $\{x_n\}_{n=0}^{\infty}$ such that $\{x_{n_k}\}_{k=0}^{\infty}$ converges to a certain $x^* \in X$ as n tends to ∞. As T is contractive, we deduce that T is continuous and that the sequence $\{\, d(x_n, x_{n+1})\}_{n=0}^{\infty}$ has strictly decreasing positive terms and hence is convergent.

Then, using the continuity of the metric, we have

$$\lim_{k \to \infty} d(x_{n_k}, T\, x_{n_k}) = d(x^*, Tx^*)$$

and therefore

$$d(x^*, Tx^*) = \lim_{n \to \infty} d(x_n, x_{n+1}) = \lim_{n \to \infty} d(x_{n+1}, x_{n+2}) = d(Tx^*, T^2 x^*).$$

If we admit $x^* \neq Tx^*$, then from the contractive condition we get the contradiction

$$d(x^*, Tx^*) = d(Tx^*, T(Tx^*)) < d(x^*, Tx^*).$$

Consequently, $x^* = Tx^*$, i.e., $F_T = \{x^*\}$.

This shows that for any $x_0 \in X$, the Picard iteration converges in X and its limit is the unique fixed point of T. $\qquad\square$

Corollary 2.2. *Let (X, d) be a complete metric space and $T : X \to X$ be a contractive operator. If there exists $x_0 \in X$ such that the Picard iteration $\{T^n x_0\}_{n=0}^{\infty}$ has a convergent subsequence, then $F_T = \{x^*\}$ and x^* is the limit of this subsequence.*

Example 2.3. Let $X = l^\infty := \{u \in l^2(\mathbb{R}) : |u_k| \leq 1/k\}$ and $T : l^\infty \to l^\infty$, defined by $T\, u_k = \dfrac{k}{k+1} \cdot u_k$. Then:

(i) l^∞ is a compact metric space;   (ii) T is not a contraction;
(iii) T is contractive;   (iv) $F_T = \{0\}$, the null sequence;
(v) The Picard iteration converges (uniformly) to the null sequence, i.e.,

$$T^n u_k^{(0)} = \left(\frac{k}{k+1} \right)^n u_k^{(0)} \to 0 \quad (\text{as } n \to \infty),$$

for any $u_k^{(0)} \in l^\infty$.

Remark. For a contractive operator, we generally have no information about the convergence rate of the Picard iteration.

2.3 Quasi-Nonexpansive Operators

In the previous two sections we have given examples of *continuous* Picard operators. The main aim of this section is to prove that a Picard operator needs not to be continuous.

Theorem 2.3. *Let (X, d) be a complete metric space and $T : X \to X$ be a mapping for which there exists $a \in \left[0, \dfrac{1}{2}\right)$ such that*

$$d(Tx, Ty) \le a[\, d(x, Tx) + d(y, Ty)\,], \quad \text{for all } x, y \in X. \tag{8}$$

Then T is a Picard operator.

Proof. First we remark that if T satisfies (8), then *card $F_T \le 1$*.

Let $x_0 \in X$, and $x_n = T^n x_0$, $n = 0, 1, 2, \ldots$ be the Picard iteration. Then by (8) we have

$$d(x_n, x_{n+1}) = d(Tx_{n-1}, Tx_n) \le a[d(x_{n-1}, x_n) + d(x_n, x_{n+1})],$$

which implies

$$d(x_n, x_{n+1}) \le \frac{a}{1-a} \cdot d(x_{n-1}, x_n), \quad n = 1, 2, \ldots \tag{9}$$

Since $0 \le \dfrac{a}{1-a} < 1$, for $a \in \left[0, \dfrac{1}{2}\right)$, we deduce, in a similar manner to that in the proof of Theorem 2.1, that $\{x_n\}_{n=0}^{\infty}$ is a Cauchy sequence, and hence a convergent sequence, too. Let $x^* \in X$ be its limit. Then we have

$$d(x^*, Tx^*) \le d(x^*, x_n) + d(x_n, Tx^*) \le d(x^*, x_n) + a[d(x^*, x_{n-1}) + d(x^*, Tx^*)],$$

and hence

$$d(x^*, Tx^*) \le \frac{1}{a} \cdot d(x^*, x_n) + \frac{a}{1-a} \cdot d(x_{n-1}, x_n), \forall n \in \mathbb{N}$$

which, together with (9), gives

$$d(x^*, Tx^*) \le \frac{1}{a} \cdot d(x^*, x_n) + \left(\frac{a}{1-a}\right)^n \cdot d(x_0, x_1), \quad n = 1, 2, \ldots \tag{10}$$

Now, letting $n \to \infty$ in (10), we obtain

$$d(x^*, Tx^*) = 0 \iff x^* = Tx^*, \text{ that is, } F_T = \{x^*\}$$

and therefore,

$$x_n \to x^* (n \to \infty), \text{ for each } x_0 \in X. \qquad \square$$

Example 2.4. Let $X = \mathbb{R}$ and $T : X \to X$, $T(x) = 0$, if $x \in (-\infty, 2]$ and $Tx = -\dfrac{1}{2}$, if $x > 2$. Then: (i) T is not continuous; (ii) T fulfills (8) (with $a = \dfrac{1}{5}$) and hence, by Theorem 2.3, T is a Picard mapping; (iii) T is not nonexpansive (to show this, take $x = 2$ and $y = 9/4$).

Corollary 2.3. *Let the assumptions in Theorem 2.3 be satisfied. Then the error estimates of the Picard iteration are given by*

$$d(x_n, x^*) \leq \frac{\alpha^n}{1-\alpha} \cdot d(x_0, x_1), \quad n = 0, 1, 2, \dots \tag{11}$$

$$d(x_n, x^*) \leq \frac{\alpha}{1-\alpha} \cdot d(x_n, x_{n-1}), \quad n = 0, 1, 2, \dots , \tag{12}$$

where $\alpha = \dfrac{a}{1-a}$.

Remarks.

1) If there exists $k \in \mathbb{N}^*$ such that T^k is a contraction (or is contractive, or satisfies (8)), then $F_T = \{x^*\}$.

The class of contractive operators is included in the class of nonexpansive operators. For a nonexpansive operator T, however, the conclusion $F_T \neq \emptyset$ is not generally true. A generalization of a nonexpansive operator, with at least one fixed point, is that of the quasi nonexpansive operators.

An operator $T : X \to X$ is said to be *quasi nonexpansive* if T has at least one fixed point in X and, for each fixed point p, we have

$$d(Tx, p) \leq d(x, p), \quad \forall x \in X. \tag{*}$$

The class of quasi-nonexpansive operators is strongly connected to the Newton's iterative method. Other examples of quasi-nonexpansive operators can be found in the class of generalized φ-contractions.

2) A contractive definition which is included in the class of quasi-nonexpansive mappings was obtained by Zamfirescu in 1972. Zamfirescu's theorem is a generalization of Banach's, Kannan's and Chatterjea's fixed point theorems.

Theorem 2.4. *Let (X, d) be a complete metric space and $T : X \to X$ be a mapping for which there exist the real numbers α, β and γ satisfying $0 \leq \alpha < 1$, $0 \leq \beta < 0.5$ and $0 \leq \gamma < 0.5$, such that, for each $x, y \in X$, at least one of the following is true:*
(z_1) $d(Tx, Ty) \leq \alpha\, d(x, y)$;
(z_2) $d(Tx, Ty) \leq \beta[d(x, Tx) + d(y, Ty)]$;
(z_3) $d(Tx, Ty) \leq \gamma[d(x, Ty) + d(y, Tx)]$.
Then T is a Picard operator.

Proof. We first fix $x, y \in X$. At least one of (z_1), (z_2) or (z_3) is true. If (z_2) holds, then we have

$$d(Tx, Ty) \leq \beta[d(x, Tx) + d(y, Ty)] \leq$$
$$\leq \beta\{d(x, Tx) + [d(y, x) + d(x, Tx) + d(Tx, Ty)]\}.$$

So

$$(1 - \beta)\, d(Tx, Ty) \leq 2\beta\, d(x, Tx) + \beta\, d(x, y),$$

which yields

$$d(Tx, Ty) \leq \frac{2\beta}{1 - \beta}\, d(x, Tx) + \frac{\beta}{1 - \beta}\, d(x, y). \tag{13}$$

If (z_3) holds, then similarly we get

$$d(Tx, Ty) \leq \frac{2\gamma}{1 - \gamma}\, d(x, Tx) + \frac{\gamma}{1 - \gamma}\, d(x, y). \tag{14}$$

Therefore, denoting

$$\delta = \max\left\{ \alpha, \frac{\beta}{1 - \beta}, \frac{\gamma}{1 - \gamma} \right\},$$

we have $0 \leq \delta < 1$ and then, for all $x, y \in X$, the following inequality

$$d(Tx, Ty) \leq 2\delta \cdot d(x, Tx) + \delta \cdot d(x, y) \tag{15}$$

holds. In a similar manner we obtain

$$d(Tx, Ty) \leq 2\delta \cdot d(x, Ty) + \delta \cdot d(x, y), \tag{16}$$

valid for all $x, y \in X$.

From (15) it follows that $card\ F_T \leq 1$. We will show that T has a (unique) fixed point. Let $x_0 \in X$ be arbitrary and $\{x_n\}_{n=0}^{\infty}$,

$$x_n = T^n x_0, \quad n = 0, 1, 2, \ldots$$

be the Picard iteration associated to T.

If $x := x_n$, $y := x_{n-1}$ are two successive approximations, then by (16) we have

$$d(x_{n+1}, x_n) \leq \delta \cdot d(x_n, x_{n-1}).$$

From this we deduce that $\{x_n\}_{n=0}^{\infty}$ is a Cauchy sequence, and hence a convergent sequence, too. Let $x^* \in X$ be its limit. In particular we have

$$\lim_{n \to \infty} d(x_{n+1}, x_n) = 0.$$

By triangle rule and (15) we get

$$d(x^*, Tx^*) \leq d(x^*, x_{n+1}) + d(Tx_n, Tx^*) \leq$$

$$\leq d(x^*, x_{n+1}) + \delta\, d(x^*, x_n) + 2\,\delta d(x_n, Tx_n),$$

which, by letting $n \to \infty$, yields

$$d(x^*, Tx^*) = 0 \iff x^* = Tx^*,$$

since $d(x_n, Tx_n) = d(x_n, x_{n+1}) \to 0$, and therefore

$$F_T = \{x^*\} \text{ and } x_n \to x^*(n \to \infty),$$

for each $x_0 \in X$.   □

Remarks.

1) The error estimate of the Picard iteration associated to a Zamfirescu
mapping is given by the same estimates (11) and (12) in the case of a Kannan
mapping, but with α replaced by

$$\delta = \max\left\{ \alpha, \frac{\beta}{1-\beta}, \frac{\gamma}{1-\gamma} \right\};$$

2) A generalization of Zamfirescu's contractiveness definition was obtained
by Ciric in 1974. It will be treated in a unified manner in Section 2.6.

Example 2.5. If T is a Kannan (or Zamfirescu) mapping, then T is a
(strictly) quasi nonexpansive operator.

Indeed, if T is a Kannan operator, then from (8) with $y = p \in F_T$ we get

$$d(Tx, p) \le a\, d(x, Tx) \le a\,[d(x, p) + d(p, Tx)]$$

and hence

$$d(Tx, p) \le \frac{a}{1-a}\, d(x, p) < d(x, p).$$

For a Zamfirescu operator, we put $x := p$ and $y := x$ in (15) and obtain

$$d(Tx, p) \le \delta\, d(x, p) < d(x, p).$$

2.4 Maia's Fixed Point Theorem

Definition 2.2. Let (X, d) be a nonempty set. A map $T : X \to X$ is said
to be a *Bessaga mapping* if there exists $x^* \in X$ such that

$$F_{T^n} = \{x^*\}, \quad \text{for all } n \in \mathbb{N}. \tag{17}$$

Example 2.6. It is easy to check that any Picard mapping is a Bessaga
mapping but the reverse is not true. This shows that any mapping satisfying
one of the Theorems 2.1-2.4 is a Bessaga mapping. On the other hand, if T
is a Bessaga mapping on the set X, then X can be organized as a complete
metric space, such that T should be a contraction on X.

Theorem 2.5. *Let X be a nonempty set, $T : X \to X$ a mapping satisfying (17) and $a \in (0,1)$ a given number. Then there exists a metric d on X such that*
  (a) (X,d) *is a complete metric space;*
  (b) T *is an a-contraction with respect to d.*

By combining Theorem 2.5 and Example 2.6 it results that, for any mapping T satisfying one of the contractive conditions in Theorem of Kannan or Zamfirescu (and many other similar conditions), it may be possible to find another complete metric on X with respect to which the operator T is a contraction.

Example 2.7. The linear map

$$T : \mathbb{R}^2 \to \mathbb{R}^2, \quad T(x,y) = \left(\frac{8x + 8y}{10}, \frac{x+y}{10} \right)$$

is not a contraction with respect to the Euclidean metric, but is a $\frac{9}{10}$-contraction with respect to the (equivalent) metric δ defined in Example 1.3, 2).

However, for a certain Bessaga mapping, it is practically not an easy task to construct this equivalent and complete metric. An alternative to this attempt is to transfer a part of the assumptions from the metric d to a second metric ρ, as shown by the Maia's fixed point theorem.

Theorem 2.6. *Let X be a nonempty set, d and ρ two metrics on X and $T : X \to X$ a mapping. Assume that*
  (i) $d(x,y) \le \rho(x,y)$, *for all $x,y \in X$;*
  (ii) (X,d) *is a complete metric space;*
  (iii) $T : (X,d) \to (X,d)$ *is continuous;*
  (iv) $T : (X,\rho) \to (X,\rho)$ *is an a-contraction with $a \in [0,1)$.*
  Then T is a Picard mapping.

Proof. Let $x_0 \in X$ be arbitrary and $\{x_n\}_{n=0}^{\infty}$, $x_n = T^n x_0$, $n = 0, 1, 2, \ldots$, be the Picard iteration associated to T.
  From (iv), using the same arguments as in the proof of Theorem 2.1, we deduce that $\{x_n\}_{n=0}^{\infty}$ is a Cauchy sequence in (X, ρ).
  By (i), it results that $\{x_n\}_{n=0}^{\infty}$ is a Cauchy sequence in (X,d) as well, and by (ii), we deduce that it converges to a certain x^* in X.
  Now, by (iii), $x^* \in F_T$ and, by (iv), $F_T = \{x^*\}$. $\square$

Remarks.

1) Assumption *(i)* in Theorem 2.6 may be weakened to

  (i') *There exists $c > 0$ such that $d(x,y) \le c \cdot \rho(x,y)$, for all $x,y \in X$,*

or to

(i'') *There exists $c > 0$ such that $d(Tx, Ty) \leq c \cdot \rho(x, y)$, for all $x, y \in X$,*

which is particularly useful when dealing with integral equations;

2) Condition *(iv)* in Theorem 2.6 may be replaced by one of the following conditions: "$T : (X, \rho) \to (X, \rho)$ is a Kannan mapping" or "$T : (X, \rho) \to (X, \rho)$ is a Zamfirescu mapping" or "$T : (X, \rho) \to (X, \rho)$ is a φ−contraction", see the next Section 2.5 etc.

2.5 φ-Contractions

Let $\varphi : \mathbb{R}_+ \to \mathbb{R}_+$ be a function. In connection with the function φ we consider the following properties:

(i_φ) φ is monotone increasing, i.e., $t_1 \leq t_2$ implies $\varphi(t_1) \leq \varphi(t_2)$;
(ii_φ) $\varphi(t) < t$ for all $t > 0$;
(iii_φ) $\varphi(0) = 0$;
(iv_φ) φ is continuous;
(v_φ) $\{\varphi^n(t)\}$ converges to 0 for all $t \geq 0$;
(vi_φ) $\sum\limits_{n=0}^{\infty} \varphi^n(t)$ converges for all $t > 0$;
(vii_φ) $t - \varphi(t) \to \infty$ as $t \to \infty$;
($viii_\varphi$) φ is subadditive.

The next lemma shows some relationships existing between the above conditions.

Lemma 2.1.
1) (i_φ) and (ii_φ) imply (iii_φ);
2) (ii_φ) and (iv_φ) imply (iii_φ);
3) (i_φ) and (v_φ) imply (ii_φ).

Definition 2.3. 1) A function φ satisfying (i_φ) and (v_φ) is said to be a *comparison function*;
2) A function φ satisfying (i_φ) and (vi_φ) is said to be a *(c)-comparison function*;
3) A comparison function satisfying (vii_φ) is called *strict comparison function*.

Lemma 2.2.
1) Any (c)-comparison function is a comparison function ;
2) Any strict comparison function is a comparison function;
3) Any comparison function satisfies (iii_φ);
4) Any comparison function satisfying ($viii_\varphi$) satisfies (iv_φ), too;
5) If φ is a comparison function, then, for any $k \in \mathbb{N}^*$, φ^k is a comparison function, too;
6) If φ is a (c)-comparison function, then the function

$$s : \mathbb{R}_+ \to \mathbb{R}_+, \quad s(t) = \sum_{k=0}^{\infty} \varphi^k(t), \quad t \in \mathbb{R}_+ \tag{18}$$

satisfies (i_φ) and (iii_φ).

Example 2.8.
1) $\varphi(t) = at$, $t \in \mathbb{R}_+$, $a \in [0,1)$ satisfies all the conditions (i_φ)-$(viii_\varphi)$;
2) $\varphi(t) = \dfrac{t}{1+t}$, $t \in \mathbb{R}_+$ is a (strict) comparison function but not a (c)-comparison function;
3) $\varphi(t) = \dfrac{1}{2}t$, if $0 \le t \le 1$ and $\varphi(t) = t - \dfrac{1}{2}$, if $t > 1$ is a (c)-comparison function but it is not a strict comparison function.

Definition 2.3. Let (X,d) be a metric space. A mapping $T : X \to X$ is said to be a φ-*contraction* if there exists a comparison function $\varphi : \mathbb{R}_+ \to \mathbb{R}_+$ such that

$$d(Tx, Ty) \le \varphi(d(x,y)), \quad \text{for all } x, y \in X. \tag{19}$$

Theorem 2.7. *Let (X,d) be a complete metric space and $T : X \to X$ a φ-contraction. Then T is a Picard mapping.*

Proof. Let $x_0 \in X$ and let $\{x_n\}_{n=0}^{\infty}$, $x_n = Tx_{n-1} = T^n x_0$, $n = 1, 2, \ldots$, be the Picard iteration associated to T. Then

$$d(x_n, x_{n+1}) \le \varphi^n(d(x_0, x_1))$$

and by (v_φ), we obtain that $d(x_n, x_{n+1}) \to 0$ as $n \to \infty$, that is,

$$d(T^n x_0, T^{n+1} x_0) \to 0, \text{ as } n \to \infty, \tag{20}$$

which means that x_0 is *asymptotically regular under T.*
   In fact, any $x_0 \in X$ is asymptotically regular under T, which means that T is asymptotically regular.
   We show now that $\overline{B}(x; \varepsilon)$, with $\varepsilon > 0$, is an invariant set with respect to T. Indeed, for $\varepsilon > 0$, let $\delta(\varepsilon) = \varepsilon - \varphi(\varepsilon)$ and $y \in \overline{B}(x; \varepsilon)$. Then

$$d(Ty, x) \le d(Ty, Tx) + d(Tx, x) \le \varphi(d(y,x)) + d(x, Tx) \le \varphi(\varepsilon) + d(x, Tx).$$

Hence

$$d(x, Tx) < \delta(\varepsilon) \implies d(Ty, x) \le \varphi(\varepsilon) + \varepsilon - \varphi(\varepsilon) = \varepsilon,$$

which shows that $Ty \in \overline{B}(x, \varepsilon)$, that is, $\overline{B}(x, \varepsilon)$ is invariant with respect to T.
   By (19), $\{T^n x_0\}_{n \in \mathbb{N}}$ is a Cauchy sequence for any $x_0 \in X$. For any given $\varepsilon > 0$, there exists $n_0 \in \mathbb{N}$ such that

$$d(T^n x_0, T^{n+1} x_0) < \delta(\varepsilon), \quad \text{for all } n \ge n_0$$

and this implies that $T^n x_0 \in \overline{B}(T^n x_0; \varepsilon)$, for all $n \ge n_0$.

As (X, d) is a complete metric space, $\{T^n x_0\}_{n \in \mathbb{N}}$ is convergent.

Let $x^* = \lim\limits_{n \to \infty} T^n(x_0)$. Since any comparison function satisfies (ii$_\varphi$), any φ-contraction is continuous. Hence

$$x^* = T \left(\lim_{n \to \infty} T x_{n-1} \right) = T x^*,$$

which shows that $x^* \in F_T$.

Assume there exists $y^* \in F_T$, $y^* \neq x^*$. Then $d(x^*, y^*) \neq 0$ and the condition of φ-contractiveness implies

$$0 < d(x^*, y^*) = d(T x^*, T y^*) \leq \varphi(d(x^*, y^*)) < d(x^*, y^*),$$

which is a contradiction.   □

Corollary 2.4. *Let (X, d) be a complete metric space and $T : X \to X$ be a mapping with the property that there exists $k \in \mathbb{N}^*$ such that T^k is a φ-contraction. Then $F_T = \{x^*\}$.*

Remarks.

1) The metrical fixed point theory is very rich in fixed point theorems given for various classes of φ-contractions, which are obtained for different collections of properties of the comparison function φ;

2) As Theorem 2.7 illustrates, almost all of them prove only the convergence of the Picard iteration to the unique fixed point of T. Only a few of these fixed point theorems are able to provide information on the convergence rate of the Picard iteration;

3) As we have shown, condition (vi$_\varphi$) is equivalent to the following one:

(c) There exist k_0 and α, $0 < \alpha < 1$, and a convergent series of nonnegative terms $\sum v_n$, such that

$$\varphi^{\kappa+1}(t) \leq \alpha \cdot \varphi^k(t) + v_k \qquad (21)$$

holds for all $k \geq k_0$ and $t \in \mathbb{R}_+$.

Condition (21) is in fact the generalized ratio test for series of positive terms which, for the particular case of series of *decreasing positive terms*, gives a necessary and sufficient condition of convergence, since any comparison series $\sum \varphi^k(t)$ consists of decreasing positive terms, see Berinde [Be97a].

The next theorem transposes all the conclusions in Banach's contraction mapping principle (Theorem 2.1) to a class of φ−contractions.

Theorem 2.8. *Let (X, d) be a complete metric space and $T : X \to X$ be a φ−contraction with φ a (c)-comparison function. Then*

(i) $F_T = \{x^*\}$;

(ii) The Picard iteration $\{x_n\} = \{T^n x_0\}_{n \in \mathbb{N}}$ converges to x^* (as $n \to \infty$), for each $x_0 \in X$;

(iii) The following estimation holds

$$d(x_n, x^*) \leq s(d(x_n, x_{n+1})), \quad n = 0, 1, 2, \ldots, \tag{22}$$

where $s(t) = \sum\limits_{k=0}^{\infty} \varphi^k(t)$ is the sum of the comparison series.

Proof. By Theorem 2.7 we get (i) and (ii).

Let $x_n = T^n x_0$, $n = 0, 1, 2, \ldots$ be the Picard iteration associated to T. In order to prove (iii), we use the φ-contractiveness condition and get

$$d(x_{n+k}, x_{n+k+1}) \leq \varphi^k(d(x_n, x_{n+1})), \quad n = 0, 1, 2, \ldots, \; k \geq 1.$$

So

$$d(x_{n+p}, x_n) \leq \sum_{k=0}^{p-1} \varphi^k(d(x_n, x_{n+1}))$$

and, letting $p \to \infty$, we obtain the estimate (22). $\qquad\square$

Remarks.

1) For $\varphi(t) = a\,t$, $0 \leq a < 1$, by Theorem 2.8 we obtain Theorem 2.1. The *a posteriori* estimate in Theorem 2.1 can be obtained directly by (22), while the *a priori* estimate is obtained by means of the inequality

$$d(x_n, x_{n+1}) \leq \varphi^n(d(x_0, x_1));$$

2) A result similar to Theorem 2.8 may be obtained for the class of φ-contractions with φ a strict comparison function.

In this case, the error estimate for the Picard iteration is given by

$$d(x_n, x^*) \leq \varphi^n(t_{x_0}), \quad n = 0, 1, 2, \ldots,$$

where

$$t_{x_0} := \sup\left\{ t \in \mathbb{R}_+ \,|\, t - \varphi(t) \leq d(x_0, x_1) \right\},$$

see Rus [Rus83];

3) We end this section by stating a fixed point theorem of Maia type, whose proof requires only standard arguments.

Theorem 2.9. *Let X be a nonempty set, d and ρ two metrics on X and $T : X \to X$ a mapping. Suppose that:*
(i) there exists $c > 0$ such that

$$d(Tx, Ty) \leq c\,\rho(x, y), \quad \text{for all } x, y \in X;$$

(ii) (X, d) is a complete metric space;
(iii) $T : (X, d) \to (X, d)$ is continuous;
(iv) $T : (X, \rho) \to (X, \rho)$ is a φ-contraction.
Then $T : (X, d) \to (X, d)$ is a Picard mapping.

Example 2.9.

1) If φ is right continuous and satisfies (i$_\varphi$) and (ii$_\varphi$), then from Theorem 2.8 we obtain the fixed point theorem of Browder;

2) If φ is upper semicontinuous and satisfies (ii$_\varphi$), then from Theorem 2.8 we obtain the fixed point theorem of Boyd-Wong;

3) If φ satisfies (ii$_\varphi$) and (iv$_\varphi$), then by Theorem 2.8 we obtain as a particular case the fixed point theorem of Krasnoselskij-Stechenko.

2.6 Generalized φ-Contractions

Many interesting generalizations of the contraction mapping principle have been obtained by considering contraction conditions which involve not only the distance $d(x,y)$ on the right-hand side, but also the displacements of x and y under the mapping T: $d(x,Tx)$, $d(x,Ty)$, $d(y,Tx)$ and $d(y,Ty)$.

Typical fixed point theorems in this class are Kannan's, Zamfirescu's and Ciric's fixed point theorems. The main aim of this section is to unify all these results in a single theorem, by using the concepts of multivariable comparison function and generalized φ-contraction.

Definition 2.4. A map $\varphi : \mathbb{R}_+^5 \to \mathbb{R}_+$ is called *(5-dimensional) comparison function (strict comparison function, (c)-comparison function)* if $\varphi(u) \leq \varphi(v)$, for any $u,v \in \mathbb{R}_+^5$, $u \leq v$ and

$$\psi : \mathbb{R}_+ \to \mathbb{R}_+ , \quad \psi(t) = \varphi(t,t,t,t,t), \ t \in \mathbb{R}_+ \qquad (23)$$

satisfies (v$_\varphi$) (and (vii$_\varphi$), respectively, (vi$_\varphi$)).

Example 2.10. The following functions $\varphi : \mathbb{R}_+^5 \to \mathbb{R}_+$ are 5-dimensional comparison functions:

1) $\varphi(t) = a \cdot \max\{t_1, t_2, t_3, t_4, t_5\}$, for each $t = (t_1, t_2, \ldots, t_5) \in \mathbb{R}_+^5$, where $a \in [0,1)$ is a constant;

2) $\varphi(t) = a \cdot \max\left\{t_1, t_2, t_3, t_4, \dfrac{t_4 + t_5}{2}\right\}$, $a \in [0,1)$;

3) $\varphi(t) = a(t_2 + t_3)$, $a \in [0, 1/2)$;

4) $\varphi(t) = at_1 + b(t_2 + t_3)$, $a, b \in \mathbb{R}_+$ such that $a + 2b < 1$;

5) $\varphi(t) = a \cdot \max\{t_2, t_3\}$, $a \in (0,1)$;

6) $\varphi(t) = \left(\sum\limits_{i=1}^{5} a_i t_i^p\right)^{1/p}$, where $a_i \in \mathbb{R}_+$ such that $\sum\limits_{i=1}^{5} a_i < 1$ and $p \geq 1$;

7) $\varphi(t) = \max\{at_1, b(t_2 + t_4), c(t_3 + t_5)\}$, where $a \in [0,1)$, $b, c \in [0, 1/2)$.

Definition 2.5. Let (X, d) be a metric space. A mapping $T : X \to X$ is called *generalized φ-contraction* if there exists a 5-dimensional comparison function $\varphi : \mathbb{R}^5 \to \mathbb{R}_+$ such that

$$d(Tx, Ty) \leq \varphi(d(x,y), d(x, Tx), d(y, Ty), d(x, Ty), d(y, Tx)), \qquad (24)$$

for all $x, y \in X$.

Lemma 2.3. *Let (X, d) be a metric space and $T : X \to X$ be a generalized φ-contraction. Then, for all $x_0 \in X$ and all $i, j \in \{1, 2, \ldots n\}$ we have*

$$d(T^i x_0, T^j x_0) \leq \psi \left(\delta \left(O_T \left(x_0; n \right) \right) \right).$$

Proof. Let us denote as usually $x_n = T^n x_0$, $n = 0, 1, 2, \ldots$.
Since for each $i, j \in \{1, 2, \ldots, n\}$ we have

$$\{i-1, j-1, i, j\} \subset \{0, 1, 2, \ldots n\},$$

we deduce that

$$x_{i-1}, x_i, x_{j-1}, x_j \in O_T(x_0; n) = \{x_0, Tx_0, \ldots, T^n x_0\}.$$

Hence, from the generalized contraction condition, we obtain

$$d(x_p, x_q) \leq \delta \left(O_T(x_0; n) \right) \text{ for each } p, q \in \{i-1, j-1, i, j\},$$

where $\delta \left(O_T(x_0; n) \right)$ denotes the diameter of $O_T(x_0; n)$. Then

$$\begin{aligned} d(x_i, x_j) = d(Tx_{i-1}, Tx_{j-1}) &\leq \\ &\leq \varphi(d(x_{i-1}, x_{j-1}), d(x_{i-1}, x_j), d(x_{j-1}, x_j), d(x_{i-1}, x_j), d(x_{j-1}, x_i)) \leq \\ &\leq \psi \left(\delta \left(O_T \left(x_0; n \right) \right) \right), \end{aligned}$$

due to the monotonicity of φ. $\square$

Remark. For each $n \in \mathbb{N}^*$, there exists $k \leq n$ such that

$$d(x_0, T^k x_0) = \delta \left(O_T(x_0; n) \right),$$

since $\psi(r) = \varphi(t, t, t, t, t) \leq t$, for all $t \geq 0$.

Lemma 2.4. *If $T : X \to X$ is a generalized φ-contraction with respect to a comparison function φ for which the function $h : \mathbb{R}_+ \to \mathbb{R}_+$,*

$$h(t) = t - \varphi(t, t, t, t, t), \quad t \in \mathbb{R}_+, \qquad (25)$$

is an increasing bijection, then, for any $n \in \mathbb{N}$, we have

$$\delta \left(O_T \left(x_0; n \right) \right) \leq h^{-1}(d(x_0, Tx_0)), \ \forall \ x_0 \in X.$$

Proof. Let $n \in \mathbb{N}^*$ be arbitrarily taken. The previous remark suggests that there exists $k \leq n$ such that

$$d(x_0, T^k x_0) = \delta \left(O_T \left(x_0; n \right) \right),$$

and hence, by applying Lemma 2.3, we obtain

$$\delta(O_T(x_0; n)) = d(x_0, T^k x_0) \leq d(x_0 T x_0) + d(T x_0, T^k x_0) \leq$$
$$\leq d(x_0, T x_0) + \psi(\delta(O_T(x_0; n))),$$

which leads to

$$\delta(O_T(x_0; n)) - \psi(\delta(O_T(x_0; n))) \leq d(x_0, T x_0), \quad x_0 \in X, \ n \in \mathbb{N}.$$

But h is bijective and monotone increasing, hence h^{-1} is increasing, too, and the conclusion follows from the last inequality. □

The main result of this section is given by the following theorem.

Theorem 2.10. *Let (X, d) be a complete metric space and $T : X \to X$ be a φ-contraction with φ such that the function ψ given by (23) is continuous and the function h given by (25) is an increasing bijection. Then*
  (i) T is a Picard mapping (let $F_T = \{x^*\}$);
  (ii) the following estimate

$$d(T^n x_0, x^*) \leq \psi^n(h^{-1}(d(x_0, T x_0))), \quad n = 0, 1, 2 \ldots,$$

holds, for all $x_0 \in X$.

Proof. Let $x_0 \in X$, $m, n \in \mathbb{N}$, $m > n$. Put

$$i = 1, \ j = m - n + 1, \ x = T^{n-1} x_0 = x_{n-1}$$

and apply Lemma 2.3. It results

$$d(x_n, x_m) = d(T x_{n-1}, T x_{m-1}) \leq \psi(r_1), \tag{26}$$

where
$$r_1 = \delta(O_T(x_{n-1}; m - n + 1)).$$

Now, by the Remark before Lemma 2.4, there exists k_1, $1 \leq k_1 \leq m - n + 1$, such that
$$\delta(O_T(x_{n-1}; m - n + 1)) = d(x_{n-1}, T^{k_1} x_{n-1}). \tag{27}$$

Using again Lemma 2.3 we have

$$d(x_{n-1}, T^{k_1} x_{n-1}) = d(T x_{n-2}, T^{k_1+1} x_{n-2}) \leq \psi(r_2), \tag{28}$$

where
$$r_2 = \delta(O_T(x_{n-2}; k_1 + 1)).$$

Since ψ is monotone increasing and $k_1 + 1 \leq m - n + 2$, from (26)-(28) we obtain
$$d(x_n, x_m) \leq \psi^2(\delta(O_T(x_{n-2}; m - n + 2))),$$

and, inductively,

$$d(x_n, x_m) \leq \psi^n \left(\delta \left(O_T(x_0; m) \right) \right).$$

Now, using Lemma 2.4, it results

$$d(x_n, x_m) \leq \psi^n(r_3), \qquad (29)$$

where

$$r_3 = h^{-1}(d(x_0, x_1)).$$

As φ is a comparison function, that is

$$\psi^n(r) \rightarrow 0 \ (n \rightarrow \infty), \text{ for each } r \in \mathbb{R}_+,$$

from (29) we deduce that $\{x_n\}_{n=0}^{\infty}$ is a Cauchy sequence and hence it is convergent. Let $x^* = \lim_{n \rightarrow \infty} x_n$. We will show that $x^* \in F_T$. Indeed, for each $n \in \mathbb{N}$,

$$d(x^*, Tx^*) \leq d(x^*, x_{n+1}) + d(Tx_n, Tx^*) \leq d(x^*, x_{n+1}) +$$

$$+\varphi(d(x_n, x^*), d(x_n, x_{n+1}), d(x^*, Tx^*), d(x_n, Tx^*), d(x_{n+1}, x^*)). \qquad (30)$$

Assume first that

$$\max\{d(x_n, x^*), d(x_n, x_{n+1}), d(x^*, Tx^*), d(x_n, Tx^*), d(x_{n+1}, x^*)\} = d(x^*, Tx^*).$$

Then, using the monotonicity of φ, from (30) we obtain

$$d(x^*, Tx^*) \leq d(x^*, x_{n+1}) + \psi(d(x^*, Tx^*)),$$

which is equivalent to

$$d(x^*, Tx^*) \leq h^{-1}(d(x^*, x_{n-1})). \qquad (31)$$

Since h^{-1} is monotone increasing, positive and $h^{-1}(0) = 0$, it results that h^{-1} is continuous at zero. Letting $n \rightarrow \infty$ in (31), we get

$$d(x^*, Tx^*) = 0,$$

which means $x^* \in F_T$. Now, if

$$\max\{d(x_n, x^*), d(x_n, x_{n+1}), d(x^*, Tx^*), d(x_n Tx^*), d(x_{n+1}, x^*)\} = d(x_n, x^*),$$

then, by (30), we obtain

$$d(x^*, Tx^*) \leq d(x_{n+1}, x^*) + \psi(d(x_n, x^*)),$$

which, in view of the continuity of ψ at 0, and by letting $n \rightarrow \infty$, yields

$$d(x^*, Tx^*) \leq 0$$

that is, again, $d(x^*, Tx^*) = 0$.

If the maximum takes one of the values $d(x_{n+1}, x^*)$, $d(x_n, x_{n+1})$ or $d(x_n, Tx^*)$, the proof is similar to the previous cases.

Let us discuss the last possibility, i.e.,

$$\max\{d(x_n, x^*), d(x_n, x_{n+1}), d(x^*, Tx^*), d(x_n, Tx^*), d(x_{n+1}, x^*)\} = d(x_n, Tx^*).$$

Then, by (30), it results

$$d(x^*, Tx^*) \leq d(x_{n+1}, x^*) + \psi(d(x_n, Tx^*)).$$

Letting $n \to \infty$ in the previous inequality and using the continuity of ψ, we obtain

$$d(x^*, Tx^*) - \psi(d(x^*, Tx^*)) \leq 0,$$

that is,

$$h^{-1}(d(x^*, Tx^*)) \leq 0$$

which leads to

$$h^{-1}(d(x^*, Tx^*)) = 0 \iff d(x^*, Tx^*) = 0.$$

In order to prove the uniqueness of the fixed point we proceed as follows. Let $x^*, y^* \in F_T$, $x^* \neq y^*$. Then $d(x^*, y^*) > 0$ and

$$d(x^*, y^*) = d(T^n x^*, T^n y^*) \leq \psi^n(\delta(O_T(x^*; m))) = \psi^n(\delta(\{x^*\})) = \psi^n(0) = 0,$$

a contradiction. Now (i) is proved.

In order to obtain the estimate (ii), we take $m \to \infty$ in (29). $\square$

Particular cases.

1) For φ as in Example 2.10., part 1), from Theorem 2.10 we obtain the Ciric's fixed point theorem [Cir74];

2) For φ as in Example 2.10., 3), from Theorem 2.10 we obtain Kannan's fixed point theorem, i.e., Theorem 2.3;

3) For φ as in Example 2.10., 4), from Theorem 2.10 we get a fixed point theorem obtained by Reich (1971) and Rus (1971), see Taskovic [Tas86];

4) For φ as in Example 2.10., 5), from Theorem 2.10 we obtain a fixed point theorem given by Bianchini (1972) and Dugundji (1976), see Rus [Ru79c];

5) For φ as in Example 2.10., 7), from Theorem 2.10 we obtain the very interesting Zamfirescu's fixed point theorem, i.e., Theorem 2.4 in this Chapter;

6) By considering other particular expressions for φ, we may find many other interesting fixed point theorems.

2.7 Weak Contractions

Definition 2.5. Let (X, d) be a metric space. A map $T : X \to X$ is called *weak contraction* if there exist a constant $\delta \in (0, 1)$ and some $L \geq 0$ such that

$$d(Tx, Ty) \leq \delta \cdot d(x, y) + Ld(y, Tx), \quad \text{for all } x, y \in X. \tag{32}$$

Remark. Due to the symmetry of the distance, the weak contractive condition (32) implicitly includes the following dual one

$$d(Tx, Ty) \leq \delta \cdot d(x, y) + L \cdot d(x, Ty), \quad \text{for all } x, y \in X, \tag{33}$$

obtained from (32) by formally replacing $d(Tx, Ty)$ and $d(x, y)$ by $d(Ty, Tx)$ and $d(y, x)$, respectively, and then interchanging x and y.

Consequently, in order to check the weak contractiveness of T, it is necessary to check both (32) and (33);

Obviously, any strict contraction satisfies (32), with $\delta = a$ and $L = 0$, and hence is a weak contraction (that possesses a unique fixed point).

Other examples of weak contractions are given by the next propositions.

Proposition 2.2. *Any Kannan mapping, i.e., any mapping satisfying the contractive condition (8) in Theorem 2.3, is a weak contraction.*

Proof. By condition (8) and triangle rule, we get

$$d(Tx, Ty) \leq b[d(x, Tx) + d(y, Ty)] \leq$$
$$\leq b\Big\{ [d(x, y) + d(y, Tx)] + [d(y, Tx) + d(Tx, Ty)] \Big\}$$

which yields

$$(1 - b)d(Tx, Ty) \leq bd(x, y) + 2b \cdot d(y, Tx)$$

and which implies

$$d(Tx, Ty) \leq \frac{b}{1 - b} d(x, y) + \frac{2b}{1 - b} d(y, Tx), \quad \text{for all } x, y \in X,$$

and hence, in view of $0 < b < \dfrac{1}{2}$, (32) holds with $\delta = \dfrac{b}{1 - b}$ and $L = \dfrac{2b}{1 - b}$.

Since (8) is symmetric with respect to x and y, (33) also holds. $\qquad\square$

Proposition 2.3. *Any mapping T satisfying the contractive condition: there exists $c \in \left[0, \dfrac{1}{2}\right)$ such that*

$$d(Tx, Ty) \leq c[d(x, Ty) + d(y, Tx)], \quad \text{for all } x, y \in X, \tag{34}$$

is a weak contraction.

Proof. Using $d(x, Ty) \leq d(x, y) + d(y, Tx) + d(Tx, Ty)$ by (34) we get after simple computations,

$$d(Tx, Ty) \leq \frac{c}{1-c} d(x, y) + \frac{2c}{1-c} d(y, Tx),$$

which is (32), with $\delta = \dfrac{c}{1-c} < 1$ (since $c < 1/2$) and $L = \dfrac{2c}{1-c} \geq 0$.
The symmetry of (34) also implies (33). $\qquad\qquad\qquad\qquad\qquad\qquad\square$

An immediate consequence of Propositions 2.2 and 2.3 is the following.

Corollary 2.5. *Any Zamfirescu mapping, i.e., any mapping satisfying the assumptions (z_1)-(z_3) in Theorem 2.4, is a weak contraction.*

In a similar way we can prove that any quasi contraction with $0 \leq h < 1/2$ is a weak contraction.

Having in view the fact that the class of weak contractions properly includes large classes of quasi contractions and weak contractions and quasi contractions are independent, see Example 2.12, on the one hand, and the extensive literature related to quasi contractions, on the other hand, it is the aim of this section to prove two fixed points theorems in the class of weak contractions: an existence theorem (Theorem 2.11) as well as an existence and uniqueness theorem (Theorem 2.12). Their merit is that they extend all results in Section 2.3 and offer a method for approximating fixed points, for which both a priori and a posteriori estimates are available.

Theorem 2.11. *Let (X, d) be a complete metric space and $T : X \to X$ be a weak contraction, i.e., a mapping satisfying (32) with $\delta \in (0, 1)$ and some $L \geq 0$. Then*
1) *$Fix\,(T) = \{x \in X : Tx = x\} \neq \emptyset$;*
2) *For any $x_0 \in X$, the Picard iteration $\{x_n\}_{n=0}^{\infty}$ given by (2) converges to some $x^* \in Fix\,(T)$;*
3) *The following estimates*

$$d(x_n, x^*) \leq \frac{\delta^n}{1-\delta} d(x_0, x_1), \quad n = 0, 1, 2, \dots \tag{35}$$

$$d(x_n, x^*) \leq \frac{\delta}{1-\delta} d(x_{n-1}, x_n), \quad n = 1, 2, \dots \tag{36}$$

hold, where δ is the constant appearing in (32).

Proof. We shall prove that T has at least one fixed point in X. To this end, let $x_0 \in X$ be arbitrary and let $\{x_n\}_{n=0}^{\infty}$ be the Picard iteration defined by (2). Take $x := x_{n-1}$, $y := x_n$ in (32) to obtain

$$d(Tx_{n-1}, Tx_n) \leq \delta \cdot d(x_{n-1}, x_n),$$

which shows that

$$d(x_n, x_{n+1}) \leq \delta \cdot d(x_{n-1}, x_n). \tag{37}$$

Using (37), we obtain by induction

$$d(x_n, x_{n+1}) \le \delta^n d(x_0, x_1), \quad n = 0, 1, 2, \ldots$$

and then

$$d(x_n, x_{n+p}) \le \delta^n \left(1 + \delta + \cdots + \delta^{p-1}\right) d(x_0, x_1) =$$

$$= \frac{\delta^n}{1 - \delta} (1 - \delta^p) \cdot d(x_0, x_1), n, p \in \mathbb{N}, p \ne 0. \tag{38}$$

Since $0 < \delta < 1$, (38) shows that $\{x_n\}_{n=0}^{\infty}$ is a Cauchy sequence and hence is convergent. Denote

$$x^* = \lim_{n \to \infty} x_n. \tag{39}$$

Then

$$d(x^*, Tx^*) \le d(x^*, x_{n+1}) + d(x_{n+1}, Tx^*) = d(x_{n+1}, x^*) + d(Tx_n, Tx^*).$$

By (32) we have

$$d(Tx_n, Tx^*) \le \delta \, d(x_n, x^*) + L \, d(x^*, Tx_n)$$

and hence

$$d(x^*, Tx^*) \le (1 + L)d(x^*, x_{n+1}) + \delta \cdot d(x_n, x^*), \tag{40}$$

valid for all $n \ge 0$. Letting $n \to \infty$ in (40) we obtain

$$d(x^*, Tx^*) = 0$$

i.e., x^* is a fixed point of T.

The estimate (35) can be obtained from (38) by letting $p \to \infty$.

In order to obtain (36), observe that by (37) we inductively obtain

$$d(x_{n+k}, x_{n+k+1}) \le \delta^{k+1} \cdot d(x_{n-1}, x_n), \quad k, n \in \mathbb{N},$$

and hence, similarly to deriving (38) we obtain

$$d(x_n, x_{n+p}) \le \frac{\delta(1 - \delta^p)}{1 - \delta} d(x_{n-1}, x_n), \quad n \ge 1, \, p \in \mathbb{N}^*. \tag{41}$$

Now letting $p \to \infty$ in (41), (36) follows.   □

Remarks.

1) Theorem 2.11 is a significant extension of Theorem 2.1, Theorem 2.3, Theorem 2.4 and many other related results;

2) Note that, although the three particular fixed point theorems mentioned at 1) actually forces the uniqueness of the fixed point, the weak contractions need not have a unique fixed point, as shown by Example 2.11;

3) Recall that an operator $T : X \to X$ is said to be a *weakly Picard operator* if the sequence $\{T^n x_0\}_{n=0}^{\infty}$ converges for all $x_0 \in X$ and the limits are fixed points of T, see Definition 2.1 in Section 2.1.

The fixed point x^* attained by the Picard iteration depends on the initial guess $x_0 \in X$. Therefore, Theorem 2.11 provides a large class of weakly Picard operators;

4) It is easy to see that condition (32) implies the so called Banach orbital condition

$$d(Tx, T^2x) \le a\, d(x, Tx)\,, \quad \text{for all } x \in X,$$

studied by various authors in the context of fixed point theorems.

It is possible to force the uniqueness of the fixed point of a weak contraction, by imposing an additional contractive condition, quite similar to (32), as shown by the next theorem.

Theorem 2.12. *Let (X, d) be a complete metric space and $T : X \to X$ a weak contraction for which there exist $\theta \in (0, 1)$ and some $L_1 \ge 0$ such that*

$$d(Tx, Ty) \le \theta \cdot d(x, y) + L_1 \cdot d(x, Tx)\,, \quad \text{for all } x, y \in X\,. \qquad (42)$$

Then

1) T has a unique fixed point, i.e. $F(T) = \{x^*\}$;
2) The Picard iteration $\{x_n\}_{n=0}^{\infty}$ given by (2) converges to x^*, for any $x_0 \in X$;
3) The a priori and a posteriori error estimates

$$d(x_n, x^*) \le \frac{\delta^n}{1 - \delta}\, d(x_0, x_1)\,, \quad n = 0, 1, 2, \dots$$

$$d(x_n, x^*) \le \frac{\delta}{1 - \delta}\, d(x_{n-1}, x_n)\,, \quad n = 1, 2, \dots$$

hold;
4) The rate of convergence of the Picard iteration is given by

$$d(x_n, x^*) \le \theta\, d(x_{n-1}, x^*)\,, \quad n = 1, 2, \dots \qquad (43)$$

Proof. Assume T has two distinct fixed points $x^*, y^* \in X$. Then by (42), with $x := x^*$, $y := y^*$, we get

$$d(x^*, y^*) \le \theta \cdot d(x^*, y^*) \iff (1 - \theta)\, d(x^*, y^*) \le 0\,,$$

so contradicting $d(x^*, y^*) > 0$.

Letting $y := x_n$, $x := x^*$ in (42), we obtain the estimate (43). The rest of the proof follows by Theorem 2.11. $\qquad\square$

Remarks.

1) Note that, by the symmetry of the distance, (42) is satisfied for all $x, y \in X$ if and only if

$$d(Tx, Ty) \le \theta\, d(x, y) + L_1 d(y, Ty)\,, \qquad (44)$$

also holds, for all $x, y \in X$.

So, similarly to the case of the dual conditions (32) and (33), in concrete applications it is necessary to check that both conditions (42) and (44) are satisfied;

2) Note that condition (42) has been used to prove stability results for certain fixed point iteration procedures, see Chapter 7;

3) It is known that condition (42) alone does not ensure that T has a fixed point. But if T satisfying (42) has a fixed point, it is certainly unique;

4) It is a simple task to prove that any operator T satisfying one of the conditions (1), (8), (34), or the conditions (z_1)-(z_3) in Theorem 2.4, also satisfies the uniqueness conditions (42) and (44).

Therefore, in view of Example 2.11, Theorem 2.12 (and also Theorem 2.11) properly generalizes Zamfirescu's fixed point theorem.

Moreover, any quasi contraction with $0 \le h < \dfrac{1}{2}$ also satisfies (42) and (44). This shows that Theorem 2.12 unifies and generalizes the fixed point theorems of Banach, Kannan, Chatterjea and Zamfirescu and partially covers the Ciric's fixed point theorem;

5) As it can be seen, Theorem 2.12 (as well as Theorem 2.11, except for the uniqueness of the fixed point) preserves all conclusions in the Banach contraction principle in its complete form, given in Theorem 2.1, under significantly weaker contractive conditions. Indeed, the metrical contractive conditions known in literature that involve in the right-hand size the displacements

$$d(x,y), d(x,Tx), d(y,Ty), d(x,Ty), d(y,Tx)$$

with the nonnegative coefficients

$$a(x,y), b(x,y), c(x,y), d(x,y), e(x,y),$$

respectively, are commonly based on a restrictive assumption of the form

$$0 < a(x,y) + b(x,y) + c(x,y) + d(x,y) + e(x,y) < 1,$$

while, our condition (32) do not require $\delta + L$ be less than 1, thus providing a large class of contractive type mappings.

Example 2.11. Let $T : [0,1] \to [0,1]$ be the identity map, i.e., $Tx = x$, for all $x \in [0,1]$. Then

1) T does not satisfy the Ciric's contractive condition

$$d(Tx,Ty) \le h \cdot \max\left\{ d(x,y), d(x,Tx), d(y,Ty), d(x,Ty), d(y,Tx) \right\}$$

since $\max\left\{ d(x,y), d(x,Tx), d(y,Ty), d(x,Ty), d(y,Tx) \right\} = |x - y|$ and

$$|x - y| > h \cdot |x - y|, \quad \text{for all } x \ne y \quad \text{and } 0 \le h < 1.$$

2) T satisfies condition (32) with $\delta \in (0,1)$ arbitrary and $L \ge 1-\delta$. Indeed, conditions (32) and (33) lead to

$$|x - y| \leq \delta|x - y| + L \cdot |y - x|$$

which is true for all $x, y \in [0, 1]$ if we take $\delta \in (0, 1)$ arbitrary and $L \geq 1 - \delta$.

3) The set of fixed points of T is the interval $[0, 1]$, i.e., $Fix\,(T) = [0, 1]$.

It was an open problem whether any quasi contraction is a weak contraction. The next example, together with Example 2.11, shows that Ciric's quasi-contractive condition and weak contractive condition are independent.

Example 2.12. Let $X = [0, 1] \cup \left[\dfrac{3}{2}, \dfrac{5}{3}\right]$ with the usual norm and

$T : X \to X$ be given by $Tx = 0$, if $x \in [0, 1]$ and $Tx = 1$, if $x \in \left[\dfrac{3}{2}, \dfrac{5}{3}\right]$. Then:

(a) T does not satisfy the Banach orbital condition and, therefore, it is not a weak contraction;

(b) T is a quasi contraction with $h = 2/3$.

Indeed, for $x \in [0, 1], Tx = 0, d(x, Tx) = 0, T^2x = 0, d(Tx, T^2x) = 0$ and since $0 \leq x$ we have $d(Tx, T^2x) \leq d(x, Tx)$.

If $x \in \left[\dfrac{3}{2}, \dfrac{5}{3}\right]$, then $d(x, Tx) = d(x, 1) \leq \dfrac{2}{3}$ and $T(Tx) = 0$, hence

$$d(Tx, T^2x) > d(x, Tx)$$

and so T does not satisfy the Banach orbital condition.

If $x, y \in [0, 1]$ or $x, y \in \left[\dfrac{3}{2}, \dfrac{5}{3}\right]$, then $d(Tx, Ty) = 0$, when the quasi contractive condition is obviously satisfied.

If $x \in [0, 1]$ and $y \in \left[\dfrac{3}{2}, \dfrac{5}{3}\right]$, then $d(Tx, Ty) = 1, d(x, y) \geq \dfrac{1}{2}, d(x, Tx) = d(x, 0) = x, d(y, Ty) = d(y, 1) = |y - 1|, d(y, Tx) = d(y, 0) = y, d(x, Ty) = d(x, 1) = |x - 1|$ and therefore

$$\max\left\{d(x, y), d(x, Tx), d(y, Ty), d(x, Ty), d(y, Tx)\right\} = y \geq \dfrac{3}{2}$$

and so the Ciric's quasi contractive condition

$$d(Tx, Ty) \leq h \cdot \max\left\{d(x, y), d(x, Tx), d(y, Ty), d(x, Ty), d(y, Tx)\right\}$$

is satisfied with $h = 2/3$.   □

Using the notions and results we introduced in Section 2.5, we now can extend the main results obtained in the present section, in the following way.

Definition 2.6. Let (X, d) be a metric space. A self operator $T : X \to X$ is said to be a *weak φ-contraction* or *(φ, L)-weak contraction*, provided that there exist a comparison function φ and some $L \geq 0$, such that

$$d(Tx, Ty) \leq \varphi\big(d(x, y)\big) + L\,d(y, Tx), \quad \text{for all } x, y \in X. \tag{45}$$

Clearly, any weak contraction is a weak φ-contraction, with $\varphi(t) = \delta t$, $t \in \mathbb{R}_+$ and $0 < \delta < 1$. There exist weak φ-contractions which are not weak contractions with respect to the same metric. Also, all φ-contractions are weak φ-contractions with $L \equiv 0$ in (32).

Similarly to the case of weak contractions, the fact that T satisfies (45), for all $x, y \in X$, does imply that the following dual inequality

$$d(Tx, Ty) \leq \varphi\big(d(x, y)\big) + L\, d(x, Ty)\,, \tag{46}$$

obtained from (45) by formally replacing $d(Tx, Ty)$ and $d(x, y)$ by $d(Ty, Tx)$ and $d(y, x)$, respectively, and then interchanging x and y, is also satisfied.

Consequently, in order to prove that a certain operator T is a weak φ-contraction, we must check the both inequalities (45) and (46).

Theorem 2.11 and Theorem 2.12 could now be easily extended to weak φ-contractions.

Theorem 2.13. *Let (X, d) be a complete metric space and $T : X \to X$ a weak φ-contraction with φ a (c)-comparison function. Then*
1) *$F(T) = \{x \in X : Tx = x\} \neq \emptyset$;*
2) *For any $x_0 \in X$, the Picard iteration $\{x_n\}_{n=0}^{\infty}$ defined by $x_0 \in X$ and*

$$x_{n+1} = Tx_n\,, \quad n = 0, 1, 2, \ldots$$

converges to a fixed point x^* of T;
3) *The following estimate*

$$d(x_n, x^*) \leq s\big(d(x_n, x_{n+1})\big)\,, \quad n = 0, 1, 2, \ldots \tag{47}$$

holds, where $s(t)$ is given by (18).

Theorem 2.14. *Let X and T be as in Theorem 2.13. Suppose T also satisfies the following condition: there exist a comparison function ψ and some $L_1 \geq 0$ such that*

$$d(Tx, Ty) \leq \psi\big(d(x, y)\big) + L_1 d(x, Tx)\,,$$

holds, for all $x, y \in X$.
Then
1) *T has a unique fixed point, i.e., $F(T) = \{x^*\}$;*
2) *The estimate (47) holds;*
3) *The rate of convergence of the Picard iteration is given by*

$$d(x_n, x^*) \leq \varphi\big(d(x_{n-1}, x^*)\big)\,, \quad n = 1, 2, \ldots .$$

The proofs of Theorem 2.13 and 2.14 are essentially similar to those of Theorem 2.11 and 2.12 and, therefore, are omitted here.

2.8 Bibliographical Comments

§2.1.

Based probably on ideas of Cauchy and Liouville, Picard [Pic90] developed the method of successive approximations in a series of papers on the existence of solutions of initial value problems for ordinary differential equations.

For the case of complete normed linear spaces, nowadays called Banach spaces, Theorem 2.1 was first formulated and proved by Banach [Ban22] in his famous dissertation from 1922.

Since then, numerous generalizations or extensions of Theorem 2.1 have been obtained which, together with their various applications, still form a very dynamical field of research, circumscribed by the fixed point theory. The interested readers could find very diversified topics in any of the monographs in the reference list.

The material included in this Section is classical. The method of successive approximations is also called Picard iteration by many authors, a terminology that we adopted in this book. The concept of Picard operator was introduced by Rus [Rus83] and intensively studied, see Rus, Petrusel, A. and Petrusel, G. [RPP02], for the main results and problems on this topic as well as for a comprehensive bibliographical list.

§2.2.

The content of Section 2.2 is taken from Rus [Rus01]. Theorem 2.2 is due to Nemytzki [Nem36] and Edelstein [Ede82]. An extension of this theorem was obtained by Edelstein [Ede82] who has replaced the compactness of the space by a weaker assumption of the same kind: "there is a Picard iteration containing a convergent subsequence", see Exercise 2.5 at the end of this chapter.

§2.3.

Theorem 2.3 was given by Kannan [Knn68] in 1968, while Theorem 2.4 was obtained by Zamfirescu [Zam72] in 1972. For some applications of quasi-nonexpansive operators to the study and convergence of Newton and Newton type methods, see for example Berinde [Be95a], [Be95c], [Be95d], [Be97b] and [Be00a].

Condition ($*$), generally called "of quasi nonexpansiveness" was introduced by Tricomi [Trc16] for real functions, and later studied by Diaz and Metcalf [DiM67], [DiM69] and by Dotson [Dot70] for mappings in Banach spaces.

The convergence of Picard iteration for the whole class of quasi-nonexpansive mappings was established under several additional assumptions on T, i.e., T is continuous and asymptotically regular, in Petryshyn and Williamson [PWi73], see Exercises 2.16 and 2.17.

§2.4.

The content of this Section is basically taken from Rus [Rus83]. Theorem 2.5 was given by Bessaga [Bes59] in 1959. Example 2.7 is taken from Dugundji and Granas [DuG82], p. 24. Theorem 2.6 is due to Maia [Maa68]. For various applications of Maia's fixed point theorem to concrete problems, see Rus [Ru79c], [Rus01].

§2.5.

The results presented in this Section are taken from Rus [Rus83], [Rus01] and Berinde [Be97a]. For the proofs of the Lemmas 2.1 and 2.2, see Rus [Rus83] and Berinde [Be97a]. Theorem 2.7 rewrites Theorem 3.3.3 in Rus [Rus01], while Theorem 2.8 adapts Theorem 1.5.1 in Berinde [Be97a]. Theorem 2.9 is Theorem 3.3.6 in Rus [Rus83]. For other fixed point theorems in this class of φ-contractions, including the ones mentioned in Example 2.9, see Rus [Ru79c], [Rus01] and Taskovic [Tas86].

§2.6.

The results in this Section are mainly taken from the monograph Berinde [Be97a]. Lemmas 2.3 and 2.4 are Lemma 1.5.1 and Lemma 1.5.2, respectively, while Theorem 2.10 is Theorem 1.5.4, in Berinde [Be97a]. A similar result to that in Theorem 2.10 is obtained in Rus [Rus83] for generalized strict φ-contractions, where an error estimate is also given.

§2.7.

The results in this Section are taken from the papers Berinde [Be04d], [Be03a]. Theorem 2.11 and Theorem 2.12 are, respectively, Theorem 1 and Theorem 2 in Berinde [Be04d], while Theorem 2.13 and Theorem 2.14 are, respectively, Theorem 3 and Theorem 4 in Berinde [Be03a]. For a more detailed treatment and comparison of weak contractions to other contractive conditions, see also Berinde [Be03c]. Condition (34) appears to have been first involved in a fixed point theorem by Chatterjea [Cha72].

For the extensive literature related to quasi contractions, a class of operators in some way related to that of weak contractions see, for example, Ciric [Cir03] and references therein.

The notion of weakly Picard operator was introduced and intensively studied by Rus and his collaborators, see [Rus87], [Rus88], [Rus93], [Rus96], [Rus01], [Ru03a], [Ru03b], [RMu98], [RPS01] and [RPS03].

The so called Banach orbital condition has been studied by various authors in the context of fixed point theorems, see for example Hicks and Rhoades [HiR79], Ivanov [Iva76], Rus [Ru79c] and Taskovic [Tas86].

The general condition (42) has been used by Osilike [Os95c], [Os97a] and [Os99b] to prove stability results for certain fixed point iteration procedures.

For other metrical contractive conditions known in literature related to weak contractions and their comparison we refer to the papers by Rhoades [Rh77b] and Meszaros [Mes92].

Exercises and Miscellaneous Results

2.1. Bryant (1968)
If T is a selfmapping of a complete metric space and, if, for some positive integer k, T^k is a contraction, then T has a unique fixed point.

2.2. Weissinger (1952)
Let (X, d) be a complete metric space and $\{\alpha_n\}$ a sequence of nonnegative numbers with $\sum_{n=1}^{\infty} \alpha_n < \infty$. Let $T : X \to X$ be such that

$$d(T^n x, T^n y) \leq \alpha_n d(x, y), \text{ for all } x, y \in X.$$

Prove that T is a Picard operator.

2.3.
Let (X, d) be a complete metric space. A map $T : X \to X$ is *expanding* if $d(Tx, Ty) \geq \beta d(x, y)$, for all $x, y \in X$ and some $\beta > 1$. Prove that if T is surjective and expanding, then
(a) T is bijective;
(b) T is a Picard operator.

2.4.
Let (X, d) be complete and $T : X \to X$ a map satisfying

$$d(Tx, Ty) \leq \alpha(x, y) d(x, y), \text{ for all } x, y \in X,$$

where $\alpha : X \times X \to \mathbb{R}^+$ has the following property: for any closed interval $[a, b] \subset \mathbb{R}^+ \setminus \{0\}$,

$$\sup\{\alpha(x, y) : a \leq d(x, y) \leq b\} = \lambda(a, b) < 1.$$

Then T is a Picard operator.

2.5. Edelstein (1962)
Let (X, d) be a metric space and $T : X \to X$ be contractive. If there exists a point $x_0 \in X$ such that its sequence of iterates $\{T^n x_0\}$ contains a convergent subsequence $\{T^{n_i} x_0\}$, then $\{T^n x_0\}$ converges and $u = \lim_{n \to \infty} T^n x_0$ is the unique fixed point of T.

2.6. Converse of Banach's Fixed Point Theorem (Janos, 1967)

Let (X, d) be a compact metric space and $T : X \to X$ a continuous mapping. Assume

$$\bigcap_{n \in \mathbb{N}} T^n(X) = \{x^*\}.$$

Then for each $a \in (0, 1)$ there exists a metric ρ on X such that
(a) The metrics d and ρ are equivalent;
(b) $T : (X, \rho) \to (X, \rho)$ is an a-contraction.

2.7. Rakotch (1962)

Let (X, d) be a complete metric space. A map $T : X \to X$ is said to be *weakly contractive* if there exists a function $\lambda : (0, \infty) \to [0, 1)$ with $\sup\{\lambda(r) : 0 < p \le r \le q\} < 1$ and such that

$$d(Tx, Ty) \le \lambda[d(x, y)]d(x, y), \text{ for all } x, y \in X.$$

Prove that T has a unique fixed point.

2.8. Boyd-Wong (1969)

Let (X, d) be a complete metric space and let $T : X \to X$ satisfy

$$d(Tx, Ty) \le \varphi(d(x, y)), \quad \text{for all } x, y \in X,$$

where $\varphi : \mathbb{R}^+ \to \mathbb{R}^+$ is a real function, upper semicontinuous from the right, satisfying $\varphi(t) < t$ for $t > 0$. Then T is a Picard operator.

2.9. Meir-Keeler (1969)

Let (X, d) be a complete metric space and let $T : X \to X$ satisfy the following condition: given $\epsilon > 0$ there exists $\delta > 0$ such that

$$\epsilon \le d(x, y) < \epsilon + \delta \Rightarrow d(Tx, Ty) < \epsilon.$$

Then T is a Picard operator.

2.10. Hardy and Rogers (1973)

Let (X, d) be a metric space and T a self-mapping of X satisfying the condition: for $x, y \in X$

$$d(Tx, Ty) \le ad(x, Tx) + bd(y, Ty) + cd(x, Ty) + ed(y, Tx) + fd(x, y), \quad (\#)$$

where a, b, c, d, e, f are nonnegative and we set $\alpha = a + b + c + d + e + f$. Then
(a) If (X, d) is complete and $\alpha < 1$, then T has a unique fixed point.
(b) If $(\#)$ is modified to the condition: $x \ne y$ implies

$$d(Tx, Ty) < ad(x, Tx) + bd(y, Ty) + cd(x, Ty) + ed(y, Tx) + fd(x, y),$$

and in this case we assume (X, d) is compact, T is continuous and $\alpha = 1$, then T has a unique fixed point.

2.11. Using $T : [0,1] \to [0,1], T(x) = 1/2$, for $0 \le x < 1$ and $T(1) = 1$, show that
(a) T satisfies the Kannan contractive condition (8);
(b) T does not satisfy the Banach contraction condition (1).

2.12. Browder and Petryshyn (1966)
Let E be a Banach space and T a nonexpansive self map of E. T is said to be *asymptotically regular* if, for each point $x \in E$, $\lim(T^{n+1}x - T^n x) = 0$. Let T be nonexpansive asymptotically regular such that $I - T$ maps bounded closed subsets of E into closed subsets of E. Suppose T has a fixed point. Then, for each $x_0 \in E$, $\{T^n x_0\}$ converges to a fixed point of T in E.

2.13. Petryshyn and Williamson (1973)
Let X be a Banach space. If A and B are two sets in X, we denote the distance between A and B by

$$d(A,B) = \inf\{\|a - b\| : a \in A, b \in B\}$$

and the distance between a point p and A by $d(p, A)$.
Let D be a closed subset of a Banach space X and let T map D continuously into X such that
(a) $F_T \ne \emptyset$;
(b) For each $x \in D$ and every $p \in F_T$, $(*)$ holds, i.e., T is quasi nonexpansive;
(c) There exists an $x_0 \in D$ such that $x_n = T^n(x_0) \in D$, for each $n \ge 1$.
Then $\{x_n\}$ converges to a fixed point of T in D if and only if

$$\lim d(x_n, F_T) = 0.$$

2.14. Petryshyn and Williamson (1973)
Let D be a closed subset of a Banach space X and let T map D continuously into X such that
(a) $F_T \ne \emptyset$;
(b) T is quasi nonexpansive;
(c) There exists an $x_0 \in D$ such that $x_n = T^n(x_0) \in D$, for each $n \ge 1$;
(d) T is asymptotically regular at x_0;
(e) If $\{y_n\} \subseteq D, n \ge 1$, and $\|(I - T)y_n\| \to 0$ as $n \to \infty$, then

$$\liminf_n d(y_n, F_T) = 0.$$

Then $\{x_n\}$ converges to a fixed point of T in D.

2.15. Dotson (1970)
Let X be the real line with the usual metric and let T be defined as follows:

$$T(x) = \begin{cases} 0, & \text{if } x = 0 \\ \dfrac{x}{2} \sin \dfrac{1}{x}, & \text{if } x \ne 0 \end{cases}$$

(a) Show that T is not a nonexpansive function;
(b) Show that T is quasi nonexpansive.

Solution.
(a) Take $x = \dfrac{2}{\pi}$ and $y = \dfrac{2}{3\pi}$ to obtain $|Tx - Ty| = \dfrac{8}{3\pi} > \dfrac{4}{3\pi} = |x - y|$;
(b) Since $p = 0$ is the only fixed point of T, we have to show that $|T(x)| \le |x|$, which is immediate.

2.16. Petryshyn and Williamson (1973)

Let $B = B(0, 1)$ be the unit ball in $\mathbb{R}^2$ with the usual (Euclidean) norm. Define $T : B \to B$ by

$$T(x, y) = \left(-\frac{x}{2}, -y\right)$$

where (x, y) denote the usual coordinates for $\mathbb{R}^2$. Show that:
(a) T is nonexpansive;
(b) $F_T \ne \emptyset$;
(c) At all points z in B on the line $y = 0$, T is asymptotically regular at z, but T is not asymptotically regular at any other points in B.

2.17. Petryshyn and Williamson (1973)

Let D be a closed convex subset of a real Banach space X and let T be conditionally quasi nonexpansive mapping of D into itself. Suppose further that T satisfies the following conditions:
(a) There exists a compact set $K \subset X$ and a constant $k < 1$ such that

$$d(T(x), K) \le kd(x, K) \quad \text{for each } x \in D.$$

(b) T is conditionally quasi nonexpansive, that is, T is quasi nonexpansive whenever $F_T \ne \emptyset$.
Then the sequence $\{T^n(x_0)\}$ converges to a fixed point of T for each x_0 in D.

2.18. Ciric (1981)

Let (X, d) be a complete metric space and let $T : X \to X$ be contractive, that is $d(Tx, Ty) < d(x, y)$ for all $x, y \in X, x \ne y$, and satisfies the following condition: given $\epsilon > 0$ there exists $\delta > 0$ such that

$$\epsilon < d(x, y) < \epsilon + \delta \Rightarrow d(Tx, Ty) < \epsilon.$$

Then T is a Picard operator.

2.19. Show that if T satisfies (#) in Exercise 2.10, with $a = b$ and $c = d$ and $\alpha \le 1$, then T is a quasi nonexpansive operator.

3

The Krasnoselskij Iteration

It is well known that if T is assumed to be only a nonexpansive map, then the Picard iterations $\{T^n x_0\}_{n \geq 0}$ need no longer converge (to a fixed point of T). In fact, in general, T need not have a fixed point, as shown by Exercises 1.15, 1.16 and 1.19.

It is the purpose of this chapter to survey some old and new results on the approximation of fixed points for nonexpansive and pseudocontractive type operators by means of Krasnoselskij iteration.

The key idea in introducing Krasnoselskij iteration is the fact that, if T_λ is the averaged mapping associated to T, then if T is nonexpansive, so is T_λ, and both have the same fixed point set, see Exercise 3.3. Furthermore, T_λ has much more asymptotic behavior than the original mapping T.

Krasnoselskij was the first to notice the regularizing effect of T_λ in the case of a uniformly convex Banach space, see also the Bibliographical Comments at the end of this chapter.

3.1 Nonexpansive Operators in Hilbert Spaces

We begin this section by proving the Browder-Gohde-Kirk fixed point theorem (Theorem 1.2), which is a basic fixed point existence result for nonexpansive operators. The proof will be given in a Hilbert space setting, suitable to many convergence theorems for the Krasnoselskij iteration.

Theorem 3.1. *Let C be a closed bounded convex subset of the Hilbert space H and $T : C \to C$ be a nonexpansive operator. Then T has at least one fixed point.*

Proof. For a fixed element v_0 in C and a number s with $0 < s < 1$, we denote

$$U_s(x) = (1 - s)v_0 + s\, Tx\,, \quad x \in C.$$

Since C is convex and closed, we deduce that $U_s : C \to C$ is a s−contraction and, in virtue of Theorem 1.1, it has a unique fixed point, say u_s. On the other hand, since C is closed, convex and bounded in the Hilbert space H, it is weakly compact. Hence we may find a sequence $\{s_j\}$ in $(0,1)$ such that $s_j \to 1$ (as $j \to \infty$) and $u_j = u_{s_j}$ converges weakly to an element p of H.

Since C is weakly closed, p lies in C. We shall prove that p is a fixed point of T. If u is any arbitrary point in H, we have

$$\|u_j - u\|^2 = \|(u_j - p) + (p - u)\|^2 = \|u_j - p\|^2 + \|p - u\|^2 + 2\langle u_j - p, p - u \rangle,$$

where

$$2\langle u_j - p, p - u \rangle \to 0 \quad (\text{as } j \to \infty),$$

since $u_j - p$ converges weakly to zero in H. Setting $u = Tp$ above, we obtain

$$\lim_{j \to \infty} \left(\|u_j - Tp\|^2 - \|u_j - p\|^2 \right) = \|p - Tp\|^2.$$

Moreover, since $s_j \to 1$ and $U_{s_j} u_j = u_j$, we have

$$Tu_j - u_j = [s_j T u_j + (1 - s_j) v_0] - u_j + (1 - s_j)[Tu_j - v_0] =$$

$$= (U_{s_j} u_j - u_j) + (1 - s_j)(Tu_j - v_0) = 0 + (1 - s_j)(Tu_j - v_0) \to 0,$$

as $j \to \infty$, and therefore $\lim\limits_{j \to \infty} \|Tu_j - u_j\| = 0$.

On the other hand, since T is nonexpansive, we have

$$\|Tu_j - Tp\| \le \|u_j - p\|$$

and hence

$$\|u_j - Tp\| \le \|u_j - Tu_j\| + \|Tu_j - Tp\| \le \|u_j - Tu_j\| + \|u_j - p\|.$$

Thus

$$\limsup \left(\|u_j - Tp\| - \|u_j - p\| \right) \le \lim_{j \to \infty} \|u_j - Tu_j\| = 0$$

and, due to the boundedness of C, we have also

$$\limsup \left(\|u_j - Tp\|^2 - \|u_j - p\|^2 \right) =$$

$$= \limsup \left(\|u_j - Tp\| - \|u_j - p\| \right) \left(\|u_j - Tp\| + \|u_j - p\| \right) \le 0,$$

which yields

$$\lim_{j \to \infty} \left(\|u_j - Tp\|^2 - \|u_j - p\|^2 \right) = 0$$

and hence

$$\|p - Tp\|^2 = 0,$$

that is, p is a fixed point of T. $\qquad\square$

Remark. Even if the proof of Theorem 3.1 is more constructive than the corresponding version of this result in uniformly convex Banach spaces (Theorem 1.2), it does not provide a method for computation of fixed points.

Definition 3.1. Let H be a Hilbert space and C a subset of H. A mapping $T : C \to H$ is called *demicompact* if it has the property that whenever $\{u_n\}$ is a bounded sequence in H and $\{Tu_n - u_n\}$ is strongly convergent, then there exists a subsequence $\{u_{n_k}\}$ of $\{u_n\}$ which is strongly convergent.

We can give now a result on approximating fixed points of nonexpansive mappings by means of the Krasnoselskij iteration. To this end, we start by proving the next Lemma.

Lemma 3.1. *Let C be a bounded closed convex subset of a Hilbert space H and $T : C \to C$ be a nonexpansive and demicompact operator. Then the set F_T of fixed points of T is a nonempty convex set.*

Proof. Since T is nonexpansive, by Theorem 3.1, T has fixed points in C, that is, $F_T \neq \emptyset$. Furthermore, F_T is convex, i.e., when $x, y \in F_T$ and $\lambda \in [0,1]$ we have

$$u_\lambda = (1 - \lambda)x + \lambda y \in F_T.$$

Indeed,

$$\| Tu_\lambda - x \| = \| Tu_\lambda - Tx \| \leq \| u_\lambda - x \| \quad \text{and} \quad \| Tu_\lambda - y \| \leq \| u_\lambda - y \|,$$

which imply that

$$\| x - y \| \leq \| x - Tu_\lambda \| + \| Tu_\lambda - y \| \leq \| x - y \|.$$

This shows that for some a, b with $0 \leq a, b \leq 1$, we have

$$x - Tu_\lambda = a(x - u_\lambda) \quad \text{and} \quad y - Tu_\lambda = b(y - u_\lambda)$$

from which it follows that $Tu_\lambda = u_\lambda \in F_T$. $\qquad\square$

Theorem 3.2. *Let C be a bounded closed convex subset of a Hilbert space H and $T : C \to C$ be a nonexpansive and demicompact operator. Then the set F_T of fixed points of T is a nonempty convex set and for any given x_0 in C and any fixed number λ with $0 < \lambda < 1$, the Krasnoselskij iteration $\{x_n\}_{n=0}^\infty$ given by*

$$x_{n+1} = (1 - \lambda)x_n + \lambda T x_n, \quad n = 0, 1, 2, \ldots \tag{1}$$

converges (strongly) to a fixed point of T.

Proof. The first part follows by Lemma 3.1.

For any $x_0 \in C$, the sequence $\{x_n\}_{n=0}^\infty$ given by (1) lies in C and is bounded. Let p be a fixed point of T, and, so of the averaged map U_λ, given by

$$U_\lambda = (1 - \lambda)I + \lambda T \quad (I = \text{the identity map}). \tag{2}$$

We first prove that the sequence $\{x_n - Tx_n\}_{n \in \mathbb{N}}$ converges strongly to zero. Indeed

$$x_{n+1} - p = (1 - \lambda)\, x_n + \lambda Tx_n - p = (1 - \lambda)(x_n - p) + \lambda(Tx_n - p).$$

On the other hand, for any constant a,

$$a(x_n - Tx_n) = a(x_n - p) - a(Tx_n - p).$$

Then

$$\| x_{n+1} - p\|^2 = (1 - \lambda)^2 \| x_n - p\|^2 + \lambda^2 \| Tx_n - p\|^2 +$$
$$+ 2\lambda(1 - \lambda)\, \langle Tx_n - p, x_n - p \rangle$$

and

$$a^2 \| x_n - Tx_n\|^2 = a^2 \| x_n - p\|^2 + a^2 \| Tx_n - p\|^2 - 2a^2\, \langle Tx_n - p, x_n - p \rangle .$$

Hence, summing up the corresponding sides of the preceding two inequalities and using the fact that T is nonexpansive and $Tp = p$, we get

$$\| x_{n+1} - p\|^2 + a^2 \| x_n - Tx_n\|^2 \le [2a^2 + \lambda^2 + (1 - \lambda)^2] \cdot \| x_n - p\|^2 +$$
$$+ 2[\lambda(1 - \lambda) - a^2] \cdot \langle Tx_n - p, x_n - p \rangle .$$

If we choose now an a such that $a^2 \le \lambda(1 - \lambda)$, then from the last inequality we obtain

$$\| x_{n+1} - p\|^2 + a^2 \| x_n - Tx_n\|^2 \le$$
$$\le \left(2a^2 + \lambda^2 + (1 - \lambda)^2 + 2\lambda(1 - \lambda) - 2a^2 \right) \|x_n - p\|^2 = \|x_n - p\|^2$$

(we used the Cauchy-Schwarz inequality,

$$\langle Tx_n - p, x_n - p \rangle \le \| Tx_n - P\| \cdot \| x_n - p\| \le \| x_n - p\|^2 \Big) .$$

Letting now $a^2 = \lambda(1 - \lambda) > 0$ and summing up the obtained inequality

$$a^2 \|x_n - Tx_n\|^2 \le \| x_n - p\|^2 - \| x_{n+1} - p\|^2$$

for $n = 0$ to $n = N$ we get

$$\lambda(1 - \lambda) \sum_{n=0}^{N} \| x_n - Tx_n\|^2 \le \sum_{n=0}^{N} \left[\| x_n - p\|^2 - \| x_{n+1} - p\|^2 \right] =$$

$$= \| x_0 - p\|^2 - \| x_{N+1} - p\|^2 \le \| x_0 - p\|^2 ,$$

which shows that $\sum_{n=0}^{\infty} \| x_n - Tx_n\|^2 < \infty$ and hence $\| x_n - Tx_n\| \to 0$, as $n \to \infty$.

As T is demicompact, it results that there exists a strongly convergent subsequence $\{x_{n_i}\}$ such that $x_{n_i} \to p \in F_T$.

Since T is nonexpansive, $Tx_{n_i} \to Tp$ and $Tp = p$.

The convergence of the entire sequence $\{x_n\}_{n=0}^{\infty}$ to p now follows from the inequality $\| x_{n+1} - p \| \leq \| x_n - p \|$, which can be deduced from the nonexpansiveness of T and is valid for each n. □

Remarks.

1) The class of demicompact operators contains the compact operators, therefore by Theorem 3.2 we obtain, in particular, the result of Krasnoselskij [Kra55], and that of Schaefer [Sch57], established there in the more general context of uniformly convex Banach spaces;

2) From the proof of Theorem 3.2 it results that U_λ given by (2) is *asymptotically regular* , i.e., $\| U_\lambda^n x - U_\lambda^{n+1} x \| \to 0$, as $n \to \infty$, for any $x \in C$, that is,

$$x_n - x_{n+1} \to 0, \quad \text{as} \quad n \to \infty, \tag{3}$$

for any $x_0 \in C$.

The existence of the previous limit alone does not imply generally the convergence of the sequence $\{x_n\}_{n=0}^{\infty}$ to a fixed point of T (in Theorem 3.2 one additional assumption was the demicompactness of T). There are other possible additional assumptions to ensure the convergence of $\{x_n\}_{n=0}^{\infty}$ under the hypothesis of asymptotic regularity. For example, in the case of the real line, $C = [a, b]$ the closed bounded interval and $T : C \to C$ a continuous function, Hillam [Hil76] showed that the Picard iteration associated to T converges if and only if it is asymptotically regular;

3) Let us notice that the Krasnoselskij iteration is in fact the Picard iteration corresponding to the "averaged operator" U_λ associated to T and defined by (2);

4) The demicompactness on the whole D may be weakened to 0 by simultaneously adding an other assumption, to obtain the next result. A map T of $D \subset X$ into X is *demicompact at f* if, for any bounded sequence $\{x_n\}$ in D such that $x_n - T(x_n) \to f$ as $n \to \infty$, there exists a subsequence $\{x_{n_j}\}$ and an x in D such that $x_{n_j} \to x$ as $j \to \infty$ and $x - T(x) = f$. Clearly, when T is demicompact on D, it is demicompact at 0 but the converse is not true.

Corollary 3.1. *Let X be a uniformly convex Banach space, D a closed bounded convex set in X, and T a nonexpansive mapping of D into D such that T satisfies any one of the following two conditions:*
(i) (I-T) maps closed sets in D into closed sets in X;
(ii) T is demicompact at 0.

Then, for any given x_0 in C and any fixed number λ with $0 < \lambda < 1$, the Krasnoselskij iteration $\{x_n\}_{n=0}^{\infty}$ given by (1) converges (strongly) to a fixed point of T.

Proof. It suffices to show that the averaged map T_λ satisfies all conditions $(a) - (e)$ in Exercise 2.14. □

Remarks.
1) Conditions (i) and (ii) in Corollary 3.1 are independent;
2) If in Theorem 3.2 we remove the assumption that T is demicompact, then the Krasnoselskij iteration does not longer converge strongly, in general, but it converges (at least) weakly to a fixed point, as shown by the next theorem.

Theorem 3.3. *Suppose T is a nonexpansive operator that maps a bounded closed convex set C of H into C and that $F_T = \{p\}$. Then the Krasnoselskij iteration converges weakly to p,*

$$U_\lambda^n x_0 \rightharpoonup p,$$

for any $x_0 \in C$.

Proof. It suffices to show that if $\{x_{n_j}\}_{j=0}^\infty$, $x_{n_j} = U_\lambda^{n_j} x$ converges weakly to a certain p_0, then p_0 is a fixed point of T or of U_λ and therefore $p_0 = p$. Suppose that $\{x_{n_j}\}_{j=0}^\infty$ does not converge weakly to p. Then

$$\left\| x_{n_j} - U_\lambda p_0 \right\| \leq \left\| U_\lambda x_{n_j} - U_\lambda p_0 \right\| + \left\| x_{n_j} - U_\lambda x_{n_j} \right\| \leq$$

$$\leq \left\| x_{n_j} - p_0 \right\| + \left\| x_{n_j} - U_\lambda x_{n_j} \right\|$$

and, using the arguments in the proof of Theorem 3.2, it results

$$\left\| x_{n_j} - U_\lambda x_{n_j} \right\| \to 0, \quad \text{as} \quad n \to \infty,$$

and so the last inequality implies that

$$\lim \sup \left(\left\| x_{n_j} - U_\lambda p_0 \right\| - \left\| x_{n_j} - p_0 \right\| \right) \leq 0. \tag{4}$$

But, like in the proof of Theorem 3.2, we have

$$\left\| x_{n_j} - U_\lambda p_0 \right\|^2 = \left\| (x_{n_j} - p_0) + (p_0 - U_\lambda p_0) \right\|^2 =$$

$$= \left\| x_{n_j} - p_0 \right\|^2 + \left\| p_0 - U_\lambda p_0 \right\|^2 + 2 \left\langle x_{n_j} - p_0, p_0 - U_\lambda p_0 \right\rangle,$$

which shows, together with $x_{n_j} \rightharpoonup p_0$ (as $j \to \infty$), that

$$\lim_{n \to \infty} \left[\left\| x_{n_j} - U_\lambda p_0 \right\|^2 - \left\| x_{n_j} - p_0 \right\|^2 \right] = \left\| p_0 - U_\lambda p_0 \right\|^2. \tag{5}$$

On the other hand, we have

$$\left\| x_{n_j} - U_\lambda p_0 \right\|^2 - \left\| x_{n_j} - p_0 \right\|^2 = \left(\left\| x_{n_j} - U_\lambda p_0 \right\| - \left\| x_{n_j} - p_0 \right\| \right) \cdot$$

$$\cdot \left(\left\| x_{n_j} - U_\lambda p_0 \right\| + \left\| x_{n_j} - p_0 \right\| \right). \tag{6}$$

Since C is bounded, the sequence $\left\{ \left\| x_{n_j} - U_\lambda p_0 \right\| + \left\| x_{n_j} - p_0 \right\| \right\}$ is bounded, too, and by the relations (4)-(6) we get

$$\|p_0 - U_\lambda p_0\| \leq 0, \quad \text{i.e.} \quad U_\lambda p_0 = p_0 \Leftrightarrow p_0 \in F_T = \{p\},$$

which ends the proof. $\qquad\qquad\qquad\qquad\qquad\qquad\qquad\qquad\qquad\qquad\qquad\qquad$ $\square$

Remark. The assumption $F_T = \{p\}$ in Theorem 3.3 may be removed in order to obtain a more general result.

Theorem 3.4. *Let C be a bounded closed convex subset of a Hilbert space and $T : C \to C$ be a nonexpansive operator. Then, for any x_0 in C, the Krasnoselskij iteration converges weakly to a fixed point of T.*

Proof. Let F_T be the set of all fixed points of T in C (which is nonempty, by Theorem 3.1, and convex, by Lemma 3.1). As T is nonexpansive, for each $p \in F_T$ and each n we have

$$\| x_{n+1} - p\| \leq \| x_n - p\| ,$$

which shows that the function $g(p) = \lim_{n \to \infty} \| x_n - p\|$ is well defined and is a lower semicontinuous convex function on F_T. Let

$$d_0 = \inf\{g(p) : p \in F_T\}.$$

For each $\varepsilon > 0$, the set

$$F_\varepsilon = \{y : g(y) \leq d_0 + \varepsilon\}$$

is closed, convex, nonempty and bounded and, hence, weakly compact. Therefore $\bigcap_{\varepsilon > 0} F_\varepsilon \neq \emptyset$, and in fact

$$\bigcap_{\varepsilon > 0} F_\varepsilon = \{y : g(y) = d_0\} \equiv F_0.$$

Moreover, F_0 contains exactly one point. Indeed, since F_0 is convex and closed, for $p_0, p_1 \in F_0$, and $p_\lambda = (1 - \lambda)p_0 + \lambda p_1$,

$$g^2(p_\lambda) = \lim_{n \to \infty} \| p_\lambda - x_n\|^2 = \lim_{n \to \infty} (\|\lambda(p_1 - x_n) + (1 - \lambda)(p_0 - x_n)\|^2) =$$

$$= \lim_{n \to \infty} (\lambda^2 \|p_1 - x_n\|^2 + (1 - \lambda)^2 \| p_0 - x_n\|^2 +$$

$$+2\lambda(1 - \lambda) \langle p_1 - x_n, p_0 - x_n\rangle) = \lim_{n \to \infty} (\lambda^2 \|p_1 - x_n\|^2 +$$

$$+(1 - \lambda)^2 \| p_0 - x_n\|^2 + 2\lambda(1 - \lambda) \| p_1 - x_n\| \cdot \| p_0 - x_n\|)+$$

$$+ \lim_{n \to \infty} \{2\lambda(1 - \lambda) [\langle p_1 - x_n, p_0 - x_n\rangle - \| p_1 - x_n\| \cdot \| p_0 - x_n\|]\} =$$

$$= g^2(p) + \lim_{n \to \infty} \{2\lambda(1 - \lambda) \langle p_1 - x_n, p_0 - x_n\rangle - \| p_1 - x_n\| \cdot \| p_0 - x_n\|\}.$$

Hence

$$\lim_{n \to \infty} \{2\lambda(1 - \lambda) [\langle p_1 - x_n, p_0 - x_n\rangle - \| p_1 - x_n\| \cdot \| p_0 - x_n\|]\} = 0.$$

Since

$$\|p_1 - x_n\| \to d_0 \text{ and } \| p_0 - x_n\| \to d_0,$$

the latter relation implies that

$$\| p_1 - p_0 \|^2 = \| (p_1 - x_n) + (x_n - p_0 \|^2 = \| p_1 - x_n \|^2 +$$

$$+ \| x_n - p_0 \|^2 - 2 <p_1 - x_n, p_0 - x_n> \to d_0^2 - d_0^2 - 2d_0^2 = 0,$$

giving a contradiction.

Now, in order to show that $x_n = U_\lambda^n x_0 \rightharpoonup p_0$, is suffices to assume that $x_{n_j} \rightharpoonup p$ for an infinite subsequence and then prove that $p = p_0$. By the arguments in Theorem 3.3, $p \in F_T$. Considering the definition of g and the fact that $x_{n_j} \to p$, we have

$$\| x_{n_j} - p_0 \|^2 = \| x_{n_j} - p + p - p_0 \|^2 = \| x_{n_j} - p \|^2 + \| p - p_0 \|^2 -$$

$$-2 \langle x_{n_j} - p, p - p_0 \rangle \to g^2(p) + \| p - p_0 \|^2 = g^2(p_0) = d_0^2.$$

Since $g^2(p) \geq d_0^2$, the last inequality implies that

$$\| p - p_0 \| \leq 0,$$

which means that $p = p_0$.   □

3.2 Strictly Pseudocontractive Operators

In this section we present some convergence theorems for the Krasnoselskij iteration scheme in the class of pseudocontractive operators. The first of them is concerned with the computation of fixed points of strictly pseudocontractive operators.

Theorem 3.5. *Let C be a bounded closed convex subset of a Hilbert space and $T : C \to C$ be a strictly pseudocontractive operator, i.e., an operator for which there exists a constant $k < 1$ such that*

$$\|Tx - Ty\|^2 \leq \| x - y \|^2 + k \| (I - T) x - (I - T) y \|^2 , \quad x, y \in C. \quad (7)$$

Then, for any x_0 in C and any fixed μ such that $\mu < 1 - k$ the Krasnoselskij iteration $\{x_n\}_{n=0}^\infty$, given by $x_0 \in C$ and

$$x_{n+1} = (1 - \mu) x_n + \mu T x_n , \quad n = 0, 1, 2, \ldots, \quad (8)$$

converges weakly to a fixed point p of T.

If, additionally, we assume that T is demicompact, then $\{x_n\}_{n=0}^\infty$ converges strongly to p.

Proof. We denote as usually $T_t = (1-t)I + tT$ and show that T_t is nonexpansive. Indeed, by the pseudocontractiveness condition (7) it follows that $U = I - T$ is strongly monotone, i.e.,

$$< Ux - Uy, x - y > \geq m \parallel Ux - Uy \parallel^2, \quad \text{with } m = \frac{1-k}{2} > 0.$$

Then, for any $t > 0$

$$\parallel T_t x - T_t y \parallel^2 = \parallel (I - tU)x - (I - tU)y \parallel^2 =$$

$$= \parallel x - y \parallel^2 + t^2 \parallel Ux - Uy \parallel^2 - 2t < Ux - Uy, x - y > \leq$$

$$\leq \parallel x - y \parallel^2 + (t^2 - 2tm) \parallel Ux - Uy \parallel^2.$$

Now, if we take $t \leq 2m = 1 - k$, then from the preceding inequality we obtain

$$\parallel T_t x - \lambda_t y \parallel \leq \parallel x - y \parallel, \quad x, y \in C,$$

which shows that T_t is nonexpansive.

Now, by Theorem 3.4, T_t (and therefore T) has a fixed point p_0 in C and for any fixed λ with $0 < \lambda < 1$, the Krasnoselskij iteration $x_n = (T_t)_\lambda^n (x_0)$ associated to T_t converges weakly to some fixed point p of T in C.

But the iteration function $(T_t)_\lambda$ is in fact

$$(T_t)_\lambda = (1 - \lambda)I + \lambda T_t = (1 - \lambda)I + \lambda[(1-t)I + tT] = (1 - \lambda t)I + \lambda tT = T\mu,$$

with $\mu = \lambda t < t \leq 1 - k$.

In order to prove the second part of the theorem, based on Theorem 3.3, it suffices to show that T_μ is demicompact. But this follows immediately from the demicompactness of T using the equality

$$T_\mu x - x = \mu(Tx - x),$$

valid for every x in C. $\qquad\qquad\qquad\qquad\qquad\qquad\qquad\qquad\qquad\qquad\qquad\square$

3.3 Lipschitzian and Generalized Pseudocontractive Operators

Even though there is a rather strong connection between strictly pseudo-contractive operators and generalized pseudocontractive operators, these two classes are however independent each other.

This is the motivation why, in addition to the short previous section, we consider here generalized pseudocontractions which are also Lipschitzian, a class for which we can use the Krasnoselskij iteration in order to approximate their fixed points.

Definition 3.2. Let H be a Hilbert space with inner product $\langle \cdot, \cdot \rangle$ and norm $\|\cdot\|$. An operator $T : H \to H$ is said to be a *generalized pseudo-contraction* if there exists a constant $r > 0$ such that, for all x, y in H,

$$\| Tx - Ty \|^2 \leq r^2 \| x - y \|^2 + \| Tx - Ty - r(x - y) \|^2 . \qquad (9)$$

Remarks.

1) Condition (9) is equivalent to

$$\langle Tx - Ty, \ x - y \rangle \ \leq \ r \| x - y \|^2 , \quad \text{for all} \quad x, y \in H, \qquad (10)$$

or to

$$\langle (I - T) x - (I - T) y \rangle \ \geq \ (1 - r) \| x - y \|^2 . \qquad (11)$$

Relation (11) implies that $U = I - T$ is strongly monotone for $r < 1$.

2) If $r = 1$, then a generalized pseudo-contraction reduces to a pseudo-contraction;

3) By the Cauchy-Schwarz inequality

$$| \langle Tx - Ty, \ x - y \rangle | \leq \| Tx - Ty \| \cdot \| x - y \| ,$$

we obtain that any Lipschitzian operator T, that is, any operator for which there exists $s > 0$ such that

$$\| Tx - Ty \| \leq s \cdot \| x - y \| , \quad x, y \in H, \qquad (12)$$

is also a generalized pseudo-contractive operator, with $r = s$.

This, however, does not exclude the possibility that a certain operator T be simultaneously Lipschitzian with constant s, and generalized pseudo-contractive with constant r, and $r < s$. The existence of the last inequality is, in fact, the only reason of considering together Lipschitzian and generalized pseudo-contractive operators.

4) On the other hand, Theorem 3.6 below is obtained under the essential assumptions $r < 1$ and $s \geq 1$. Consequently, in the following, we shall assume that the Lipschitzian constant s and the generalized pseudo-contractivity constant r fulfill the conditions

$$0 < r < 1 \quad \text{and} \quad r \leq s. \qquad (13)$$

Example 3.1. Let H be the real line $\mathbb{R}$ endowed with the Euclidean inner product and norm, $K = \left[\frac{1}{2}, 2 \right]$ and $T : K \to K$ a function given by $Tx = \frac{1}{x}$, for all x in K.

Then T is Lipschitzian with constant $s = 4$ (so T is also generalized pseudo-contractive with constant $r = 4$).

Moreover, T is generalized pseudocontractive with any constant $r > 0$. It is easy to see that T has a unique fixed point, $F_T = \{1\}$, and that, for any initial choice $x_0 = a \neq 1$, the Picard iteration yields the oscillatory sequence

3.3 Lipschitzian and Generalized Pseudocontractive Operators 73

$$a, \frac{1}{a}, a, \frac{1}{a}, \dots$$

Theorem 3.6. *Let K be a non-empty closed convex subset of a real Hilbert space and $T : K \to K$ a generalized pseudocontractive and Lipschitzian operator with the corresponding constants r and s fulfilling (13). Then*
(i) T has an unique fixed point p;
(ii) for each x_0 in K, the Krasnoselskij iteration $\{x_n\}_{n=0}^{\infty}$, given by

$$x_{n+1} = (1 - \lambda)x_n + \lambda T x_n, \quad n = 0, 1, 2, \dots, \tag{14}$$

converges (strongly) to p, for all $\lambda \in (0, 1)$ satisfying

$$0 < \lambda < 2(1 - r)/(1 - 2r + s^2). \tag{15}$$

(iii) Both the a priori

$$\| x_n - p \| \leq \frac{\theta^n}{1 - \theta} \cdot \| x_1 - x_0 \|, \quad n = 1, 2, \dots \tag{16}$$

and a posteriori

$$\| x_n - p \| \leq \frac{\theta}{1 - \theta} \cdot \| x_n - x_{n-1} \|, \quad n = 1, 2, \dots \tag{17}$$

estimates hold, with

$$\theta = \left((1 - \lambda)^2 + 2\lambda(1 - \lambda)\,r + \lambda^2 s^2 \right)^{1/2}. \tag{18}$$

Proof. We consider the averaged operator F associated to T,

$$F x = (1 - \lambda)x + \lambda \cdot T x, \quad x \in K, \tag{19}$$

for all $\lambda \in [0, 1]$. Since K is convex, we have that $F(K) \subset K$ for each $\lambda \in [0, 1]$.
As a closed subset of a Hilbert space, K is a complete metric space. We claim that F is a θ-contraction with θ given by (18).
Indeed, since T is generalized pseudo-contractive and Lipschitzian, we have

$$\| F x - F y \|^2 = \| (1 - \lambda)\, x + \lambda T x - (1 - \lambda)\, y - \lambda T y \|^2 =$$

$$= \| (1 - \lambda)(x - y) + \lambda(T x - T y) \|^2 = (1 - \lambda)^2 \cdot \| x - y \|^2 +$$

$$+ 2\lambda(1 - \lambda) \cdot \langle T x - T y, x - y \rangle + \lambda^2 \cdot \| T x - T y \|^2 \leq$$

$$\leq \left((1 - \lambda)^2 + 2\lambda(1 - \lambda)r + \lambda^2 s^2 \right) \cdot \| x - y \|^2,$$

which yields

$$\| F x - F y \| \leq \theta \cdot \| x - y \|, \quad \text{for all} \quad x, y \in K.$$

In view of condition (15), it results that $0 < \theta < 1$, so the mapping F is a θ−contraction. In order to obtain the conclusion we now apply the contraction mapping principle (Theorem 2.1) for the operator F and the complete metric space K. □

Remarks.
1) The *a priori* estimate (16) in Theorem 3.6 shows that the Krasnoselskij iteration converges to p at least as fast as the geometric series of ratio θ;
2) The Krasnoselskij iteration solves several situations when the Picard iteration does not converge.

Example 3.2. Let K be as in Example 3.1. Here $s = 4$ and $r > 0$ arbitrary. Taking, for example, $r = 0.5$ we get

$$2(1-r)/(1-2r+s^2) = 1/16,$$

and so, by Theorem 3.6, the sequence $\{x_n\}_{n=0}^{\infty}$ given by

$$x_{n+1} = (1-\lambda)\cdot x_n + \lambda\cdot\frac{1}{x_n}, \qquad n = 0,1,2,\ldots \qquad (20)$$

converges strongly to the fixed point $p = 1$ of T, for all values of λ in the interval $\left(0,\dfrac{1}{16}\right)$.

Remark. It is of interest to answer the following question: amongst all the Krasnoselskij iterations $\{x_n\}_{n=0}^{\infty}$ in the family (14), obtained when λ ranges the interval $(0,a)$, with

$$a = \frac{2(1-r)}{(1-2r+s^2)},$$

is there a certain iteration to be the fastest one (in that family) ?

To answer this question, we shall adopt a suitable concept of convergence rate.

Let $\{x_n\}$ and $\{y_n\}$ be two sequences that converge to p (as $n \to \infty$), satisfying the estimate (16) with $\theta = \theta_1$ and $\theta = \theta_2$, respectively, and such that $\theta_1, \theta_2 \in (0,1)$. We shall say that $\{x_n\}$ *converges faster than* $\{y_n\}$ if

$$\theta_1 < \theta_2.$$

Equipped now with this concept of rate of convergence, Theorem 3.7 below answers in the affirmative the previous question.

Theorem 3.7. *Let all assumptions in Theorem 3.6 be satisfied. Then the fastest iteration* $\{x_n\}_{n=0}^{\infty}$ *in the family (14), with* $\lambda \in (0,a)$, *is the one obtained for*

$$\lambda_{\min} = (1-r)/(1-2r+s^2). \qquad (21)$$

Proof. We have to find the minimum of the quadratic function

$$f(x) = (1 - x)^2 + 2x(1 - x) r + x^2 s^2,$$

with respect to x, that is to minimize the function

$$f(x) = (1 - 2r + s^2) x^2 - 2(1 - r) x + 1, \quad x \in (0, a),$$

with a given by
$$a = 2(1 - r)/(1 - 2r + s^2). \tag{22}$$

This is an elementary task. Indeed from (13) we have that

$$1 - 2r + s^2 \geq (1 - r)^2 > 0,$$

and hence f does admit a minimum, which is attained for

$$x = \lambda_{\min},$$

with $\lambda_{\min}$ given by (21). The minimum value of $f(x)$ is then

$$f_{\min} = (s^2 - r^2)/(1 - 2r + s^2),$$

which shows that the minimum value of θ given by (18) is

$$\theta_{\min} = \left((s^2 - r^2) / (1 - 2r + s^2) \right)^{1/2},$$

that completes the proof. $\qquad\qquad\qquad\qquad\qquad\qquad\qquad\qquad\quad\Box$

Remarks.

1) It is important to notice that if $s < 1$, that is, T is actually a s−contraction, then $a > 1$ and hence $\lambda = 1 \in (0, a)$. This shows that among all Krasnoselskij iterations (14) that converge to the fixed point of T, we also find the Picard iteration associated to T, which is obtained from (14) for $\lambda = 1$. (This of course does not happen if $s \geq 1$);

2) As for the Picard iteration we have a similar a priori estimation, we can compare the Picard iteration to the fastest Krasnoselskij iteration in the family (14), with $\lambda \in (0, a)$:

a) If $r = s^2 < 1$, then we have

$$\theta_{\min} = s,$$

which means that the fastest Krasnoselskij iteration in the family (14) coincides with the Picard iteration itself;

b) If $r \neq s^2$, then it is easy to check that we have

$$\theta_{\min} < s,$$

(since $s < 1$), which shows that the Krasnoselskij iteration (14) with $\lambda = \lambda_{\min}$ is *faster* than the Picard iteration associated to T.

In this case, the fastest iteration from (14) may be regarded as an *accelerating* procedure of the Picard iteration.

Example 3.3. For T and K as in Examples 3.1 and 3.2, and for a certain $r \in (0,1)$, we obtain the fastest Krasnoselskij iteration for

$$\lambda = (1 - r) / (1 - 2r + 16).$$

If we take $r = 0.5$, then (14) converges for each $\lambda \in \left(0, \dfrac{1}{16}\right)$. The fastest Krasnoselskij iteration $\{x_n\}_{n=0}^{\infty}$ in this family is then obtained for $\lambda = \dfrac{1}{32}$, and is given by

$$x_{n+1} = \frac{1}{32}\left(31\, x_n + \frac{1}{x_n}\right), \quad n = 0, 1, 2, \ldots.$$

The averaged operator F,

$$F(x) = \frac{1}{32}\left(31\, x + \frac{1}{x}\right),$$

associated to T is a contraction and has the contraction coefficient

$$\theta_{\min} = \frac{\sqrt{63}}{8} = 0.992,$$

which is very close to 1.

The fastest Krasnoselskij iteration obtained in this way, converges very slowly to $p = 1$, the fixed point of T, as shown by the next Example.

Example 3.4. Starting with $x_0 = 1.5$, and $x_0 = 1.25$, respectively, the first 32 iterations are the following:

n	x_n	n	x_n	n	x_n	n	x_n
0	1.5	16	1.203	0	1.25	16	1.0960
1	1.473	17	1.191	1	1.2359	17	1.0902
2	1.449	18	1.180	2	1.2226	18	1.0848
3	1.425	19	1.170	3	1.2100	19	1.0797
4	1.402	20	1.160	4	1.1980	20	1.0749
5	1.381	21	1.151	5	1.1866	21	1.0704
6	1.360	22	1.142	6	1.1759	22	1.0662
7	1.341	23	1.133	7	1.1657	23	1.0584
8	1.322	24	1.126	8	1.1561	24	1.0515
9	1.304	25	1.118	9	1.1470	25	1.0484
10	1.287	26	1.111	10	1.1384	26	1.0454
11	1.271	27	1.105	11	1.1303	27	1.0426
12	1.256	28	1.098	12	1.1226	28	1.0400
13	1.242	29	1.087	13	1.1153	29	1.0376
14	1.228	30	1.082	14	1.1085	30	1.0353
15	1.215	31	1.077	15	1.1021	31	1.0331

3.4 Pseudo φ-Contractive Operators

In this section we want to show how we can unify in a single concept various notions as nonexpansive, Lipschitzian, pseudo-contractive type operators etc. For this new class of operators, called pseudo φ-contractive, we shall prove a convergence theorem for the Krasnoselskij fixed point procedure.

Let H be a Hilbert space with the inner product $\langle \cdot, \cdot \rangle$ and the norm $\|\cdot\|$. For the operators $T : H \to H$, let us denote by

1) C_0, the class of a−contractions, $0 \le a < 1$;
2) C_1, the class of nonexpansive operators;
3) C_2, the class of strictly pseudo-contractive operators;
4) C_3, the class of pseudo-contractive operators;
5) C_4, the class of generalized pseudo-contractive operators.

The next lemmas are immediate consequences of the results given in the previous sections and chapters.

Lemma 3.2.
1) $T \in C_3$ if and only if

$$\langle Tx - Ty,\, x - y \rangle \le \| x - y \|^2, \quad \text{for all } x, y \in H;$$

2) $T \in C_3$ if and only if

$$\langle (I - T)\,x - (I - T)\,y,\, x - y \rangle \ge 0, \text{ for all } x, y \in H.$$

Lemma 3.3.
1) $T \in C_4$ if and only if there exists $r > 0$ such that

$$\langle Tx - Ty,\, x - y \rangle \le r \cdot \| x - y \|^2, \quad \text{for all } x, y \in H;$$

2) $T \in C_4$ if and only if there exists $r > 0$ such that

$$\langle (I - T)\,x - (I - T)\,y,\, x - y \rangle \ge (1 - r) \cdot \| x - y \|^2, \quad \text{for all } x, y \in H.$$

Lemma 3.4. $T \in C_2$ if and only if there exists $k > 0$ such that

$$\langle (I - T)\,x - (I - T)\,y,\, x - y \rangle \ge k \cdot \| x - y \|^2, \quad \text{for all } x, y \in H.$$

Remark. It is also easy to prove the following inclusions

$$C_0 \subset C_1 \subset C_2 \subset C_3 \subset C_4.$$

Definition 3.3. An operator $T : H \to H$ is said to be (*strictly*) *pseudo φ-contractive* if, for any $a, b, c \in \mathbb{R}$ with $a + b + c = 1$, there exists a (comparison) function $\varphi : \mathbb{R}_+ \to \mathbb{R}_+$, such that

$$a \cdot \| x - y \|^2 + b \cdot \langle Tx - Ty, \, x - y \rangle + c \cdot \| Tx - Ty \|^2 \leq \varphi^2 \left(\| \, x - y \| \right),$$

$$(23)$$

holds, for all x, y in H.

Example 3.4.
   1) Any Lipschitzian operator T is pseudo φ−contractive with $a = 0, b = 0, c = 1$ and $\varphi(t) = t$;
   2) Any pseudo-contractive operator is also of pseudo φ−contractive type with $a = 0, b = 1, c = 0$ and $\varphi(t) = t$;
   3) Any generalized pseudo-contractive operator is a (strictly, if $r < 1$) pseudo φ−contractive operator, with $a = 0, b = 1, c = 0$ and $\varphi(t) = r \cdot t, r > 0$;
   4) Any strictly pseudocontractive operator is a pseudo φ−contractive operator, with $a = \dfrac{k-1}{2k}$, $b = 1$, $c = \dfrac{1-k}{2k}$ and $\varphi(t) = t$;
   5) Any strongly pseudocontractive operator is a pseudo φ−contractive operator, with $a = \dfrac{r\,t}{2(1+r)}$, $b = 1$, $c = -\dfrac{r\,t}{2(1+r)}$, $\varphi(u) = \dfrac{rt^2 + 2r + 2}{2t(r+1)} \cdot u$.

There are many convergence theorems concerning the approximation of fixed points for several classes of pseudocontractive type operators. The next theorem shows that the Krasnoselskij iteration converges to a fixed point of any strictly pseudo φ−contraction.

Theorem 3.8. *Let K be a nonempty closed convex subset of a real Hilbert space H and $T : K \to K$ a strictly pseudo φ−contractive operator. Then*
   (i) T has an unique fixed point p in K;
   (ii) For each $x_0 \in K$, the Krasnoselskij iteration $\{x_n\}_{n=0}^{\infty}$ given by (14) converges strongly to p, for all $\lambda \in (0,1)$;
   (iii) If, additionally, φ is a (c)−comparison function, then

$$\| \, x_n - p \| \leq s \left(\| \, x_n - x_{n+1} \| \right), \qquad n = 1, 2, \ldots$$

(where $s(t) = \sum\limits_{k=0}^{\infty} \varphi^k(t)$ denotes the sum of the comparison series).
   Proof. The proof is similar to that of Theorem 3.6. We consider the associated operator

$$Fx = (1 - \lambda)\,x + \lambda\,Tx, \quad x \in K$$

and show that $F : K \to K$ is a φ−contraction. Indeed, by (23) we get

$$\| \, Fx - Fy \|^2 \leq \varphi^2 \left(\| \, x - y \| \right), \quad \text{for all } \; x, y \in K,$$

which shows that F is a φ−contraction.
   Now, by Theorems 2.7 and 2.8, the conclusion immediately follows.   □

Remarks.

1) If T is not a strictly pseudo φ-contraction, then Theorem 3.8 is no longer valid;

2) We can obtain a result similar to the one given by Theorem 2.10 by considering in the right hand side of (23) the expression

$$\varphi^2 \left(\| \, x - y \, \| \, , \| \, x - Tx \, \| \, , \| \, y - Ty \, \| \, , \| \, x - Ty \, \| \, , \| \, y - Tx \, \| \right),$$

given by a 5-dimensional comparison function rather than a one-dimensional function;

3) If T is Lipschitzian and generalized pseudocontractive (with $r < 1$), then by Theorem 3.8 we obtain exactly Theorem 3.6, by taking the most used comparison function, i.e.,

$$\varphi(t) = r \cdot t;$$

4) The next two examples illustrate why we needed to consider special classes of pseudocontractive operators and not simply pseudocontractive operators in some of the convergence theorems stated in this chapter.

Example 3.5. Let $\mathbb{R}$ denote the reals with the usual norm, $K = [0, 1]$ and define $T : K \to \mathbb{R}$ by $Tx = \dfrac{1}{2}x + 1$. Then T is a $\dfrac{1}{2}$-contraction and hence is strongly pseudocontractive, but T has no fixed points in K.

Example 3.6. Let $\mathbb{R}$ denote the reals with the usual norm, $K = \{1, 2\}$ and define $T : K \to K$ by $T(1) = 2, T(2) = 1$. Then T is strongly pseudocontractive, but T has no fixed point in K.

3.5 Quasi Nonexpansive Operators

The convergence of Picard iteration for two classes of particular quasi nonexpansive operators was studied in Section 2.3, see also Exercise 2.14, which gives a convergence theorem for the whole class of quasi nonexpansive operators, when some additional assumptions are satisfied.

In the case of Hilbert spaces, see Exercise 3.5, it is known that nonexpansive operators are asymptotically regular. Since quasi nonexpansive operators strictly include the nonexpansive ones, even though a quasi nonexpansive operator is generally not asymptotically regular, however, its averaged operator is asymptotically regular in the case of uniformly Banach spaces, as the next Lemma shows.

Lemma 3.5. *Let X be a uniformly convex Banach space, D a subset of X, and T a mapping of D into X such that $F_T \neq \emptyset$ and T is quasi nonexpansive. Let T_λ be the averaged operator associated to T, i.e.,*

$$T_\lambda(x) = (1 - \lambda)x + \lambda Tx, \ x \in D.$$

If there exists $x_0 \in D$ and $\lambda \in (0,1)$ such that the Krasnoselskij iteration $\{T_\lambda^n(x_0)\}$ is defined and lies in D for each $n \geq 1$, then T_λ is asymptotically regular at x_0, that is,

$$\lim_{n \to \infty} [T_\lambda^n(x_0) - T_\lambda^{n+1}(x_0)] = 0.$$

Proof. Let p be any element in F_T and let x_0 be a point in D satisfying the conditions above. T_λ is also quasi nonexpansive since $F_{T_\lambda} = F_T \neq \emptyset$ and for all x in D we have

$$\|T_\lambda(x) - p\| = \|\lambda x - \lambda p + (1-\lambda)(Tx - p)\| \leq \lambda \|x - p\| + (1-\lambda) \|x - p\| =$$

$$= \|x - p\|.$$

This implies

$$\|x_{n+1} - p\| = \|T_\lambda x_n - p\| \leq \|x_n - p\|, \text{ for each } n \geq 1,$$

and therefore $\{\|x_n - p\|\}$ converges to some $d_0 \geq 0$.

If $d_0 = 0$, then $\lim_{n \to \infty} x_n = p$ and so in this case $x_n - x_{n+1} = T_\lambda^n(x_0) - T_\lambda^{n+1}(x_0) \to 0$ as $n \to \infty$, as required. In the case $d_0 > 0$, since $\|x_n - p\| \to d_0$, $\|T_\lambda x_n - p\| \leq \|x_n - p\|$ for each n, and

$$\lim_{n \to \infty} \|T_\lambda x_n - p\| = \lim_{n \to \infty} \|x_n - p\| = d_0,$$

it follows from the uniform convexity of X that

$$\lim_{n \to \infty} \|(x_n - p) - (T_\lambda x_n - p)\| = 0,$$

i.e.,

$$\lim_{n \to \infty} \|(x_n - T_\lambda x_n)\| = \lim_{n \to \infty} \|T_\lambda^n(x_0) - T_\lambda^{n+1}(x_0)\| = 0. \qquad \square$$

The following Lemma will be also useful to prove the main result of this section and is important by itself.

Lemma 3.6 *Let X be a strictly convex Banach space and D a closed convex subset of X. If T is a continuous mapping of D into X such that $F_T \neq \emptyset$ and*

$$\|Tx - p\| \leq \|x - p\|, \text{ for } x \in D \setminus F_T \text{ and } p \in F_T, \qquad (24)$$

then F_T is a convex set.

Proof. Let x and y be any two distinct points of F_T and, for $t \in (0,1)$, denote $z_t = tx + (1-t)y$. Since D is convex, $z_t \in D$. Suppose, contrary to our assertion, that $z_t \notin F_T$ for some $t \in (0,1)$. This means $z_t \in D \setminus F_T$. Then, it follows by (24) that

$$\|x - y\| \le \|x - T(z_t)\| + \|T(z_t) - y\| \le \|x - z_t\| + \|z_t - y\|.$$

Since X is strictly convex, we have that

$$x - T(z_t) = a(T(z_t) - y), \text{ for some } a > 0,$$

from which we obtain

$$T(z_t) = \frac{1}{1+a}x + \frac{a}{1+a}y,$$

which shows that $T(z_t)$ lies on the line determined by x and y. On the other hand,

$$\|x - T(z_t)\| \le \|x - z_t\| \text{ and } \|T(z_t) - y\| \le \|z_t - y\|.$$

Thus $T(z_t)$ must coincide with z_t. □

In the last part of this section we are interested to obtain convergence theorems for Krasnoselskij iteration under the basic assumption that T or T_λ is strictly quasi nonexpansive and that T satisfies the so-called Frum-Ketkov contractive condition. To this end we also need the following lemma.

Lemma 3.7. *Let D be a closed convex subset of X and T a selfmap of D such that*

$$d(T(x), K) \le kd(x, K), \text{ for all } x \in D \qquad (25)$$

for some convex compact set K in X and constant $k < 1$. If $T_\lambda = \lambda I + (1-\lambda)T$ is the averaged mapping and $\lambda \in (0, 1)$, then

$$d(T_\lambda(x), K) \le k_\lambda d(x, K), \text{ for each } x \in D, \qquad (26)$$

where $k_\lambda = \lambda + (1 - \lambda)k < 1$.

Proof. Let λ be fixed in $(0, 1)$, and $x \in D$, fixed. Since clearly $0 < k_\lambda < 1$, it suffices to prove (26).

For a given $\delta > 0$, there exist $y_\delta \in K$ and $z_\delta \in K$ such that

$$\|x - y_\delta\| \le d(x, K) + \delta/(2\lambda), \quad \|Tx - z_\delta\| \le d(Tx, K) + \delta/(2(1 - \lambda)).$$

Let $w_\lambda = \lambda y_\delta + (1 - \lambda)z_\delta$. Since K is convex, we have $w_\lambda \in K$. Then

$$d(T_\lambda x, K) \le \|T_\lambda x - w_\lambda\| = \|\lambda(x - y_\delta) + (1 - \lambda)(Tx - z_\delta)\| \le$$
$$\le \lambda \|x - y_\delta\| + (1 - \lambda) \|Tx - z_\delta\| \le k_\lambda d(x, K) + \delta,$$

and since $\delta > 0$ was chosen arbitrarily, the conclusion follows. □

The main result of this section is given by the next Theorem.

Theorem 3.9. *Let D be a closed convex set in a strictly convex Banach space X and let $T : D \to D$ be a conditionally quasi-nonexpansive operator. Suppose further that there exists a convex compact set K in X and a number $k < 1$ such that (25) holds.*

Then, for any $x_0 \in D$ and any $\lambda \in (0,1)$, the Krasnoselskij iteration $\{T_\lambda^n(x_0)\}$ converges to a fixed point of T.

Proof. By the convexity of D it follows that T_λ maps D into itself. Since T satisfies (25), by Lemma 3.7, T_λ satisfies (26) and hence, in view of Frum-Ketkov fixed point theorem, see Exercise 3.20, $Fix(T_\lambda) \neq \emptyset$. Moreover, since X is strictly convex and T is conditionally quasi-nonexpansive, it results that T_λ is conditionally strictly quasi nonexpansive, i.e.,

$$\|T_\lambda x - T_\lambda\| < \|x - y\|$$

for all $x \neq y$ in D, whenever $Fix(T_\lambda) \neq \emptyset$.

In fact, as $Fix(T_\lambda) \neq \emptyset$, T_λ is strictly nonexpansive.

On the other hand, by the same Frum-Ketkov contractive condition, it results

$$d(T_\lambda^n(x_0), K) \leq k_\lambda^n d(x_0, K)$$

and since $k_\lambda < 1$, this implies $\lim_{n \to \infty} d(T_\lambda^n(x_0), K) = 0$, and since K is compact, this forces $\{x_n \equiv T_\lambda^n(x_0)\}$ to contain a convergent subsequence $\{x_{n_j}\}_{j \geq 1}$ with $\lim_{j \to \infty} = x^*$.

The quasi nonexpansiveness condition implies that

$$\lim_{n \to \infty} d(x_n, Fix(T_\lambda)) = d \geq 0$$

exists. Therefore, it suffices to prove that $d = 0$. If $x^* \in Fix(T_\lambda)$, then $d = 0$. If $x^* \notin Fix(T_\lambda)$, then by the strictly quasi nonexpansiveness property, for every $x \in D \setminus Fix(T_\lambda)$, there exists $p = p_x \in Fix(T_\lambda)$ such that

$$\|T_\lambda x - T_\lambda\| < \|x - y\|.$$

This implies that T_λ is continuous at x^*, and hence

$$\|T_\lambda x^* - p\| = \left\|T_\lambda\left(\lim_{j \to \infty} x_{n_j}\right) - p\right\| = \lim_{n \to \infty} \|T_\lambda^n(x_0) - p\| =$$

$$\lim_{j \to \infty} \left\|T_\lambda^{n_j}(x_0) - p\right\| = \lim_{j \to \infty} \left\|x_{n_j} - p\right\| = \left\|\lim_{j \to \infty} x_{n_j} - p\right\| = \|x^* - p\|, \quad (27)$$

(where the middle equalities hold since, T_λ quasi nonexpansive implies that $\lim_{n \to \infty} \|T_\lambda^n(x_0) - p\|$ exists).

But the equality (27) is a contradiction, hence always $d = 0$.

Now, by $\lim_{n \to \infty} d(x_n, Fix(T_\lambda)) = 0$ we can prove that $\{x_n\}$ is a Cauchy sequence and, as it contains a convergent subsequence, it is convergent in the whole and $x^* \in Fix(T_\lambda)$. $\square$

3.6 Bibliographical Comments

§3.1.

The first result on the convergence of averaged sequences involving two successive terms of the Picard iteration, i.e., the expression

$$\frac{1}{2}(x_n + Tx_n),$$

has been obtained by Krasnoselskij [Kra55]. There, it was shown that if K is a closed bounded convex subset of a uniformly convex Banach space and $T : K \to K$ is a nonexpansive and compact operator (i.e., T is continuous and $T(K)$ is relatively compact), then the sequence $\{x_n\}_{n=0}^{\infty}$ defined by

$$x_{n+1} = \frac{1}{2}(x_n + Tx_n), n \geq 0$$

converges strongly to a fixed point of T.

Krasnoselskij gave no estimation of the rate of convergence of $\{x_n\}_{n=0}^{\infty}$ and, in fact, it is typical of iteration methods involving nonexpansive mappings that their convergence may be arbitrarily slow. Actually, Oblomskaja [Obl68] gave a linear example where convergence is slower that $n^{-\alpha}$ for all $\alpha \in (0,1)$. In this context, we also mention the monograph Patterson [Pat74, Chapter 4] which contains a thorough discussion of successive approximation method for linear operators, and an extensive bibliography.

Schaefer [Sch57] extended Krasnoselskij's result to the case when the constant $1/2$ is replaced by a $\lambda \in (0,1)$, obtaining in this way the first result for the general Krasnoselskij iteration, defined by (1). Then, Edelstein [Ede66] extended the previous result to the case when E is strictly convex.

Petryshyn [Pt66a] extended the results of Krasnoselskij and Schaefer to demicompact nonexpansive mappings $T : K \to E$ that satisfy a Leray-Schauder condition on the boundary ∂K of K, using the so-called iteration-retraction method, that can work only in Hilbert spaces, while the results of Krasnoselskij and Schaefer were derived in the more general setting of a uniformly convex Banach space.

A new technique, based on a generalization of the projection method to Banach spaces was recently developed by Alber [Alb96] and his collaborators.

Browder and Petryshyn [BrP66], [BrP67] carried further the results of Krasnoselskij and Schaefer, investigating the convergence of the Krasnoselskij (and Picard) iterations for nonexpansive operators $T : E \to E$ which are asymptotically regular and for which $I - T$ maps bounded closed sets into closed sets. Further extensions were obtained by Diaz and Metcalf [DiM67], [DiM69], Dotson [Dot70], Outlaw [Out69] and Petryshyn [Pet67], [Pet71].

The weak convergence of the Krasnoselskij iteration process was first proved by Schaefer [Sch57], for the class of continuous nonexpansive operators. The extension of this result to general nonexpansive operators was carried out in two stages by Browder and Petryshyn [BrP66] and Opial [Op67a], respectively.

The results included in this Section are taken from the following sources: Theorem 3.1, which is the well known Browder-Gohde-Kirk fixed point theorem in a Hilbert space setting, is Theorem 4 in Browder and Petryshyn [BrP67]; Theorem 3.2 is Theorem 6 of Petryshyn [Pt66a], reformulated in Browder and Petryshyn [BrP67], while Theorem 3.3 is Theorem 7 and Theorem 3.4 is Theorem 8, both taken from the same paper by Browder and Petryshyn [BrP67], where many other interesting results for approximating fixed points are given. Corollary 3.1 is Corollary 2.1 in Petryshyn and Williamson [PWi73], where several results from Browder and Petryshyn [BrP67] are extended and improved.

§3.2.

Theorem 3.5 in this Section rewrites Theorem 12 in Browder and Petryshyn [BrP67]. Theorem 14 in the same paper concerns the convergence of a modified Krasnoselskij iteration, obtained by fixing the first term of the linear convex combination, i.e., the iterative sequence is defined by means of the iteration function $F_\lambda x := \lambda T x + (1 - \lambda) u_0$, $\lambda \in (0, 1)$, where u_0 is fixed.

Several other results for this iteration procedure have been also obtained independently by Browder [Br67b] and respectively by Halpern [Hal67], in a Hilbert space setting. Their results say that: if x_λ is the fixed point of F_λ (which is a λ-contraction), then the sequence $\{x_\lambda\}$ converges strongly to a fixed point of T as $\lambda \to 1$. Later, Reich [Rei80] extended this result to uniformly smooth Banach spaces. Thereafter, Singh, S.P. and Watson, B. [SWa93] extended the result of Browder and Halpern to nonexpansive nonself operators satisfying Rothe's boundary condition.

Recently Xu, H.K. and Yin [XYi95] proved the convergence in the case of nonexpansive nonself operators defined on a nonempty closed convex (not necessarily bounded) subset of a Hilbert space. By adding the inwardness condition, Xu, H.K. [XuH97] extended the latter to uniformly smooth Banach spaces. For other related results, see also Jaggi [Ja77a], [Ja77b], Rhoades, B.E., Sessa, S., Khan, M.S., Swaleh, M. [RSK87], Jung and Kim, S.S. [JKS95], [JK98a] and [JK98b] and Section 6.5.

§3.3.

The content of Section 3.3 is taken from Berinde [Be02e], [Be02a]. Theorem 3.6, without part (iii) regarding error estimates, has been proved by Verma, R.U. [Ve97a], but the proof given here is at least formally different.

Theorem 3.7 has the merit to find the fastest Krasnoselskij iteration, under the assumptions of Theorem 3.6. The argument we exploited in order to do this was mentioned in passing in Browder and Petryshyn [BrP67].

§3.4.

The results in Section 3.4 are taken from Berinde [Be03a]. Various parts of them were communicated, in different stages of evolution, at some international conferences. Examples 3.5 and 3.6 are taken from Osilike [Os97c].

§3.5.

All results in this section are taken from Petryshyn and Williamson [PWi73]: Lemmas 3.5, 3.6 and 3.7 are respectively Lemma 2.1, Lemma 2.2 and Lemma 3.1, while Theorem 3.9 is Theorem 3.3 there. Exercise 3.21 is Example 3.1. Condition (25) was first used in Frum-Ketkov [FrK67], see Exercise 3.20, but a correct proof of this result was given by Nussbaum [Nus72]. For a recent result involving a Frum-Ketkov condition see Binh [Bin04].

Exercises and Miscellaneous Results

3.1. (a) Prove that if H is a Hilbert space then for any $u, v \in H$ we have

$$\|u + v\|^2 + \|u - v\|^2 = 2\left(\|u\|^2 + \|v\|^2\right). \tag{$*$}$$

(b) Show that a Banach space X is a Hilbert space if and only if the identity $(*)$ is satisfied for all $u, v \in X$.

3.2. Let H be a Hilbert space, $C \subset H$ a closed bounded convex subset. For a fixed element v_0 in C and a number $s \in (0, 1)$, define U_s by
$U_s(x) = (1 - s)v_0 + sTx, \quad x \in C.$
Show that: (a) U_s maps C into C; (b) U_s is a s-contraction.

3.3. Let H be a Hilbert space, $C \subset H$ a closed bounded convex subset, $T : C \to C$ and for $\lambda \in (0, 1)$, define the averaged map
$T_\lambda(x) = (1 - \lambda)x + \lambda Tx, \quad x \in C.$ Show that:
(a) T_λ maps C into C;
(b) If T is nonexpansive then T_λ is nonexpansive as well;
(c) T and T_λ have the same fixed point set, i.e., $Fix\,(T) = Fix\,(T_\lambda)$.

3.4. Browder and Petryshyn (1967)
Let H be a Hilbert space, $C \subset H$ a closed bounded convex subset, $T : C \to C$ nonexpansive and, for $\lambda \in (0, 1)$, define the averaged map

$$T_\lambda(x) = (1 - \lambda)x + \lambda Tx, \quad x \in C.$$

Show that if $\{x_n\}$ is the Picard iteration associated to T_λ and $x_0 \in C$, that is, the Krasnoselskij iteration associated to T and x_0, then

$$\sum_{n=0}^{\infty} \|x_{n+1} - x_n\|^2 < \infty.$$

Deduce from the above result that T_λ is asymptotically regular.

3.5. Let H be a Hilbert space, $C \subset H$ a closed bounded convex subset. If $T : C \to C$ is nonexpansive, then T is asymptotically regular, i.e., for any $x \in C$,

$$\left\|T^{n+1}x - T^n x\right\| \to 0 \text{ as } n \to \infty.$$

3.6. Let H be a Hilbert space and $C \subset H$ be a closed bounded convex subset. For each $x \in H$ define $R_C x$ as the nearest point to x in C.
(a) If $C = B(x_0, r)$, show that $R_C : H \to C$ is given by

$$R_C x = \begin{cases} x, & \text{if } \|x - x_0\| \leq r \\ \dfrac{r(x - x_0)}{\|x - x_0\|}, & \text{if } \|x - x_0\| \geq r; \end{cases}$$

(b) Show that R_C is nonexpansive.

3.7. Figueiredo-Karlovitz
If the mapping R_C defined in Exercise 3.6 for $C = B(0,1)$ is nonexpansive for a Banach space X of dimension > 2, then X is a Hilbert space.

3.8. Let H be a Hilbert space, $C \subset H$ a closed bounded convex subset and $T : C \to C$ a strictly pseudo-contractive operator. Show that there exist values of $\lambda \in (0,1)$ such that the averaged operator

$$T_\lambda(x) = (1 - \lambda)x + \lambda T x, \quad x \in C,$$

is nonexpansive.

3.9. Let H be a Hilbert space, $K \subset H$ a closed bounded convex subset. Show that any Lipschitzian operator $T : K \to K$ is also generalized pseudo-contractive with the same constant but the reverse is not true.

3.10. If K is a closed convex subset of a strictly convex Banach space X and $T : K \to K$ is nonexpansive, then F_T is closed and convex.

3.11. Let $X = \mathbb{R}^2$ be endowed with the norm $\|(x,y)\|_\infty = \max\{|x|, |y|\}$ and define $T : \mathbb{R}^2 \to \mathbb{R}^2$ by $T(x,y) = (x, |x|)$. Then
(a) T is nonexpansive;
(b) F_T is not convex.

3.12. Consider the unit ball in the space C_0 of all sequences of real numbers with limit 0 endowed with the sup norm and define $T : C_0 \to C_0$ by

$$Tx = (x_1, 1 - |x_1|, x_2, x_3, \dots), \quad x = (x_1, x_2, x_3, \dots).$$

Show that
(a) T is nonexpansive;
(b) $F_T = \{u, -u\}$, where $u = (1, 0, 0, 0, \dots)$ (hence F_T is disconnected).

3.13. Let $C[0,1]$ be endowed with the Chebyshev's norm and let B be given by

$$B = \{x : [0,1] \to \mathbb{R} \mid x(0) = 0, x(1) = 1 \text{ and } 0 \leq x(t) \leq 1, t \in (0,1)\}.$$

Define T on B by $Tx(t) = tx(t), t \in [0,1]$. Then
(a) T has no fixed points in B;
(b) If $\{x_n(t)\}$ is the Krasnoselskij iteration with $x_0(t) = 0$, we have

$$\lim_{n \to \infty} \|T x_n - x_n\| = 0.$$

3.14. Alspach (1981)

Let $X = L^1[0,1]$ and $K = \left\{ f \in X | \int_0^1 f = 1, 0 \leq f \leq 2 \text{ a.e.} \right\}$. Then
(a) K is a closed convex subset of $[0,2]$ (and hence it is weakly compact);
(b) The mapping $T : K \to K$ given by

$$Tf(t) = \begin{cases} \min\{2f(2t), 2\}, & \text{if } 0 \leq t \leq \dfrac{1}{2} \\ \max\{2f(2t-1) - 2, 0\}, & \text{if } \dfrac{1}{2} < t < 1 \end{cases}$$

is isometric on K but has no fixed points. (This shows that a weakly compact convex set in a Banach space does not have the fixed point property for nonexpansive operators)

3.15. Let K be a subset of a Banach space X and $T : K \to K$ be nonexpansive and $x_0 \in K$. Show that

$$\lim_{n \to \infty} \left\| T^n x_0 - T^{n+1} x_0 \right\|$$

always exists but this limit may be nonzero.

3.16. Baillon, Bruck and Reich (1978)
Let X be a Banach space, K a bounded, closed and convex subset of X, $T : K \to K$ nonexpansive and T_λ the averaged operator, i.e.,

$$T_\lambda(x) = (1 - \lambda)x + \lambda Tx, \quad x \in K \text{ and } \lambda \in (0,1).$$

Then, for any $x \in K$,

$$\lim_{n \to \infty} \left\| T_\lambda^{n+1} x - T_\lambda^n x \right\| = \frac{1}{k} \lim_{n \to \infty} \left\| T_\lambda^{n+k} x - T_\lambda^n x \right\| = \lim_{n \to \infty} \frac{1}{n} \left\| T_\lambda^n x \right\|.$$

3.17. Ishikawa (1976)
Let X be a Banach space, K a bounded, closed and convex subset of X and $T : K \to K$ be nonexpansive. For $\lambda \in (0,1)$, let T_λ be the averaged operator associated to T, i.e.,

$$T_\lambda(x) = (1 - \lambda)x + \lambda Tx, \quad x \in K$$

and define the sequences $\{x_n\}$ and $\{y_n\}$ as follows

$$x_{n+1} = T_\lambda x_n; \quad y_n = Ty_n, \ n = 0,1,2,\dots$$

Then
(a) For each $i, n \in \mathbb{N}$,

$$\|y_{i+n} - x_i\| \geq (1 - \lambda)^{-n} [\|y_{i+n} - x_{i+n}\| - \|y_i - x_i\|] + (1 + n\lambda)\|y_i - x_i\|;$$

and
(b) $\lim_{n \to \infty} \|x_n - Tx_n\| = 0$.

3.18. Opial (1967)
Let X be a uniformly Banach space having a weakly continuous duality map and let x^* be the weak limit of a weakly convergent sequence $\{x_n\}$. Then

$$\limsup_{n\to\infty} \|x_n - x^*\| < \limsup_{n\to\infty} \|x_n - x\|, \text{ for all } x \neq x^*.$$

(Opial's condition)

3.19. Browder and Petryshyn (1967)
If X is uniformly convex, C is bounded and $T : C \to C$ is asymptotically regular, then the weak sequential limits of $\{T^n x\}$ are fixed points of T, i.e., $\omega_w(x) \subset F_T$.

3.20. Frum-Ketkov (1967)
Let D be a closed convex subset of a Banach space X and $T : D \to D$ a continuous map. Assume that there exist a compact set $K \subset X$ and a constant $k < 1$ such that

$$d(Tx, K) \leq kd(x, K), \text{ for each } x \in D.$$

Then T has a fixed point.

3.21. Petryshyn and Williamson (1973)
Let $X = l^p, 1 < p < \infty$ the space of infinite sequences of real numbers $x = (x_1, x_2, \dots)$ whose norm, $\|x\| \equiv \left(\sum_{i \geq 1} |x_i|^p \right)^{1/p}$ is finite. Show that

(a) l^p is uniformly convex;
(b) The collection $\{e_i | i \geq 1\}$ forms a Schauder basis for l_p, where e_i are the unit vectors in l^p of the form $e_j = \{\delta_{ij}\}_{j \geq 1}$, that is, each $x \in l^p$ has a unique representation in terms of this collection;
Let B be the unit ball in l_p with center 0 and let $\{f_i\}_{i \geq 1}$ be a family of nonexpansive self-mappings of the interval $[-1, 1]$ with $f_i(0) = 0$, $i \geq 1$. Define T for $x \in B$ by

$$Tx \equiv f_1(x_1)e_1 + \frac{1}{2} \sum_{i > 1} f_i(x_i)e_i, \ x = (x_1, x_2, \dots) \in B.$$

(c) Show that T is well defined, $T(B) \subset B$ and T is nonexpansive;
(d) Show that $K \equiv \{x \in l^p | x_i = 0, i > 1; |x_1| \leq 1\}$ is convex and compact and for any $x \in B$, T satisfies the Frum-Ketkov contractive condition:

$$d(Tx, K) \leq \frac{1}{2}d(x, K);$$

(e) Apply Theorem 3.8 to show that the Krasnoselskij iteration associated to T converges for any $x_0 \in B$ and any $\lambda \in (0, 1)$ to a fixed point of T in B.

4

The Mann Iteration

Although, chronologically, it was introduced two years earlier than the Krasnoselskij iteration, the Mann iteration is formally a generalization of the latter and, in its normal form, is obtained by replacing the parameter λ in the Krasnoselskij iteration formula by a sequence of real numbers $\{a_n\} \subset [0,1]$. Since in many cases the convergence of the normal Mann iteration could be obtained from the corresponding results proved for the Ishikawa iteration procedure, the aim of this chapter is to present merely some representative sample results regarding the Mann iteration, in general, without a (complete) proof.

4.1 The General Mann Iteration

Definition 4.1. Let E be a linear space, C a convex subset of E and let $T : C \to C$ be a mapping and $x_1 \in C$, arbitrary. Let $A = [a_{n\,j}]$ be an infinite real matrix satisfying
   (A_1) $a_{n\,j} \geq 0$ for all n, j and $a_{n\,j} = 0$ for $j > n$;
   (A_2) $\sum\limits_{j=1}^{n} a_{n\,j} = 1$ for all $n \geq 1$;
   (A_3) $\lim\limits_{n\to\infty} a_{n\,j} = 0$ for all $j \geq 1$.
   The sequence $\{x_n\}_{n=1}^{\infty}$ defined by $x_{n+1} = T(v_n)$, where

$$v_n = \sum_{j=1}^{n} a_{n\,j}\, x_j,$$

is called the *Mann iterative process* or, simply, *the Mann iteration*.

Remark. The Mann iterative process $\{x_n\}_{n=1}^{\infty}$ can be briefly denoted by $M(x_1, A, T)$ to indicate the initial guess x_1, the matrix A and the operator T to whom the process is associated. Similarly, we can denote the Krasnoselskij iteration $\{x_n\}_{n=0}^{\infty}$ by $K(x_0, \lambda, T)$.

There exists a rich literature on the convergence of Mann iteration for different classes of operators considered on various spaces. We begin by stating without proof a result on the Mann iteration whose statement is very closed to the form in which was originally formulated but here in a setting that is different from that in which the original result of Mann was formulated.

Theorem 4.1. *Suppose E is a locally convex Hausdorff linear topological space, C is a closed convex subset of E, $T : C \to C$ is continuous, $x_1 \in C$ and $A = [a_{n\,j}]$ satisfies $(A_1), (A_2)$ and (A_3). If either of the sequences $\{x_n\}$ or $\{v_n\}$ in the Mann iterative process $M(x_1, A, T)$ converges to a point p, then the other sequence also converges to p, and p is a fixed point of T.*

Definition 4.2. A Mann process $M(x_1, A, T)$ is said to be *normal* provided that $A = [a_{n\,j}]$ satisfies $(A_1), (A_2), (A_3), (A_4)$ $a_{n+1,j} = (1 - a_{n+1,n+1})a_{n\,j}$, $j = 1, 2, \ldots, n$; $n = 1, 2, 3, \ldots$ and (A_5) either $a_{n\,n} = 1$ for all n, or $a_{n\,n} < 1$ for all $n > 1$.

Theorem 4.2. *The following are true:*

(a) In order that $M(x_1, A, T)$ be a normal Mann process, it is necessary and sufficient that $A = [a_{n\,j}]$ satisfies $(A_1), (A_2), (A_4), (A_5)$ and

(A_3') $\sum\limits_{n=1}^{\infty} a_{n\,n}$ *diverges.*

(b) The matrices $A = [a_{n\,j}]$ (other than the infinite identity matrix) in all normal Mann process $M(x_1, A, T)$ are constructed as follows:

Choose $\{c_n\}$ such that $0 \le c_n < 1$ for all n and the series $\sum\limits_{n=1}^{\infty} c_n$ diverges, and define $A = [a_{n\,j}]$ by

$$\begin{cases} a_{11} = 1, \ a_{1j} = 0 \ \text{for} \ j > 1; \\ a_{n+1,n+1} = c_n, \ n = 1, 2, 3, \ldots; \\ a_{n+1,j} = a_{j\,j} \prod\limits_{i=j}^{n} (1 - c_i), \ \text{for} \ j = 1, 2, \ldots, n \\ a_{n+1,j} = 0, \ \text{for} \ j > n+1, \ n = 1, 2, 3, \ldots \end{cases} \tag{1}$$

(c) The sequence $\{v_n\}$ in a normal Mann process $M(x_1, A, T)$ satisfies

$$v_{n+1} = (1 - c_n) v_n + c_n T v_n, \ \text{for all} \ n = 1, 2, 3, \ldots, \tag{2}$$

where

$$c_n = a_{n+1,n+1}. \tag{3}$$

Examples.

1) The simplest example of Mann iteration is obtained by choosing $c_n = 1$ for all $n \ge 1$, which corresponds to the Picard iteration.

Another one is obtained letting $c_n = 1/(n+1)$, when the obtained matrix A is the Cesaro matrix;

2) If $\lambda \in [0,1]$ and $A_\lambda = [a_{n\,j}]$ is defined by

$$a_{n\,1} = \lambda^{n-1}, a_{n\,j} = \lambda^{n-j}(1-\lambda), \text{ for } j = 2, 3, \ldots, n$$

and

$$a_{n\,j} = 0 \text{ for } j > n, \; n = 1, 2, 3, \ldots,$$

then $M(x_1, A_\lambda, T)$ is the normal Mann process. Since the diagonal sequence for A_λ is given by

$$c_n = a_{n+1,n+1} = 1 - \lambda, \quad \text{for all } n = 1, 2, 3, \ldots,$$

we see that it actually corresponds to the Krasnoselskij iteration.

Remarks.

1) The matrix A given by (1) is a regular matrix (i.e., A is a bounded linear operator on l_∞ which is limit preserving for convergent sequences);

2) Following Theorem 4.2, we shall consider in the sequel only normal Mann processes, defined by (2), which will be simply called *Mann iteration* procedures;

3) Most of the literature deals with the specialized Mann iteration method defined by $x_1 \in E$ and (2), where $\{c_n\}$ satisfies

(i) $c_1 = 1$; (ii) $0 < c_n < 1$, $n \geq 2$ and (iii) $\sum_{n=1}^{\infty} c_n = \infty$.

However, in the sequel we will start with some results for the general Mann iteration. The reason is that in the literature there are several theorems of the following type: T is a selfmap of a complete metric space E, satisfying a contractive condition that may or may not be strong enough to guarantee the convergence to a fixed point of the Picard iteration associated to T.

Under these conditions it is also assumed that the Mann iteration associated to T converges, for a certain $\{c_n\}$, and it is then shown that, under these circumstances, it converges to a fixed point of T.

All such kinds of results could be obtained as particular cases of some generic theorems of the following type.

Theorem 4.3. *Let T be a selfmap of a closed convex subset K of a real Banach space $(E, \|\cdot\|)$. Let $\{x_n\}_{n=1}^{\infty}$ be a general Mann iteration of T with A equivalent to convergence. Suppose that $\{x_n\}_{n=1}^{\infty}$ converges to a point $p \in K$. If there exist the constants $\alpha, \beta, \gamma, \delta \geq 0$, $\delta < 1$ such that*

$$\|Tx_n - Tp\| \leq \alpha \cdot \|x_n - p\| + \beta \cdot \|x_n - Tx_n\| + \gamma \cdot \|p - Tx_n\| +$$

$$+\delta \cdot \max\{\|p - Tp\|, \|x_n - Tp\|\}, \tag{4}$$

then p is a fixed point of T.

Proof. The conditions on A, that is, A equivalent to convergence, imply that it is regular, i.e., A is limit-preserving over c, the space of convergent sequences. If we define

$$\mathcal{C}_A = \left\{ x : Ax = \left(\sum_{j=1}^{n} a_{nj} x_j \right) \in c \right\}, \ x = (x_1, x_2, ..., x_n, ...),$$

then the condition that A is equivalent to convergence means that $\mathcal{C}_A = c$. Thus

$$\lim_{n \to \infty} x_n = p,$$

which implies that $\{Tx_n\} \in \mathcal{C}_A$ and hence $\{Tx_n\} \in c$. Since A is regular, we must have

$$\lim_{n \to \infty} Tx_n = p$$

and therefore

$$\lim_{n \to \infty} \|x_n - Tx_n\| = 0.$$

Taking the limit of (4) as $n \to \infty$ yields

$$\|p - Tp\| \le \delta \cdot \|p - Tp\|,$$

which implies $Tp = p$. □

Remarks.

1) It has been shown in that the general Mann iteration method can be written in the form $x = Aw$, where $x = \{x_n\}$, $w = \{Tx_n\}$, and $A = [a_{nk}]$ is the weighted mean matrix generated by $a_{nk} = p_k/P_n$, where

$$p_1 > 0, \ p_k = \frac{c_k\, p_1}{\prod_{i=2}^{k}(1 - c_i)}, \ P_n = \sum_{i=1}^{n} p_i = \frac{p_1}{\prod_{i=2}^{n}(1 - c_i)}, \ k > 1;$$

2) In all convergence theorems of the type mentioned above, the sequence $\{c_n\}$ satisfies (i), (ii) and

$$\text{(iv)} \ \underline{\lim} \ c_n > 0,$$

or a condition that implies (iv), and (4) can be deduced from a certain particular contractive condition.

It has been also shown by that condition (iv) implies A is equivalent to convergence. Therefore, in order to apply Theorem 4.3, it is sufficient only to show that the given particular condition implies (4).

Example 4.1.

One contractive condition that forces (4) is the following one: there exist the constants $a \ge 0$, $0 \le q < 1$ such that for all x, y in E,

$$\|Tx - Ty\| \le q \cdot \max\{a\, \|x - y\|,$$

$$\| x - Tx \| + \| y - Ty \| , \| x - Ty \| + \| y - Tx \| \}.$$

Indeed, replacing x by x_n and y by p in the preceding inequality, we have

$$\| Tx_n - Tp \| \le q \max \{ a \| x_n - p \| ; \| x_n - Tx_n \| + \| p - Tp \| ,$$

$$\| x_n - Tp \| + \| p - Tx_n \| \} \le qa \| x_n - p \| + q \| x_n - Tx_n \| + q \| p - Tx_n \| +$$
$$+ q \max \{ \| p - Tp \| , \| x_n - Tp \| \}$$

and so (4) is satisfied.

4.2 Nonexpansive and Quasi-Nonexpansive Operators

Let E be a strictly convex Banach space. The following lemma is an immediate consequence of strict convexity.

Lemma 4.1. *If E is a strictly convex Banach space and $u, v \in E$ such that $\| v \| \le \| u \|$ and for $0 < t < 1$, $\| (1-t)u + tv \| = \| u \|$, then $u = v$.*

In order to prove an important result concerning the convergence of the Mann iteration we also need the next lemma, which holds in any Banach space.

Lemma 4.2. *Let C be a closed convex subset of a Banach space E and $T : C \to C$ be a quasi nonexpansive operator, p a fixed point of T, and $x_1 \in C$. If $M(x_1, A, T)$ is any normal Mann process (with the sequences $\{x_n\}, \{v_n\}$), then the following are true:*
(i) $\| v_{n+1} - p \| \le \| v_n - p \|$, for each $n = 1, 2, 3, \ldots$
(ii) If $\{v_n\}$ clusters at p, then $\{v_n\}$ converges to p;
(iii) If $\{v_n\}$ clusters at y and z, then $\| y - p \| = \| z - p \|$.

Proof. From part (c) of Theorem 4.2 we deduce that

$$v_{n+1} - p = (1 - c_n)(v_n - p) + c_n(Tv_n - p),$$

where $c_n = a_{n+1,n+1}$. Since T is quasi-nonexpansive, that is

$$\| Tx - p \| \le \| x - p \| , \text{ for all } x \in C,$$

we get

$$\| v_{n+1} - p \| \le (1 - c_n) \| v_n - p \| + c_n \| v_n - p \| = \| v_n - p \| ,$$

which proves (i).

Statements (ii) and (iii) now immediately follow from (i).   □

Theorem 4.4. *Let E be a strictly convex Banach space, C be a closed convex subset of E, and $T : C \to C$ be a continuous and quasi nonexpansive operator, such that $T(C) \subset K \subset C$, where K is compact. Let $x_1 \in C$ and $M(x_1, A, T)$ be a normal Mann process such that the sequence $\{c_n\}$ given by (3) clusters at some $c \in (0,1)$.*

Then the sequences $\{x_n\}, \{v_n\}$ in the Mann process $M(x_1, A, T)$ converge strongly to a fixed point of T .

Proof. We denote by $co\, D$ the closed convex hull of the set D. Since $co\, K \subset C$, it results that

$$T(co\, K) \subset T(C) \subset K \subset co\, K,$$

and so, by Mazur's theorem, the set $co\, K$ is compact.

Since T is continuous, by Schauder's fixed point theorem there exists a point $p \in co\, K$ such that $Tp = p$.

On the other hand, there is a subsequence $\{c_{n_k}\}$ of $\{c_n\}$ such that $c_{n_k} \to c$ (as $k \to \infty$). The corresponding subsequence $\{v_k\} = \{v_{n_k}\}$ of $\{v_n\}$ is contained in $co\, (K \bigcup \{x_1\})$ which is compact again by Mazur's theorem.

Hence, there exists a subsequence of $\{v_k\}$, denoted also by $\{v_k\}$, which converges to some $y \in C$. Of course,

$$c_k \to c$$

and so by Theorem 4.2 and the continuity of T we have

$$v_{k+1} = (1 - c_k)\, v_k + c_k T\, v_k \to (1 - c)\, y + cTy.$$

Since $\{v_n\}$ clusters at both y and $(1 - c)\, y + cTy$, and p is a fixed point of T, part (iii) of Lemma 4.2 gives that

$$\|y - p\| = \|[(1 - c)\, y + cTy] - p\| \,,$$

which can be equivalently written as

$$\|(1 - c)\, (y - p) + c(Ty - p)\| = \|y - p\| \,.$$

Since $\|Ty - p\| \le \|\, y - p\|$ and $0 < c < 1$, then by Lemma 4.1 it results that $y - p = Ty - p$, that is, $Ty = y$. So y is a fixed point of T, and since $\{v_n\}$ clusters at y, part (ii) of Lemma 4.2 implies that $v_n \to y$.

Now, by Theorem 4.1, we have that $x_n \to y$.   □

Remarks.

1) If T is nonexpansive and the normal Mann process is $M(x_1, A_{1/2}, T)$, then from Theorem 4.4 we obtain a result of Edelstein, which in turn is a generalization of the result of Krasnoselskij;

2) If T is nonexpansive and the Mann iteration process is given by $M(x_1, A_\lambda, T)$, with $0 < \lambda < 1$, then from Theorem 4.3 we get as a particular case the result of Schaefer.

In fact, Schaefer's result was obtained in uniformly convex Banach spaces, while Theorem 4.4 and its special case mentioned above work under the weaker hypothesis of strict convexity;

3) We can drop the continuity assumption on T if we are working in a more particular class of Banach spaces, i.e., in uniformly convex Banach spaces (which are also strictly convex). It is well known that any uniformly convex Banach space is also reflexive by Pettis-Milman theorem, and consequently any closed bounded convex set is weakly compact in that ambient space.

We need the following lemma, which is an easy consequence of uniform convexity.

Lemma 4.3. *Let E be a uniformly convex Banach space and $\{c_n\}$ a sequence in $[a, b]$, where $0 < a < b < 1$. Suppose $\{w_n\}, \{y_n\}$ are sequences in E such that $\|w_n\| \leq 1$, $\|y_n\| \leq 1$ for all n. We define a sequence $\{z_n\}$ by*

$$z_n = (1 - c_n)\, w_n + c_n y_n.$$

If $\lim \| z_n \| = 1$, then $\lim \| w_n - y_n \| = 0$.

Remark. For the normal Mann process $M(x_1, A, T)$ with the sequence $\{c_n\}$ given by (3), we shall alternatively use the notation $M(x_1, c_n, T)$.

Theorem 4.5. *Let C be a closed convex subset of a uniformly convex Banach space E, $T : C \to C$ a quasi-nonexpansive operator on C which has at least one fixed point $p \in C$. If $x_1 \in C$ and $M(x_1, c_n, T)$ is a normal Mann process such that the sequence $\{c_n\}$ is bounded away from 0 and 1, then each of the sequences $\{v_{n+1} - v_n\}$ and $\{Tv_n - v_n\}$ converges (strongly) to $0 \in E$.*

Proof. From part (c) of Theorem 4.2 we have that

$$\|v_{n+1} - v_n\| = c_n \, \| Tv_n - v_n \|,$$

and hence, having in view that $0 < a \leq c_n \leq b < 1$, if either one of the sequences $\{v_{n+1} - v_n\}$ or $\{Tv_n - v_n\}$ converges to 0 then the other does also. If $\lim \|v_n - p\| = 0$, then obviously $\lim \|v_{n+1} - v_n\| = 0$. Otherwise, since by Lemma 4.2 the sequence $(\|v_n - p\|)$ is non-increasing, we certainly have $\lim \| v_n - p\| = d > 0$. We define now the sequences $\{w_n\}$, $\{y_n\}$ and $\{z_n\}$ by

$$w_n = (v_n - p)/ \| v_n - p\| \, , \quad y_n = (Tv_n - p)/ \| v_n - p\| \, ,$$

and, respectively,

$$z_n = (v_{n+1} - p)/ \| v_n - p\| .$$

Since, as in the proof of Lemma 4.2, we have

$$v_{n+1} - p = (1 - c_n)(v_n - p) + c_n(Tv_n - p),$$

by dividing it by $\| v_n - p\|$ it results that

$$z_n = (1 - c_n)\, w_n + c_n \, y_n.$$

Since $\|w_n\| = 1$, $\|y_n\| \leq 1$, and $\|z_n\| \to d/d = 1$, by Lemma 4.3 we have that $\lim \|w_n - y_n\| = 0$, which gives

$$\lim \|Tv_n - v_n\| = 0,$$

and this completes the proof.   □

Corollary 4.1. *Let E be a uniformly convex Banach space and $T : E \to E$ a nonexpansive operator which has at least one fixed point. Then for any $\lambda \in (0,1)$, the Krasnoselskij iteration $K(x_1, \lambda, T)$ is asymptotically regular for each $x_1 \in E$.*

Proof. The Krasnoselskij iteration is a particular case of the normal Mann iteration, with matrix A_λ, and so by Theorem 4.5 we get that for any $x_1 \in E$

$$v_{n+1} - v_n = T_\lambda^n x_1 - T_\lambda^{n-1} x_1 \to 0,$$

as required.   □

Remarks.
1) We know from Chapter 3 of this book that, if T is nonexpansive, then the iteration function involved in the Krasnoselskij process, that is

$$T_\lambda = \lambda I + (1 - \lambda)T,$$

is also nonexpansive and has the same fixed points set as T.

2) Using similar arguments, we can prove the next two theorems which are generalizations of the results of Browder and Petryshyn.

Theorem 4.6. *Let C be a closed convex subset of a uniformly convex Banach space E and $T : C \to C$ be a quasi nonexpansive operator on C that has at least one fixed point $p \in C$.*

If $I - T$ is closed and $M(x_1, c_n, T)$ is a normal Mann process with $x_1 \in C$, such that $\{c_n\}$ is bounded away from 0 and 1, then for any sequence $\{v_n\}$ that clusters (strongly) at some $y \in C$, we have $Ty = y$ and the sequences $\{x_n\}, \{v_n\}$ converge (strongly) to y.

Proof. There exists a subsequence $\{v_{n_k}\}$ of $\{v_n\}$ such that $v_{n_k} \to y$. It follows by Theorem 4.4 that $(I - T)v_n \to 0$, and hence $(I - T)v_{n_k} \to 0$.

Since $I - T$ is closed, we deduce that $(I - T)y = 0$, that is $Ty = y$ and, as $\{v_n\}$ clusters at y, it follows by Lemma 4.2 that $v_n \to y$.

Since $v_n - x_{n+1} = v_n - Tv_n \to 0$, we finally get $x_n \to y$.   □

Remarks.
1) Any continuous operator T on C has the property that $I - T$ is continuous on C, and so is closed. Hence, for any nonexpansive operator T, $I - T$ is closed;

2) In the previous chapter we gave a result, namely Theorem 3.2, on the approximation of fixed points of demicompact operators by means of the

Krasnoselskij iteration. We can improve Theorem 4.5 by considering the demi-closedness property instead of the closedness of the operator $I - T$, as in Theorem 4.7 below.

Definition 4.2. A mapping $S : C \to E$ is said to be *demiclosed* provided that if $\{u_n\}$ is a sequence in C which converges weakly to $u \in C$, and if $\{Su_n\}$ converges strongly to $v \in E$, then $Su = v$.

Remark. For a closed and convex set C, every weakly continuous mapping $T : C \to C$ is weakly closed and every weakly closed mapping of $T : C \to C$ is demiclosed. We have

Theorem 4.7. *Let C be a closed convex subset of a uniformly convex Banach space E, $T : C \to C$ a nonexpansive operator on C that has at least one fixed point $p \in C$.*

Let $x_1 \in C$ and $M(x_1, c_n, T)$ be the normal Mann process such that $\{c_n\}$ is bounded away from 0 and 1. Then the following are true:

(i) There exists a subsequence of $\{v_n\}$ which converges weakly to some $y \in C$, and if $I - T$ is demiclosed then each weak subsequential limit point of $\{v_n\}$ is a fixed point of T.

(ii) If $I - T$ is demiclosed and T has only one fixed point $p \in C$, then the sequences $\{x_n\}$, $\{v_n\}$ converge weakly to p;

(iii) If $I - T$ is weakly closed, then each weak cluster point of $\{v_n\}$ is a fixed point of T.

Remarks.

1) The assumption "T has at least one fixed point" involved in Theorems 4.1 - 4.7 is very natural in this context. Indeed, if C is bounded and convex and $T : C \to C$ is weakly continuous, then T has at least one fixed point, by the Tihonov fixed point theorem, while in the case of nonexpansive operators the conclusion holds by the Browder-Gohde-Kirk fixed point theorem (Theorem 1.2 in this book, see also Theorem 3.1);

2) It is well known (see Opial [Op67a]) that if T is nonexpansive and the uniformly convex Banach space E has a weakly continuous duality mapping, then $I - T$ is necessarily demiclosed. However, there exist some uniformly convex Banach spaces that do not have weakly continuous duality mappings (e.g. L^p, $1 < p < \infty$, $p \neq 2$);

3) As T weakly continuous implies $I - T$ demiclosed, by Theorem 4.7 we obtain that, if T has only one fixed point $p \in C$, then the Krasnoselskij iteration $K(x_1, \lambda, T)$ converges to this fixed point, i.e., $v_{n+1} = T_\lambda^n x_1 \to p$, which is valid in any uniformly convex Banach space;

4) In view of Theorem 4.1 (which extends Mann's result), in order to use the Mann iterative process for nonexpansive type mappings all one needs is to establish the convergence of either $\{x_n\}$ or $\{v_n\}$. Consequently, in the following we shall consider only the sequence $\{v_n\}$ which will be denoted by $\{x_n\}$;

5) In Theorem 3.2 we used the demicompactness condition in order to obtain the convergence of Krasnoselskij iteration.

This result could be extended to Mann iteration by simultaneously weakening the demicompactness property.

Definition 4.3. Let E be a Banach space, C a convex subset of E and $T : C \to C$ an operator with F_T the set of fixed points. T is said to *satisfy condition (D) on C* if there exists a nondecreasing function $\varphi : [0, \infty) \to [0, \infty)$ with $\varphi(0) = 0$ and $\varphi(r) > 0$ for $r > 0$ such that

$$\| x - Tx \| \geq \varphi \left(\inf \{ \| x - z \| : z \in F_T \} \right)$$

for all $x \in C$.

A relationship between demicompact operators and mappings that satisfy condition (D) is shown by the next lemma.

Lemma 4.5. *Let C be a closed bounded subset of a Banach space E and $T : C \to C$ an operator with $F_T \neq \emptyset$. If $I - T$ maps closed bounded subsets of C onto closed subsets of E, then T satisfies condition (D) on C.*

Let $\{x_n\}$ be the normal Mann iteration associated to $T : C \to C$ and defined by $x_1 \in C$ and the sequence $\{c_n\}$, that is, the iteration $M(x_1, c_n, T)$ given by

$$x_{n+1} = (1 - c_n) x_n + c_n T x_n, \tag{4}$$

where $c_n \in [a, b]$ and $0 < a < b < 1$.

We state without proof the following result based on condition (D).

Theorem 4.8. *Let C be a closed, bounded, convex, nonempty subset of a uniformly convex Banach space E and $T : C \to C$ be a nonexpansive operator with the fixed point set of T in C denoted by F_T. If T satisfies condition (D), then for any $x_1 \in C$ the Mann iteration (4) converges to a point of F_T.*

Remark. The fixed point to which a certain normal Mann iterative process converges depends, in general, on the initial approximation x_1 as well as on the sequence $\{c_n\}$ that determine the Mann iteration. Moreover, the Mann iteration need not converge to the fixed point of T nearest x_1, as shown by the following example.

Example 4.2. Let E be the space $\mathbb{R}^2$ equipped with the Euclidean norm and, with (r, θ) denoting the polar coordinates. Let

$$C = \left\{ (r, \theta) : 0 \leq r \leq 1, \frac{\pi}{4} \leq \theta \leq \frac{\pi}{2} \right\}.$$

The map $T : C \to C$ defined by

$$T((r, \theta)) = \left(r, \frac{\pi}{2} \right), \quad \text{for each point } (r, \theta) \text{ in } C,$$

is nonexpansive and the set of its fixed points is the line segment

$$F_T = \left\{ \left(r, \frac{\pi}{2} \right) : \ 0 \le r \le 1 \right\}.$$

Take $U_0 = (r_0, \theta_0) = (1, \frac{\pi}{4})$ and $\alpha_n \in [0, 1]$ for $n \ge 1$, and construct the Mann sequence $\{U_n\}$ by

$$U_{n+1} = (1 - \alpha_n)U_n + \alpha_n T U_n, \ n \ge 0$$

which gives

$$U_{n+1} \equiv (r_{n+1}, \theta_{n+1}) = \left(r_n, \theta_n + \alpha_n \left(\frac{\pi}{2} - \theta_n \right) \right).$$

Hence $r_n = r_0 = 1, \ n \ge 0$ and

$$\theta_{n+1} = \alpha_n \frac{\pi}{2} + (1 - \alpha_n)\theta_n, \ n \ge 0 \text{ and } \theta_0 = \frac{\pi}{4}.$$

1) For $\alpha_n \equiv 1$ we get $\theta_n = \frac{\pi}{2}, \ n \ge 0$ and so

$$U_n \to \left(1, \frac{\pi}{2} \right) \in F_T$$

which is not the nearest fixed point of T to U_0, because the nearest one is the point $p = \left(\frac{\sqrt{2}}{2}, \frac{\pi}{2} \right)$.

2) The same happens when $\alpha_n \equiv \frac{1}{2}$, when we find

$$\theta_n = \frac{\pi}{2^{n+2}} + \frac{\pi}{2} \cdot \frac{2^n - 1}{2^n}, \ n \ge 0$$

and hence

$$U_n \to \left(1, \frac{\pi}{2} \right) \in F_T$$

which is also not the nearest fixed point of T to U_0;

3) For $\alpha_n \equiv 0$, we get $\theta_n = \frac{\pi}{4}$ and hence

$$\lim_{n \to \infty} U_n = \left(1, \frac{\pi}{4} \right) \notin C.$$

An important class of quasi-contractive mappings, which is independent of the class of strictly pseudocontractive mappings, is the class of Zamfirescu mappings. In Section 2.3 we have proved (Theorem 2.4) that for any Zamfirescu mapping T considered on a complete metric space, the Picard iteration converges to the unique fixed point of T.

It is the aim of this section to show that, in a more particular ambient space, suitable for constructing the Mann iteration, the latter iterative procedure also converges to the unique fixed point of T.

Theorem 4.9. *Let E be a uniformly convex Banach space, K a closed convex subset of E, and $T : K \to K$ be a Zamfirescu mapping. Then the Mann iteration $\{x_n\}$,*

$$x_{n+1} = (1 - \alpha_n)x_n + \alpha_n Tx_n, \ n = 1, 2, \ldots \qquad (5)$$

with $\{\alpha_n\}$ satisfying the conditions
(i) $\alpha_1 = 1$; (ii) $0 \leq \alpha_n < 1$, for $n > 1$ and (iii) $\sum \alpha_n(1 - \alpha_n) = \infty$,
converges to the unique fixed point of T.

Proof. Theorem 2.4 shows that T has a unique fixed point in K. Let us denote it by p. For any $x_1 \in K$, we have

$$\| x_{n+1} - p \| \leq (1 - \alpha_n) \| x_n - p \| + \alpha_n \| Tx_n - p \| .$$

Since any Zamfirescu mapping is quasi-contractive, we deduce that

$$\| Tx_n - p \| \leq \| x_n - p \| ,$$

which shows that the sequence $\{\| x_n - p \|\}$ is decreasing. We also have

$$\| x_n - Tx_n \| = \| (x_n - p) - (Tx_n - p) \| \leq 2 \| x_n - p \| .$$

Now let us assume that there exist a number $a > 0$ such that $\| x_n - p \| \geq a$, for all n.

Suppose $\{\| x_n - Tx_n \|\}_{n \geq 1}$ does not converge to zero. Then there are two possibilities: either there exists an $\varepsilon > 0$ such that $\| x_n - Tx_n \| \geq \varepsilon$ for all n or

$$\liminf \| x_n - Tx_n \| = 0.$$

In the first case, using Lemma of Groetsch, see Exercise 4.11, with $b = 2\delta_X \left(\varepsilon / \| x_0 - p \| \right)$ we get

$$\| x_{n+1} - p \| \leq (1 - \alpha_n(1 - \alpha_n) b) \| x_n - p \| \leq$$

$$\leq \| x_{n-1} - p \| - \alpha_{n-1}(1 - \alpha_{n-1})b \| x_n - p \| - b\alpha_n(1 - \alpha_n) \| x_n - p \| \leq$$

$$\leq \| x_{n-1} - p \| - b[\alpha_{n-1}(1 - \alpha_{n-1}) + \alpha_n(1 - \alpha_n)] \cdot \| x_n - p \| .$$

By induction one obtains

$$a \leq \| x_{n+1} - p \| \leq \| x_0 - p \| - b \sum_{k=0}^{n} \alpha_k(1 - \alpha_k) \cdot \| x_n - p \| .$$

Therefore

$$a \left[1 + b \sum_{k=0}^{n} \alpha_k(1 - \alpha_k) \right] \leq \| x_0 - p \| ,$$

which contradicts (iii).

In the second case, there exists a subsequence $\{x_{n_k}\}$ such that

$$\lim_{k} \| x_{n_k} - T x_{n_k} \| = 0. \tag{6}$$

Using similar arguments to those exploited in proving Theorem 2.4, we get

$$\| T x_{n_k} - T x_{n_l} \| \leq L \left[\| x_{n_k} - T x_{n_k} \| + \| x_{n_l} - T x_{n_l} \| \right],$$

where

$$L = \max \left\{ \frac{\alpha}{1 - \alpha}, \beta, \frac{\gamma}{1 - 2\gamma} \right\},$$

α, β, γ being the constants appearing in conditions $(z_1) - (z_3)$. The previous inequality shows that $\{T x_{n_k}\}$ is a Cauchy sequence, hence convergent.

Let u be its limit. From (6) it results that

$$\lim_{k \to \infty} x_{n_k} = \lim_{k \to \infty} T x_{n_k} = u.$$

Moreover,

$$\| u - Tu \| \leq \| u - x_{n_k} \| + \| x_{n_k} - T x_{n_k} \| + \| T x_{n_k} - Tu \|.$$

We will show that $u = Tu$, that is, u is a fixed point of T. Indeed, if x_{n_k}, u satisfy (z_1), then

$$\| T x_{n_k} - Tu \| \leq \alpha \| x_{n_k} - u \|.$$

If x_{n_k}, u satisfy (z_2), then

$$\| T x_{n_k} - Tu \| \leq \beta \left[\| x_{n_k} - T x_{n_k} \| + \| u - Tu \| \right]$$

which leads to

$$\| u - Tu \| \leq \left[\| u - x_{n_k} \| + (1 + \beta) \| x_{n_k} - T x_{n_k} \| \right] / (1 - \beta)$$

and, finally, if x_{n_k}, u satisfy (z_3), then

$$\| T x_{n_k} - Tu \| \leq \gamma \left[\| x_{n_k} - Tu \| + \| u - T x_{n_k} \| \right] \leq$$

$$\leq \gamma \left[\| x_{n_k} - T x_{n_k} \| + \| T x_{n_k} - Tu \| + \| u - T x_{n_k} \| \right],$$

or

$$\| T x_{n_k} - Tu \| \leq \gamma (1 - \gamma)^{-1} \left[\| x_{n_k} - T x_{n_k} \| + \| u - T x_{n_k} \| \right].$$

Hence $u = Tu$.

Now, since p is the unique fixed point of T, it results that $p = u$ and so the two conditions $\lim_{k} x_{n_k} = u (= p)$ and $\{\| x_n - p \|\}$ decreasing with respect to n yields $\lim_{n} x_n = p$. □

Remarks.

1) Having in view that any Kannan mapping is a Zamfirescu mapping, from Theorem 4.9 we obtain the convergence of the Mann iteration in the class of Kannan mappings;

2) If $\alpha_n = \dfrac{1}{2}$ for all n, from Theorem 4.9 we obtain Theorem 2 and Theorem 3 of Kannan [Knn71], while if $\alpha_n = \lambda$ for all n, we obtain Theorem 3 of Kannan [Knn73];

3) As both Picard iteration and Krasnoselskij iteration converge in the class of Zamfirescu mappings, it is natural to try to compare these methods in order to know which one converges faster to the (unique) fixed point of T, see Chapter 9.

Theorem 4.9 can be extended to an arbitrary Banach space, by simultaneously weakening the conditions on the sequence involved in the Mann iteration, as shown by the following theorem whose proof is very simple.

Theorem 4.10. *Let E be an arbitrary Banach space, K a closed convex subset of E, and $T : K \to K$ an operator satisfying conditions $(z_1) - (z_3)$ in Theorem 2.4 with $d(x,y) = \|x - y\|$. Let $\{x_n\}_{n=0}^\infty$ be defined by (5) and $x_0 \in K$, with $\{\alpha_n\} \subset [0,1]$ satisfying*

$$\text{(iv)} \qquad \sum_{n=0}^\infty \alpha_n = \infty.$$

Then $\{x_n\}_{n=0}^\infty$ converges strongly to the unique fixed point of T.

Proof. By Theorem 2.4, we know that T has a unique fixed point in K. Call it p and consider $x, y \in K$.

By $(z_1) - (z_3)$, with $a \equiv \alpha, b \equiv \beta, c \equiv \gamma$ similarly to the proof of Theorem 2.4 we find that, denoting

$$\delta = \max\left\{a, \frac{b}{1-b}, \frac{c}{1-c}\right\}, \tag{7}$$

we have $0 < \delta < 1$ and the inequality

$$\|Tx - Ty\| \le \delta\|x - y\| + 2\delta\|x - Tx\| \tag{8}$$

holds, for all $x, y \in K$.

Let $\{x_n\}_{n=0}^\infty$ be the Mann iteration (5), with $x_0 \in K$ arbitrary. Then

$$\|x_{n+1} - p\| = \|(1-\alpha_n)x_n + \alpha_n Tx_n - (1 - \alpha_n + \alpha_n)p\| =$$
$$= \|(1-\alpha_n)(x_n - p) + \alpha_n(Tx_n - p)\| \le$$
$$\le (1-\alpha_n)\|x_n - p\| + \alpha_n\|Tx_n - p\|. \tag{9}$$

Take $x := p$ and $y := x_n$ in (8) to obtain

$$\|Tx_n - p\| \le \delta \cdot \|x_n - p\|,$$

which together with (9) yields

$$\|x_{n+1} - p\| \le [1 - (1-\delta)\alpha_n]\|x_n - p\|, \quad n = 0, 1, 2, \dots. \tag{10}$$

Inductively we get

$$\|x_{n+1} - p\| \leq \prod_{k=0}^{n} \left[1 - (1 - \delta)\alpha_k\right] \cdot \|x_0 - p\|, \quad n = 0, 1, 2, \ldots . \tag{11}$$

As $0 < \delta < 1$, $\alpha_k \in [0, 1]$ and $\sum_{k=0}^{\infty} \alpha_k = \infty$, by a standard argument it results that

$$\lim_{n \to \infty} \prod_{k=0}^{n} \left[1 - (1 - \delta)\alpha_k\right] = 0,$$

which by (11) implies

$$\lim_{n \to \infty} \|x_{n+1} - p\| = 0,$$

i.e., $\{x_n\}_{n=0}^{\infty}$ converges strongly to p. $\qquad\square$

Remarks.

1) Condition (iv) in Theorem 4.10 is more relaxed than conditions (i) - (iii) in Theorem 4.9. Indeed, in view of

$$0 < \alpha_k(1 - \alpha_k) < \alpha_k,$$

valid for all α_k satisfying (i) - (ii), condition (iii) implies (iv).
There also exist values of $\{\alpha_n\}$, e.g., $\alpha_n \equiv 1$, such that (iv) is satisfied but (iii) is not;

2) Since the contractive condition of Kannan, i.e., condition (8) in Section 2.3, is a special case of Zamfirescu contractive conditions, Theorems 2 and 3 of Kannan [Knn71] are special cases of Theorem 4.10 or Theorem 4.9 in this section, with $\alpha_n = 1/2$.
Theorem 3 of Kannan [Knn73] is the special case of Theorem 4.10 or Theorem 4.9 with $\alpha_n = \lambda$, $0 < \lambda < 1$. However, note that all the results of Kannan are obtained in uniformly Banach spaces, like Theorem 4.9, while Theorem 4.10 is valid in arbitrary Banach spaces;

3) Because of the more restrictive assumptions (i) - (ii), the convergence of Picard iteration cannot be obtained as a particular case of Theorem 4.9, but, due to the more natural assumption (iv), it can be obtained by Theorem 4.10, by taking $\alpha_n \equiv 1$;

4) By Theorem 4.10 we can also obtain, as a particular case, a convergence theorem for Mann iteration in the class of operators that satisfy Chatterjea's contractive condition (34) in Section 2.7.

4.3 Strongly Pseudocontractive Operators

Let E be a Banach space, K a subset of E, and $T : K \to K$ a strongly pseudocontractive operator, i.e. (see Definition 1.13 and Remark 2 following it), there exists a number $t > 1$ such that the inequality

$$\| x - y \| \leq \|(1 + r)(x - y) - rt(Tx - Ty)\| \tag{12}$$

holds for all $x, y \in K$ and $r > 0$.

As mentioned in Chapter 1, a mapping T is strongly pseudocontractive if and only if $I - T$ is a strongly accretive mapping, i.e. (see Definition 1.14 in Chapter 1) there exist $j(x - y) \in J(x - y)$ and a positive number k such that

$$\langle (I - T) x - (I - T) y , \; j(x - y) \rangle \geq k \, \| x - y \|^2 \tag{12'}$$

that, in turn, is equivalent to the fact that the next inequality

$$\| x - y \| \leq \| x - y + r \left[(I - T - kI) x - (I - T - kI) y \right] \| \tag{12''}$$

holds for any $x, y \in K$ and any $r > 0$ (where $k = \dfrac{t - 1}{t}$).

Based on the form (12'') of the strong pseudo-contractiveness property, one can prove that the Mann iteration process converges strongly to the unique fixed point of a Lipschitzian and strongly pseudocontractive operator.

Theorem 4.11. *Let E be a Banach space and K a nonempty closed convex and bounded subset of E. If $T : K \to K$ is a Lipschitzian strongly pseudocontractive operator such that the fixed point set of T, F_T, is nonempty, then the Mann iteration $\{x_n\} \subset K$ generated by (5) with $x_1 \in K$ and the sequence $\{\alpha_n\} \subset (0, 1]$, with $\{\alpha_n\}$ satisfying*

$$(i) \quad \sum_{n=1}^{\infty} \alpha_n = \infty ; \qquad (ii) \quad \alpha_n \to 0 \quad (as \quad n \to \infty),$$

converges strongly to the unique fixed point of T.

Proof. Let p be a fixed point of T. Since T is a strongly pseudocontractive operator, $I - T$ is strongly accretive, i.e., the inequality (12'') holds for any $x, y \in K$ and $r > 0$. Let $L > 0$ be the Lipschitz constant. Then, from the definition of $\{x_n\}$,

$$x_{n+1} = (1 - \alpha_n) x_n + \alpha_n T x_n , \quad n = 1, 2, \ldots \tag{13}$$

and therefore we have

$$x_n = x_{n+1} + \alpha_n x_n - \alpha_n T x_n = (1 + \alpha_n) x_{n+1} +$$

$$+\alpha_n(I - T - kI)\,x_{n+1} - (2 - k)\,\alpha_n\,x_{n+1} + \alpha_n\,x_n + \alpha_n(T\,x_{n+1} - Tx_n) =$$
$$= (1 + \alpha_n)\,x_{n+1} + \alpha_n(I - T - kI)\,x_{n+1} - (2 - k)\,\alpha_n[(1 - \alpha_n)\,x_n + \alpha_n\,T\,x_n] +$$
$$+\alpha_n\,x_n + \alpha_n(T\,x_{n+1} - T\,x_n) = (1 + \alpha_n)\,x_{n+1} + \alpha_n(I - T - kI)\,x_{n+1} -$$
$$-(1 - k)\,\alpha_n\,x_n + (2 - k) \cdot \alpha_n^2(x_n - T\,x_n) + \alpha_n(T\,x_{n+1} - T\,x_n).$$

As $Tp = p$, we have

$$x_n - p = (1 + \alpha_n)(x_{n+1} - p) + \alpha_n(I - T - kI)(x_{n+1} - p) - (1 - k)\,\alpha_n(x_n - p) +$$
$$+(2 - k)\,\alpha_n^2(x_n - Tx_n) + \alpha_n(T\,x_{n+1} - T\,x_n).$$

Now, using the inequality (12"), we get

$$\|x_n - p\| \geq (1 + \alpha_n)\,\|\,x_{n+1} - p\| - (1 - k)\,\alpha_n\,\|x_n - p\| -$$
$$-(2 - k)\,\alpha_n^2\,\|\,x_n - Tx_n\| - \alpha_n\,\|\,T\,x_{n+1} - T\,x_n\|.$$

Since T is Lipschitzian, it follows that

$$\|\,T\,x_{n+1} - T\,x_n\| \leq L\,\|\,x_{n+1} - x_n\| \leq L(L + 1)\,\alpha_n\,\|\,x_n - p\|,$$

and then

$$\|x_n - p\| \geq (1 + \alpha_n)\,\|\,x_{n+1} - p\| - (1 - k)\,\alpha_n\,\|x_n - p\| -$$
$$-(2 - k)\,\alpha_n^2\,\|\,x_n - Tx_n\| - L(L + 1)\,\alpha_n^2\,\|\,x_n - p\|.$$

Hence

$$\|\,x_{n+1} - p\| \leq [1 + (1 - k)\,\alpha_n](1 + \alpha_n)^{-1}\,\|\,x_n - p\| + (2 - k)\,\alpha_n^2(1 + \alpha_n)^{-1} \cdot$$
$$\cdot \|x_n - Tx_n\| + L(L + 1)\,\alpha_n^2(1 + \alpha_n)^{-1}\,\|\,x_n - p\| \leq$$
$$\leq [1 + (1 - k)\alpha_n](1 - \alpha_n + \alpha_n^2)\,\|x_n - p\| + (2 - k)\alpha_n^2\,\|\,x_n - Tx_n\| +$$
$$+L(L + 1)\,\alpha_n^2\,\|\,x_n - p\|,$$

$$(14)$$

and, by using $\|x_n - Tx_n\| \leq (L + 1)\,\|x_n - p\|$, we obtain

$$\|\,x_{n+1} - p\| \leq (1 - k\alpha_n + M\alpha_n^2)\,\|\,x_n - p\|,$$

for some constant $M > 0$.

Since $\alpha_n \to 0$, there exists $N_0 \geq 0$ such that

$$M\alpha_n \leq k(1 - k), \forall n \geq N_0,$$

we get

$$\|\,x_{n+1} - p\| \leq (1 - k^2\alpha_n)\,\|\,x_n - p\|, \forall n \geq N_0.$$

Now using Lemma 1.2, it follows that the sequence $\{\| x_n - p\|\}$ converges to 0, that is, $\{x_n\}$ converges strongly to the (unique) fixed point p of T.   □

By the same technique of proof as above, we can obtain a convergence theorem for the Krasnoselskij iteration method in the class of Lipschitzian strictly pseudocontractive operators in a Banach space setting.

Remind that in Section 3.3 we also presented a result for Krasnoselskij iteration (Theorem 3.6) in the class of Lipschitzian generalized pseudocontractive operators, but in a Hilbert space setting. Note that, due to assumption (ii) in Theorem 4.11, the next Corollary cannot be obtained directly as a particular case of this theorem, but can be proved independently.

Corollary 4.2. *Let K and T be as in Theorem 4.11. If $\alpha_n = \dfrac{k}{2(3 + 3L + L^2)}$,*
where $k = \dfrac{t - 1}{t}$ and $Fix\,(T) = \{p\}$, then the sequence $\{x_n\}$ generated by (13) converges strongly to the unique fixed point of T and we have the estimate

$$\| x_{n+1} - p\| \leq \rho^n \, \| x_1 - p\| ,$$

where

$$\rho = 1 - k^2 / [4(3 + 3L + L^2)].$$

Proof. We have $0 < \alpha_n < 1$. As $p = Tp$, we get

$$[1 + (1 - k)\alpha_n](1 - \alpha_n + \alpha_n^2) = 1 - k\alpha_n + \alpha_n^2 - (1 - k)\alpha_n^2(1 - \alpha_n) \leq$$

$$\leq 1 - k\alpha_n + \alpha_n^2$$

and

$$\| x_n - Tx_n\| \leq (1 + L) \, \| x_n - p\| .$$

Hence, by (7) we obtain

$$\| x_{n+1} - p\| \leq (1 - k\alpha_n) \| x_n - p\| + [1 + (2 - k)(1 + L) +$$

$$+L(L + 1)\alpha_n^2] \| x_n - p\| < [1 - k\alpha_n + (3 + 3L + L^2)\alpha_n^2] \, \| x_n - p\| =$$

$$= \left[1 - k^2/(4(3 + 3L + L^2))\right] \, \| x_n - p\| = \rho \, \| x_n - p\| .$$

Therefore

$$\| x_{n+1} - p\| \leq \rho^n \, \| x_1 - p\| ,$$

as required.   □

Remark.
Even if the great majority of convergence theorems for the Mann iteration existing in literature are obtained by imposing a condition of the form (ii), this condition turned out to be artificial and unnecessary, as shown by Example 4.3 in the case of Krasnoselskij iteration, see also Chapter 9, for the more general Mann iteration.

Theorem 4.12 illustrates this fact and, moreover, points out that the boundedness of K in Theorem 4.11 is also unnecessary to get the convergence of Mann iteration. The next example shows that assumptions like (ii) in Theorem 4.11 are often artificial, being tributary to the particular technique of proof used.

Example 4.3. Let T be as in Example 3.1, i.e., $E = \mathbb{R}$ with the usual norm, $K = \left[\frac{1}{2}, 2\right]$ and $T : K \to K$ a function given by $Tx = \frac{1}{x}$, for all x in K. Then:

(a) T is Lipschitzian with constant $L = 4$;
(b) T is strongly pseudocontractive (with any constant $k \in (0, 1)$);
(c) Taking $\alpha_n = \lambda \in (0, 1)$, condition $\lim_{n \to \infty} \alpha_n = 0$ is not satisfied, but
(d) A certain Mann iteration does converge to the unique fixed point of T. Indeed, for any $t > 1$, we have

$$\|x - y\| \leq \|(1 + r)(x - y) - rt(Tx - Ty)\|$$

which is equivalent to

$$|x - y| \leq |x - y| \cdot \left|1 + r + \frac{rt}{xy}\right|$$

valid for all $x, y \in \left[\frac{1}{2}, 2\right]$ and $r > 0$. Moreover, using Theorem 3.6, which can be applied here, since T is also generalized pseudo-contractive, we deduce that Krasnoselskij iteration (which is in fact a Mann-type iteration procedure with a constant sequence $\alpha_n \equiv \lambda \in (0, 1)$), converges strongly to the unique fixed point of T, $p = 1$, for any initial approximation $x_0 \in \left[\frac{1}{2}, 2\right]$ and $\lambda \in I$, I an interval in $(0, 1)$, although $\lim_{n \to \infty} \alpha_n = \lambda \neq 0$.

Theorem 4.12. *Let E be a Banach space and K a nonempty closed convex subset of E. If $T : K \to K$ is a Lipschitzian (with constant L) and strongly pseudocontractive operator (with constant k) such that the fixed point set of T, F_T, is nonempty, then the Mann iteration $\{x_n\} \subset K$ generated by (5) with $x_1 \in K$ and the sequence $\{\alpha_n\} \subset (0, 1]$, satisfying (i) and*

$$\alpha_n \leq \frac{k - \eta}{(L + 1)(L + 2 - k)},$$

for some $\eta \in (0, k)$, converges strongly to the unique fixed point p of T.
Moreover, there exists $\{\beta_n\}_{n \geq 0}$, a sequence in $(0, 1)$ with $\beta_n \geq (\eta/(1 + k))\alpha_n$, such that for all $n \in \mathbb{N}$, the following estimate holds

$$\|x_{n+1} - p\| \leq \prod_{j=1}^{n}(1 - \beta_j)\,\|x_1 - p\|.$$

Proof. Define $\delta_n := \|x_n - p\|$, for each $n \in \mathbb{N}$. Like in the proof of Theorem 4.11, it follows that

$$\delta_n \geq (1+\alpha_n)\delta_{n+1} - (1-k)\alpha_n\delta_n - (2-k)\alpha_n^2 \|x_n - Tx_n\| - L(L+1)\alpha_n^2\delta_n. \quad (15)$$

Since T is Lipschitzian, we have

$$\|x_n - Tx_n\| \leq (L+1)\delta_n. \quad (16)$$

By denoting

$$A_n := 1 + (1-k)\alpha_n + (2-k+L)(L+1)\alpha_n^2, \; B_n := 1 + \alpha_n \text{ and } \beta_n := 1 - \frac{A_n}{B_n}$$

by (15) and (16) we obtain

$$\delta_{n+1} \leq \frac{A_n}{B_n}\delta_n. \quad (17)$$

On the other hand,

$$\beta_n = \frac{\alpha_n}{1+\alpha_n}\left[k - (L+1)(L+2-k)\alpha_n\right] \geq \frac{\alpha_n}{1+\alpha_n}\eta \geq \frac{\eta}{1+k}\alpha_n.$$

Further, from (17) we have

$$\delta_{n+1} \leq \frac{A_n}{B_n} \cdots \frac{A_1}{B_1} = \prod_{j=1}^{n}(1-\beta_j)\delta_1.$$

Now, clearly, $\sum_{n=1}^{\infty}\beta_n = \infty$, and hence $\prod_{j=1}^{\infty}(1-\beta_j) = 0$. Thus $\delta_n \to 0$, i.e., $x_n \to p$ in norm as $n \to \infty$. $\qquad\square$

Now, by Theorem 4.12, we can obtain directly a convergence theorem regarding the Krasnoselskij iteration procedure, which is given by formula (5) with $\alpha_n \equiv \lambda$.

Corollary 4.3. *Let E, K, T, L, k, p, η be as in Theorem 4.12. Then the Krasnoselskij iteration $\{x_n\} \subset K$ generated by $x_1 \in K$ and (18), where $\lambda \in (0, a)$, and*

$$a = k/[(L+1)(L+2-k)],$$

converges strongly to the (unique) fixed point p of T. Moreover, the following estimate holds

$$\|x_{n+1} - p\| \leq q^n \|x_1 - p\|,$$

where

$$q = \frac{1 + (1-k)\lambda + (L+1)(L+2-k)\lambda^2}{1+\lambda}.$$

Proof. Take $\alpha_n \equiv \lambda$ in Theorem 4.12. $\qquad\square$

If T is not a Lipschitzian operator, we can still prove the convergence of the Mann iteration, but in some particular Banach spaces as, for example, in uniformly smooth Banach spaces. A typical result of this kind is given by the next theorem, which is a particular case of Theorem 5.2 in Chapter 5.

Theorem 4.13. *Let E be a real uniformly smooth Banach space and K a bounded closed convex and nonempty subset of E. Let $T : K \to K$ be a strongly pseudocontractive operator such that $Tp = p$ for some $p \in K$, and let $\{x_n\}$ be the Mann iteration process generated by $x_1 \in K$ and the sequence $\{\alpha_n\}$ satisfying the following conditions:*
(i) $0 \le \alpha_n < 1$ for all $n \ge 1$;
(ii) $\lim\limits_{n \to \infty} \alpha_n = 0$;
(iii) $\sum\limits_{n=0}^{\infty} \alpha_n = \infty$.
Then, for arbitrary $x_1 \in K$, the sequence $\{x_n\}$ given by (5) converges strongly to p and p is unique.

Proof. We use the fact that in a uniformly smooth Banach space Lemma 1.1 is valid. For the rest of the proof see Theorem 5.2. $\qquad\square$

4.4 Bibliographical Comments

§4.1.

The general Mann iterative process given in Definition 4.1 was introduced in 1953 by Mann [Man53]. Its convergence was stated in a Banach space setting but, as shown by Dotson [Dot70, Theorem 1], it is valid in the more general context of a locally convex Haussdorf linear topological space, as stated in Theorem 4.1.

Definitions 4.1 and 4.2 as well as Theorems 4.1 and 4.2 are taken from Dotson [Dot70], where they appear as Theorems 1 and 2, respectively. Hints on the proof of Theorem 4.1 are given in the same paper, Dotson [Dot70].

Theorem 4.3 is in fact Theorem 1 in Rhoades [Rh95b], here with a slight correction (the Banach space setting put instead of the metric space setting in original, obviously inappropriate for the Mann iteration).

Example 4.1 is Corollary 1 in Rhoades [Rh95b], where many other special cases belonging to these type of conditions can be also found.

The notation $M(x_1, \alpha_n, T)$ of a normal Mann iterative process appears to have been first used by Senter and Dotson [SeD74] in 1974.

§4.2.

The content of this Section is taken from Dotson [Dot70], paragraph 3. Theorem 4.4 is Theorem 3, while Lemma 4.1 and Lemma 4.2 are respectively Lemma 1 and Lemma 2 in the same paper, Dotson [Dot70].

The rest of the Section is taken mainly from paragraph 4 in Dotson [Dot70]: Theorem 4.5 being Theorem 4, while Theorem 4.6 is Theorem 5. Corollary 4.1 is a result due to Browder and Petryshyn [BrP67].

Theorem 4.7 is also taken from Dotson [Dot70], where ii) appears as Theorem 6, while Theorem 4.8 is Theorem 1 in Senter and Dotson [SeD74], where a more general result is proved, i.e., Theorem 2, which considers T a quasi-nonexpansive operator and C not necessarily bounded.

Notice that in the same paper of Dotson [Dot70] one can find similar results for the Krasnoselskij iteration in Hilbert and Banach spaces.

Example 4.2 is taken from Senter and Dotson [SeD74]. For further results on the approximation of fixed points of quasi-nonexpansive mappings and generalized nonexpansive mappings in uniformly Banach spaces satisfying Opial's condition, see Park, J.Y. and Jeong, J.U. [PJe94].

Theorem 4.9 in this section is Theorem 4 in Rhoades [Rh74a], slightly reformulated. In the same paper one can find suitable examples illustrating the relationships existing between the classes of nonexpansive, quasi-nonexpansive, strictly pseudocontractive and generalized contractive mappings, respectively.

The stability of the Mann iteration for Zamfirescu operators was studied in Harder and Hicks [HH88b].

A survey on the relevant results regarding the convergence of Mann iteration for several classes of Lipschitzian and pseudo-contractive operators in Hilbert spaces are given in Chidume and Moore [ChM99].

Theorem 4.10 is Theorem 2 in Berinde [Be03e]. The corresponding result for Ishikawa iteration, that extends Theorem 4.10, was obtained in [Be04c].

§4.3.

The equivalence of the inequalities (12') and (12") quoted at the beginning of this Section is proved in Bogin [Bog74], see lemma of Kato [Kat67] given as Exercise 4.12.

The first part of this Section, including Theorem 4.11 and Corollary 4.2, is taken from Liu, Liwei [LiW97].

Some other results related to that in Theorem 4.11, established for various particular Banach spaces, are given in Chidume [Chi87], [Ch90a], [Ch94b]; Tan and Xu, H.K. [TX93c]; Weng [We91a]; Bethke [Bet89]; Kang, Z.B. [Kng91]; Schu [Sc91f]; Xu, Z.B. and Roach [XuR92]; Osilike and Udomene [OU01a].

Theorem 4.12 and Corollary 4.3 are taken from Sastry and Babu [SaB00].

Some other extensions of Theorem 4.11 were obtained in Chidume and Osilike [ChO98]. Theorem 4.13 is a particular case of a more general result given there (and transcribed as Theorem 5.2 in Chapter 5 of this book).

Exercises and Miscellaneous Results

4.1. Hicks and Kubicek (1977)

Let H be the complex plane, $K = \{z \in H : |z| \leq 1\}$ and $T : K \to K$ given by

$$T(re^{i\theta}) = \begin{cases} 2re^{i(\theta + \frac{\pi}{3})}, & \text{if } 0 \leq r \leq \dfrac{1}{2} \\ e^{i(\theta + \frac{2\pi}{3})}, & \text{if } \dfrac{1}{2} < r \leq 1. \end{cases}$$

Then
(a) T is discontinuous and pseudocontractive;
(b) The origin is the unique fixed point of T;
(c) The Mann iteration with the sequence $\alpha_n = 1/(n+1)$ does not converge to $(0,0)$.

4.2. Chidume and Mutangadura (2001)

Let H be the real Hilbert space $\mathbb{R}^2$ endowed with the usual Euclidean inner product. If $x = (a,b) \in H$, we define $x^\perp \in H$ to be $(b,-a)$.

Let $K := \{x \in H : \|x\| \leq 1\}$ and denote

$$K_1 := \{x \in H : \|x\| \leq \frac{1}{2}\}, \ K_2 := \{x \in H : \frac{1}{2} \leq \|x\| \leq 1\}.$$

Define $T : K \to K$ as follows

$$Tx = \begin{cases} x + x^\perp, & \text{if } x \in K_1 \\ \dfrac{x}{\|x\|} - x + x^\perp, & \text{if } x \in K_2. \end{cases}$$

Then: (a) T is Lipschitzian and pseudocontractive; (b) The origin is the unique fixed point of T; (c) No Mann sequence converges to the fixed point; (d) No Mann sequence converges to any $x \neq 0$.

4.3.
(i) Prove Lemma 4.1; (ii) Prove Lemma 4.3; (iii) Prove Theorem 4.7; (iv) Prove Lemma 4.5; (v) Prove Theorem 4.8.

4.4. Chidume (2001)

Let $E = l_\infty$ and $K = \{x \in l_\infty : \|x\|_\infty \leq 1\}$. Define $T : K \to K$ by

$$Tx = (0, x_1^2, x_2^2, x_3^2, \dots), \quad \text{for } x = (x_1, x_2, x_3, \dots) \in K.$$

Then: (i) T is quasi-nonexpansive; (ii) T is not nonexpansive.

4.5. Chidume (2001)

Let $E = l_\infty$ and $K = \{x \in l_\infty : \|x\|_\infty \leq 1\}$. Define $T : K \to K$ by

$$T(x) = \begin{cases} (0, x_1^2, x_2^2, x_3^2, \dots), & \text{if } \|x\| \leq 1 \\ \|x\|_\infty^{-2} (0, x_1^2, x_2^2, x_3^2, \dots), & \text{if } \|x\| > 1, \end{cases}$$

where $x = (x_1, x_2, x_3, \dots) \in l_\infty$. Then: (i) T is a quasi-nonexpansive map with the unique fixed point $0 = (0, 0, \dots)$;

(ii) T is not uniformly asymptotically regular (to show this, prove that for all integers $n \geq 1$, there exists $x \in B(0, 1)$ such that

$$\left\| T_\lambda^{n+1} x - T_\lambda^n x \right\| > \lambda^2 (1 - \lambda^2),$$

for arbitrary $\lambda \in (0, 1)$, where $T_\lambda = (1 - \lambda)I + \lambda T$ is the averaged map associated to T).

4.6. Show that any nonexpansive map is a continuous pseudocontraction (but the reverse is not true).

4.7. Show that $T : [0, 1] \to \mathbb{R}$ defined by $Tx = 1 - x^{\frac{2}{3}}$ is a continuous pseudo-contraction which is not nonexpansive.

4.8. Osilike and Udomene (2001)
Show that a strictly pseudocontractive map is L-Lipschitzian.

4.9. Rhoades (1974)
Let H be a Hilbert space and K a nonempty compact convex subset of H. Let $T : K \to K$ be a strictly pseudo-contractive map (with a constant k) and let $\{\alpha_n\}_{n=0}^\infty$ be a sequence of real numbers satisfying the conditions: (i) $\alpha_o = 1$; (ii) $0 < \alpha_n < 1$ for all $n \geq 1$; (iii) $\sum_{n=1}^\infty \alpha_n = \infty$, and (iv) $\lim_{n \to \infty} \alpha_n = \alpha < 1 - k$. Then the Mann iteration method generated from an arbitrary $x_0 \in K$ by

$$x_{n+1} = (1 - \alpha_n)x_n + \alpha_n T x_n, n \geq 0,$$

converges strongly to a fixed point of T.

4.10. Chidume (1994)
Let $E = L_p$ or $l_p, 1 < p \leq 2$ and let K be a nonempty closed convex subset of E. Let $T : K \to K$ be a continuous strongly pseudo-contractive mapping of K into itself. Let $\{\alpha_n\}_{n=0}^\infty$ be a sequence of real numbers satisfying the conditions: (i) $0 < \alpha_n < 1$ for all $n \geq 1$; (ii) $\sum_{n=1}^\infty \alpha_n = \infty$, and (iii) $\sum_{n=1}^\infty \alpha_n^p < \infty$. Then the Mann iteration method generated from an arbitrary $x_1 \in K$ by

$$x_{n+1} = (1 - \alpha_n)x_n + \alpha_n T x_n, n \geq 1,$$

converges strongly to the unique fixed point of T.

4.11. Groetsch (1972)
Let X be a uniformly convex Banach space and $x, y \in X$ such that $\|x\| \leq 1$, $\|y\| \leq 1$ and $\|x - y\| \geq \epsilon > 0$. Then for $0 \leq \lambda \leq 1$, $\|\lambda x + (1 - \lambda)y\| \leq 1 - 2\lambda(1 - \lambda)\delta_X(\epsilon)$, where $\delta_X(.)$ is the modulus of convexity of X.

4.12. Kato (1967)
Let X be a real Banach space, J be the normalized duality mapping on X and let $x, y \in X$. Then $\|x\| \leq \|x + \lambda y\|$, $\forall \lambda > 0$ if and only if there exists $x^* \in Jx$ such that $\langle y, x^* \rangle \geq 0$.

5

The Ishikawa Iteration

As mentioned in the previous chapter, if T is continuous and the Mann iterative process converges, then it converges to a fixed point of T. But if T is not continuous, then there is no guarantee that, even if the Mann process converges, it will converge to a fixed point of T, as shown by the following example.

Example 5.1. Let $T : [0,1] \rightarrow [0,1]$ be given by $T0 = T1 = 0$ and $Tx = 1$, $0 < x < 1$. Then $F_T = \{0\}$ and the Mann iteration $M(x_1, \alpha_n, T)$ with $0 < x_1 < 1$ and $\alpha_n = \dfrac{1}{n}$, $n \geq 1$, converges to 1, which is not a fixed point of T.

If, instead of the Mann iteration we consider another iterative process, which is in some sense a two-step Mann iterative process, then it is possible to approximate the fixed point of some classes of contractive mappings T for which Mann iteration is not known to converge to a fixed point of T.

This new iterative process is called *Ishikawa iteration*, and was first introduced for the class of Lipschitzian pseudo-contractive operators. Here we consider some other classes of operators for which not only Mann iteration but also Ishikawa iteration method can be used to approximate fixed points.

It is nowadays quite clear that, for large classes of contractive type operators, it suffices to consider the simpler Mann iteration, even if Ishikawa iteration - which is more general but also computationally more complicated than Mann iteration - could be always used. Actually, having in view some recent results presented in Chapters 3 and 4, it is also evident that a simpler method than Mann iteration, i.e., the Krasnoselskij iteration - which is a particular case of Mann iteration - can be used in some cases to approximate fixed points of some classes of operators.

5.1 Lipschitzian and Pseudo-Contractive Operators in Hilbert Spaces

As we have shown in the previous chapter, the Mann iteration process converges in the special case of Lipschitzian and strongly pseudocontractive operators.

However, if T is only a pseudocontractive mapping, then generally the Mann iterative process does not converge to the fixed point, see Exercises 4.1 and 4.2.

Interest in pseudocontractive maps stems mainly from their firm connection with the class of nonlinear accretive operators, as it was pointed out in Chapter 1. It is a classical result, see Deimling [Dei74], that if T is an accretive operator, then the solutions of the equations $Tx = 0$ correspond to the equilibrium points of some evolution systems.

This explains why a considerable research effort has been devoted to iterative methods for approximating solutions of the equation above, when T is accretive or, correspondingly, to the iterative approximation of fixed points of pseudocontractions.

Results of this kind have been obtained firstly in Hilbert spaces, but only for Lipschitz operators, and then they have been extended to more general Banach spaces (thanks to several geometric inequalities for general Banach spaces developed within the past two decades) and to more general classes of operators.

There are still no results for the case of arbitrary Lipschitzian and pseudo-contractive operators, even when the domain of the operator is a compact convex subset of a Hilbert space. This explains the importance, from this point of view, of the improvement brought by the Ishikawa iteration.

It is the aim of this section to show that, under certain assumptions on the sequences $\{\alpha_n\}, \{\beta_n\}$, the Ishikawa iterative process associated to a Lipschitzian pseudocontractive operator converges strongly to a fixed point of T. The original result of Ishikawa is stated in the following.

Theorem 5.1. *Let K be a convex compact subset of a Hilbert space H and let $T : K \to K$ be a Lipschitzian pseudocontractive map and $x_1 \in K$. Then the Ishikawa iteration $\{x_n\}$, $x_n = I(x_1, \alpha_n, \beta_n, T)$, i.e., the sequence defined by*

$$x_{n+1} = (1 - \alpha_n)x_n + \alpha_n T\left[(1 - \beta_n)x_n + \beta_n T x_n\right], \tag{1}$$

where $\{\alpha_n\}, \{\beta_n\}$ are sequences of positive numbers satisfying

$$(i) \ 0 \le \alpha_n \le \beta_n \le 1, \ n \ge 1; \quad (ii) \lim_{n \to \infty} \beta_n = 0; \quad (iii) \sum_{n=1}^{\infty} \alpha_n \beta_n = \infty,$$

converges strongly to a fixed point of T.

Proof. Since T is pseudocontractive, for any $x, y \in K$ we have

$$\| Tx - Ty \|^2 \le \| x - y \|^2 + \| (I - T) x - (I - T) y \|^2, \tag{2}$$

where I is the identity map.

From the assumption that T is Lipschitzian, we deduce that there exists a positive number L such that

$$\| Tx - Ty \| \le L \| x - y \|, \quad \text{for any } x, y \in K. \tag{3}$$

Since K is a convex compact set and T is continuous (being Lipschitzian), from Schauder's fixed point theorem we obtain that the set of fixed points of T, $Fix(T)$, is nonempty. Let p denote any point of $Fix(T)$. Recall Lemma 1.8: for any x, y, z in a Hilbert space H and a real number λ, we have

$$\| \lambda x + (1 - \lambda) y - z \|^2 = \lambda \| x - z \|^2 + (1 - \lambda) \| y - z \|^2 - \lambda (1 - \lambda) \| x - y \|^2. \tag{4}$$

Using (4) we obtain the following three equalities

$$\| x_{n+1} - p \|^2 = \| \alpha_n T [\beta_n T x_n + (1 - \beta_n) x_n] + (1 - \alpha_n) x_n - p \|^2 =$$

$$= \alpha_n \| T [\beta_n T x_n + (1 - \beta_n) x_n] - p \|^2 + (1 - \alpha_n) \| x_n - p \|^2 -$$

$$-\alpha_n (1 - \alpha_n) \| T [\beta_n T x_n + (1 - \beta_n) x_n] - x_n \|^2; \tag{5}$$

$$\| \beta_n T x_n + (1 - \beta_n) x_n - p \|^2 = \beta_n \| T x_n - p \|^2 + (1 - \beta_n) \| x_n - p \|^2 -$$

$$-\beta_n (1 - \beta_n) \| T x_n - x_n \|^2, \tag{6}$$

and, respectively,

$$\| \beta_n T x_n + (1 - \beta_n) x_n - T[\beta_n T x_n + (1 - \beta_n) x_n] \|^2 =$$

$$= \beta_n \| T x_n - T[\beta_n T x_n + (1 - \beta_n) x_n] \|^2 + (1 - \beta_n) \cdot$$

$$\cdot \| x_n - T[\beta_n T x_n + (1 - \beta_n) x_n] \|^2 - \beta_n (1 - \beta_n) \| T x_n - x_n \|^2. \tag{7}$$

Applying (2) we deduce the following two inequalities

$$\| T[\beta_n T x_n + (1 - \beta_n) x_n] - p \|^2 = \| T[\beta_n T x_n + (1 - \beta_n) x_n] - Tp \|^2 \le$$

$$\le \| \beta_n T x_n + (1 - \beta_n) x_n - p \|^2 +$$

$$+ \| \beta_n T x_n + (1 - \beta_n) x_n - T[\beta_n T x_n + (1 - \beta_n) x_n] \|^2, \tag{8}$$

and

$$\| T x_n - p \|^2 = \| T x_n - Tp \|^2 \le \| x_n - p \|^2 + \| x_n - T x_n \|^2. \tag{9}$$

Now, performing the computations in $(5) + \alpha_n [(6) + (7) + (8) + \beta_n (9)]$, we get

$$\| x_{n+1} - p \|^2 \le \| x_n - p \|^2 - \alpha_n \beta_n (1 - 2\beta_n) \| T x_n - x_n \|^2 +$$

$$+\alpha_n\beta_n \left\| Tx_n - T\left[\beta_n Tx_n + (1-\beta_n)x_n\right]\right\|^2 -$$

$$-\alpha_n(\beta_n - \alpha_n)\left\| x_n - T\left[\beta_n Tx_n + (1-\beta_n)x_n\right]\right\|^2,$$

and so, in view of (i), it follows that

$$\|x_{n+1} - p\|^2 \le \|x_n - p\|^2 - \alpha_n\beta_n(1-2\beta_n)\|Tx_n - x_n\|^2 +$$

$$+\alpha_n\beta_n \left\| Tx_n - T\left[\beta_n Tx_n + (1-\beta_n)x_n\right]\right\|^2. \tag{10}$$

Since T is Lipschitzian, we have

$$\left\| Tx_n - T\left[\beta_n Tx_n + (1-\beta_n)x_n\right]\right\| < L\beta_n \left\| Tx_n - x_n\right\| \tag{11}$$

and hence, from (10) and (11) we deduce

$$\|x_{n+1} - p\|^2 \le \|x_n - p\|^2 - \alpha_n\beta_n(1-2\beta_n - L^2\beta_n^2)\left\| Tx_n - x_n\right\|^2. \tag{12}$$

By summing (12) for $n \in \{m, m+1, \ldots, n\}$ we obtain

$$\|x_{n+1} - p\|^2 \le \|x_m - p\|^2 - \sum_{k=m}^{n} \alpha_k\beta_k(1-2\beta_k - L^2\beta_k^2)\left\| Tx_k - x_k\right\|^2,$$

which can be written as

$$\sum \alpha_k\beta_k(1-2\beta_k - L^2\beta_k^2)\left\| Tx_k - x_k\right\|^2 \le \|x_m - p\|^2 - \|x_{n+1} - p\|^2.$$

Now, by exploiting the assumption (ii), we deduce that there exists a positive integer N such that

$$2\beta_k + L^2\beta_k^2 \le 1/2, \text{ for all integers } k \ge N.$$

Then, for $m > N$ we obtain

$$\frac{1}{2}\sum_{k=m}^{n} \alpha_k\beta_k \left\| Tx_k - x_k\right\|^2 \le \|Tx_m - p\|^2 - \|Tx_{n+1} - p\|^2. \tag{13}$$

Since K is bounded, the right-hand side quantity in (13) is bounded. This means that the series in the left-hand side is convergent and therefore, by (iii), it results that

$$\liminf_n \|Tx_n - x_n\| = 0,$$

which in turn implies (K is compact) that there is a subsequence $\{x_{n_k}\}_{k=1}^{\infty}$ that converges to a certain point q of $Fix(T)$.

Now, since q is a fixed point of T, from (12) we obtain for $n \ge N$

$$\|x_{n+1} - q\| \le \|x_n - q\|,$$

that is, the sequence $\{\|x_n - q\|\}$ is non-increasing.

Having in view that there is a subsequence $\{\| x_{n_k} - q \|\}$ converging to zero, it finally results that $\{x_n\}$ converges to q. $\qquad\qquad\square$

Remarks.

1) In its original form, the Ishikawa iteration does not include the Mann iteration, because of the assumption (i) in Theorem 5.1. Indeed, if one had $\beta_n = 0$ $(n \geq 1)$, then it would results $\alpha_n = 0$, as well;

2) In the effort to obtain an Ishikawa iteration which should include the Mann iteration as a special case, some authors, amongst them Naimpally and Singh, K.L. [NaS82] and Liu, Q. [LiQ87], have modified (i) to a weaker condition of the form $0 \leq \alpha_n$, $\beta_n \leq 1$;

3) Liu, Q. [LiQ87] extended Theorem 5.1 to the class of Lipschitzian hemi-contractive maps. A hemicontractive map is a pseudocontractive map with respect to a fixed point, i.e., if p is a fixed point of T, and x is a point in the space, then T satisfies

$$\| Tx - p \|^2 \leq \| x - p \|^2 + \| x - Tx \|^2 ;$$

4) However, neither the proof of Q. Liu nor that of Ishikawa can be used to establish a similar result for the Mann iterative process;

5) Since its publication in 1974, as far as we know, Theorem 5.1 has never been extended to more general Banach spaces in its original formulation.

All extensions obtained so far cover slightly more general classes of operators and are still confined to Hilbert spaces. To overcome these difficulties some authors have introduced other iterative processes, see Chapter 6, for a brief presentation of the most important of them.

5.2 Strongly Pseudo-Contractive Operators in Banach Spaces

Starting from the results established for the Mann iteration associated to several classes of Lipschitzian pseudo-contractive operators in Hilbert spaces, a considerable effort has been devoted to extending these results in Banach spaces with certain geometric properties. One of the most general results that were obtained in this class is given by the next theorem.

Theorem 5.2. *Let E be a real uniformly smooth Banach space and K a bounded closed convex and nonempty subset of E. Suppose $T : K \to K$ is a strongly pseudocontractive operator that has at least a fixed point $x^* \in F_T$. Let $\{\alpha_n\}$, $\{\beta_n\}$ be real sequences satisfying the following conditions:*
(i) $0 \leq \alpha_n$, $\beta_n < 1$, for all $n \geq 0$;
(ii) $\lim_{n\to\infty} \alpha_n = 0$; $\lim_{n\to\infty} \beta_n = 0$;

(iii) $\sum\limits_{n=0}^{\infty} \alpha_n = \infty.$

Then, for arbitrary $x_0 \in K$, the Ishikawa iteration $I(x_0, \alpha_n, \beta_n, T)$, i.e., the sequence $\{x_n\}$ defined iteratively by

$$x_{n+1} = (1 - \alpha_n)x_n + \alpha_n T y_n, \tag{14}$$

$$y_n = (1 - \beta_n)x_n + \beta_n T x_n, \ n \geq 0, \tag{15}$$

converges strongly to x^* and, moreover, x^* is unique.

Proof. Using Lemma 1.1 we obtain

$$\| x_{n+1} - x^* \|^2 = \|(1 - \alpha_n)(x_n - x^*) + \alpha_n(Ty_n - x^*)\|^2 \leq$$

$$\leq (1 - \alpha_n)^2 \|x_n - x^*\|^2 + 2\alpha_n(1 - \alpha_n) \langle Ty_n - x^*, \ j(x_n - x^*) \rangle +$$

$$+ \max\{(1 - \alpha_n) \|x_n - x^*\|, 1\} \cdot \alpha_n \|Ty_n - x^*\| \max\{\|Ty_n - x^*\|, 1\} \cdot$$

$$\cdot b(\alpha_n) \leq (1 - \alpha_n)^2 \|x_n - x^*\|^2 + M_1 \alpha_n b(\alpha_n) + 2\alpha_n(1 - \alpha_n)\delta_n, \tag{16}$$

for some constant $M_1 > 0$ (since K is bounded), where

$$\delta_n := \langle Ty_n - x^*, j(x_n - x^*) \rangle =$$

$$= \langle Ty_n - x^*, j(x_n - x^*) - j(y_n - x^*) \rangle + \langle Ty_n - Tx^*, \ j(y_n - x^*) \rangle \leq$$

$$\leq \langle Ty_n - Tx^*, \ j(x_n - x^*) - j(y_n - x^*) \rangle + k \| y_n - x^* \|^2 =$$

$$= \Delta_n + k \| y_n - x^* \|^2,$$

where we denoted $\Delta_n := \langle Ty_n - Tx^*, \ j(x_n - x^*) - j(y_n - x^*) \rangle$ and k is the strong pseudo-contractiveness constant, $0 < k < 1$.

We shall prove that $\Delta_n \to 0$ as $n \to \infty$. Indeed, note that the sequences $\{x_n - x^*\}$ and $\{y_n - x^*\}$ are bounded subsets of E, and, by (15),

$$\|(x_n - x^*) - (y_n - x^*)\| = \beta_n \| x_n - T x_n \| \leq (diam \ K)\beta_n \to 0,$$

as $n \to \infty$. Hence, by the uniform continuity of j on bounded subsets of E, and since $\{Ty_n - Tx^*\}$ is bounded, we deduce exactly $\Delta_n \to 0$ as $n \to \infty$.

Now set $M_2 := 2(1 - \alpha_n)$. Then, by (16) we obtain the following estimates

$$\| x_{n+1} - x^* \|^2 \leq (1 - \alpha_n)^2 \| x_n - x^* \|^2 + M_1 \alpha_n b(\alpha_n) +$$

$$+ 2k\alpha_n(1 - \alpha_n) \| y_n - x^* \|^2 + 2\alpha_n(1 - \alpha_n)\Delta_n \leq (1 - \alpha_n)^2 \| x_n - x^* \|^2 +$$

$$+ 2k\alpha_n(1 - \alpha_n) \| y_n - x^* \|^2 + \alpha_n[M_2 \Delta_n + M_1 b(\alpha_n)]. \tag{17}$$

Now, using (15) we have

$$\|y_n - x^*\|^2 \leq (1 - \beta_n)^2 \|x_n - x^*\|^2 + 2k\beta_n(1 - \beta_n) \|x_n - x^*\|^2 + M_3 \beta_n b(\beta_n),$$

for some constant $M_3 > 0$ (again we used the fact that K is bounded). Hence

$$\|y_n - x^*\|^2 \leq [1 - (1-k)\beta_n] \cdot \|x_n - x^*\|^2 + M_3\beta_n b(\beta_n) \leq$$

$$\leq \|x_n - x^*\|^2 + M_3\beta_n b(\beta_n).$$

Substituting this last inequality in (17) and denoting $M_4 := 2kM_3(1-\alpha_n)$ we get the following estimates

$$\|x_{n+1} - x^*\|^2 \leq (1-\alpha_n)^2 \|x_n - x^*\|^2 + 2k\alpha_n(1-\alpha_n)\|x_n - x^*\|^2 +$$

$$+M_3\beta_n b(\beta_n) + \alpha_n[M_2\Delta_n + M_1 b(\alpha_n)] \leq$$

$$\leq [1 - \alpha_n + k\alpha_n]\|x_n - x^*\|^2 + M_4\alpha_n\beta_n b(\beta_n) +$$

$$+\alpha_n[M_2\Delta_n + M_1 b(\alpha_n)]. \qquad (18)$$

Set $J_n := M_4\beta_n b(\beta_n) + M_2\Delta_n + M_1 b(\alpha_n)$. By condition (ii) and the continuity of the function $b(\cdot)$ we obtain that $J_n \to 0$ as $n \to \infty$.

So, by (18) we get

$$\|x_{n+1} - x^*\|^2 \leq [1 - (1-k)\alpha_n] \cdot \|x_n - x^*\|^2 + \alpha_n J_n.$$

Set $\lambda_n := \|x_n - x^*\|^2$, $\sigma_n := \alpha_n J_n$ and hence the last inequality yields

$$\lambda_{n+1} \leq [1 - (1-k)\alpha_n]\lambda_n + \sigma_n. \qquad (19)$$

Notice that the sequence $\{x_n\}$ is bounded below. Let $a = \inf\{\lambda_n : n \geq 1\}$. We will prove that $a = 0$. Let us suppose $a \neq 0$, i.e., $a > 0$. Then, for all $n \geq 1$, we have $\lambda_n \geq a > 0$. Note that $\sigma_n/\alpha_n \to 0$ as $n \to \infty$. Hence there exists a positive integer N_0 such that, for all $n \geq N_0$, we have

$$0 < \frac{\sigma_n}{\alpha_n} < a \leq \frac{1}{2}(1-k)\lambda_n.$$

This implies

$$\sigma_n \leq \frac{1}{2}(1-k)\alpha_n\lambda_n, \quad \text{for all } n \geq N_0.$$

We substitute this last inequality in (19) and, since $0 < k < 1$, we get

$$0 \leq \lambda_{n+1} \leq [1 - (1-k)\alpha_n]\lambda_n + \frac{1}{2}(1-k)\alpha_n\lambda_n =$$

$$= \left[1 - \frac{1-k}{2}\cdot\alpha_n\right]\lambda_n \leq \prod_{j=0}^{n}\left[1 - \frac{1-k}{2}\alpha_j\right]\lambda_j \to 0, \quad \text{as } n \to \infty,$$

since $\alpha_n \in (0,1)$, for all $n \geq 0$, $\{\lambda_n\}$ is bounded and $\sum_{n=0}^{\infty}\alpha_n = \infty$, by (iii).

This is a contradiction and hence $a = 0$.

Now we shall prove that the sequence $\{\lambda_n\}$ converges to zero, as $n \to \infty$.

As $\inf\{\lambda_n : n \geq 1\} = 0$, there exists a subsequence $\{\lambda_{n_j}\}_{j=0}^{\infty}$ of $\{\lambda_n\}_{n=0}^{\infty}$ such that $\lambda_{n_j} \to 0$ as $j \to \infty$.

Now, given any $\varepsilon > 0$, there exists a large enough integer j_0 such that $\dfrac{\sigma_n}{(1-k)\alpha_n} < \varepsilon$ and $\lambda_{n_j} < \varepsilon$, $\forall n \geq n_{j_0}$.

Inequality (19) yields now

$$\lambda_{n_{j_0}+1} \leq [1 - (1-k)\alpha_{n_{j_0}}]\varepsilon + (1-k)\alpha_{n_{j_0}}\varepsilon = \varepsilon,$$

and a simple induction yields $\lambda_{n_{j_0}+p} \leq \varepsilon$, for all $p \geq 1$.

This last inequality implies $\lambda_n \to 0$ as $n \to \infty$, that is, $x_n \to x^*$ as $n \to \infty$.

The uniqueness of the fixed point is a direct consequence of the arguments above. Indeed, the element $p \in F_T$ was arbitrarily chosen. Suppose now there is a $p^* \in F_T$, with $p^* \neq p$.

Repeating all computations relative to p^*, we obtain that the sequence $\{x_n\}$ converges to both p^* and p, so $F_T = \{p\}$.   □

Remarks.

1) Theorem 5.2 is a significant generalization of most of the related results in literature. Furthermore, the parameters $\{\alpha_n\}$ and $\{\beta_n\}$ of the Ishikawa iteration involved in Theorem 5.2 do not depend neither on the geometry of the underlying Banach space, nor on other special properties of the operator T itself;

2) Taking $\alpha_n = c_n$ and $\beta_n = 0$ for all $n \geq 0$, from Theorem 5.2 we obtain a general convergence theorem for the Mann iteration.

Corollary 5.1. *Let E be a real uniformly smooth Banach space and let $K \subset E$ be a nonempty bounded closed and convex subset. Let $T : K \to K$ be a strongly pseudocontractive map such that there exists $x^* \in F_T$. Let $\{c_n\}_{n=0}^{\infty}$ be a real sequence satisfying the following conditions:*

(i) $0 \leq c_n < 1$ for all $n \geq 0$;

(ii) $\lim\limits_{n \to \infty} c_n = 0$;   (iii) $\sum\limits_{n=0}^{\infty} c_n = \infty$.

Then, for arbitrary $x_1 \in K$, the Mann iteration $M(x_1, c_n, T)$ defined by

$$x_{n+1} = (1 - c_n)x_n + c_n Tx_n, \ n \geq 0$$

converges strongly to x^*, and x^* is unique.

Remark. In order to prove Theorem 5.2 we used a property that characterizes the uniformly smooth Banach spaces E (equivalently, E^* is a uniformly convex Banach space): the duality mapping J is single-valued and uniformly continuous on any bounded subset of E, see Exercise 5.3. Similarly to Theorem 5.2, one obtains a more general result by considering a generalization of the concept of strongly pseudocontractive operators.

Definition 5.1. Let E be a real normed space and let K be a nonempty subset of E. A single-valued map $T : K \to E$ is said to be:

1) *φ-strongly accretive* if for any $x, y \in K$, there exist $j(x - y) \in J(x - y)$ and a strictly increasing function $\varphi : [0, \infty) \to [0, \infty)$ with $\varphi(0) = 0$ such that

$$\langle Tx - Ty, \; j(x - y) \rangle \geq \varphi(\|x - y\|) \cdot \|x - y\| ;$$

2) *φ-strongly pseudocontractive* if $I - T$ is a φ-strongly accretive mapping.

Remark. Obviously, every strongly accretive operator is φ-strongly accretive, and every strongly pseudo-contractive operator is also φ-strongly pseudo-contractive with $\varphi(t) = kt$, $0 < k < 1$ and $t \geq 0$.

The next result generalizes Theorem 5.2 to φ-strongly pseudo-contractive operators, also removing the boundedness of K. We state it without proof.

Theorem 5.3. *Let E be a uniformly smooth Banach space and K be a nonempty closed convex subset of E. Let $T : K \to K$ be a Lipschitzian φ-strongly pseudocontractive operator, with Lipschitz constant $L \geq 1$ and $F_T \neq \emptyset$.*

If $\{\alpha_n\}$, $\{\beta_n\}$ are two sequences in $[0, 1]$ satisfying

$$(i) \; \alpha_n \to 0, \; \beta_n \to 0 \; (as \; n \to \infty); \; (ii) \; \sum_{n=0}^{\infty} \alpha_n = \infty,$$

then for any given $x_0 \in K$ the Ishikawa iterative process $\{x_n\}$,

$$x_n = I(x_0, \alpha_n, \beta_n, T), \; n \geq 0,$$

converges to the unique fixed point of T in K.

Remark. Since any strongly pseudocontractive operator is also φ-strongly pseudocontractive, Theorem 5.3 improves and extends several related results in literature.

5.3 Nonexpansive Operators in Banach Spaces Satisfying Opial's Condition

The aim of this section is to show that rich (topological) properties of the ambient Banach space together with weak properties of the operator itself could still ensure the convergence of the Ishikawa iteration. More specifically, we will show that if E is a uniformly convex Banach space which satisfies Opial's condition or whose norm is Frechet differentiable, K is a bounded closed convex subset of E, and $T : K \to K$ is a nonexpansive operator, then the Ishikawa iteration $I(x_0, \alpha_n, \beta_n, T)$ converges weakly to a fixed point of T, provided that the sequences $\{\alpha_n\}$, $\{\beta_n\}$ fulfill some appropriate conditions.

Recall that a Banach space E is said to satisfy Opial's condition, see also Exercise 3.18, if for any sequence $\{x_n\}$ in E the condition $x_n \rightharpoonup x_0$ (weakly) implies

$$\lim_n \sup \|x_n - x_0\| < \lim_n \sup \|x_n - y\|, \text{ for all } y \in E, \ y \neq x_0.$$

It is known, see, for example, Opial [Op67a], that all l^p spaces for $1 < p < \infty$ satisfy Opial's condition but the L^p spaces do not, unless $p = 2$.

It is also known, see van Dulst [Dul82] that any separable Banach space can be equivalently re-normed so that it satisfies Opial's condition. Consequently, this class of Banach spaces is large enough.

Recall also that E is said to have a Frechet differentiable norm if, for each $x \in S(E)$, the unit sphere of E, the limit

$$\lim_{t \to 0} \frac{\|x + ty\| - \|x\|}{t}$$

exists and is attained uniformly in $y \in S(E)$. In this case we have

$$\frac{1}{2} \|x\|^2 + \langle h, \ J(x) \rangle \leq \frac{1}{2} \|x + h\|^2 \leq \frac{1}{2} \|x\|^2 + \langle h, \ J(x) \rangle + g(\|h\|), \quad (20)$$

for all bounded x, h in E, where $J(x) = \partial \frac{1}{2} \|x\|^2$ is the Frechet derivative of the functional $\frac{1}{2} \|x\|^2$ at $x \in E$, $\langle \cdot, \cdot \rangle$ is the duality pairing and the function $g : [0, \infty) \to [0, \infty)$ satisfies

$$\lim_{t \to 0^+} \frac{g(t)}{t} = 0.$$

For a bounded closed convex subset K of a uniformly convex Banach space E and an operator $T : K \to K$, we consider the Ishikawa iterative process $I(x_0, \alpha_n, \beta_n, T)$, that can be written as

$$x_{n+1} = T_n x_n, \ \ n = 0, 1, 2, \ldots \quad (21)$$

where

$$T_n(x) = (1 - \alpha_n)x + \alpha_n T[\beta_n T x + (1 - \beta_n)x]. \quad (22)$$

We know that if T is nonexpansive, then T_n is also nonexpansive and that $F_{T_n} \supseteq F_T$, for all $n \geq 0$, where F_T denotes the set of all fixed points of T.

We will need the next lemmas.

Lemma 5.1. *If T is nonexpansive and $p \in F_T$, then*

$$\lim_{n \to \infty} \|x_n - p\| \ \text{ exists.}$$

Proof. We have $\|x_{n+1} - p\| = \|Tx_n - Tp\| \leq \|x_n - p\|$, which shows that the sequence $\{\|x_n - p\|\}$ is non-increasing. $\square$

Lemma 5.2. *Let* $\{\alpha_n\}$ *and* $\{\beta_n\} \subset [0,1]$ *be such that*

(i) $\displaystyle\sum_{n=0}^{\infty} \alpha_n(1-\alpha_n) = \infty$; *(ii)* $\displaystyle\sum_{n=0}^{\infty} \beta_n(1-\alpha_n) < \infty$; *(iii)* $\displaystyle\lim_{n} \sup \beta_n < 1$.

Then $\displaystyle\lim_{n\to\infty} \|Tx_n - x_n\| = 0$, *provided that* T *is nonexpansive.*

Proof. Set

$$y_n = \beta_n T x_n + (1 - \beta_n) x_n.$$

Then

$$x_{n+1} = \alpha_n T y_n + (1 - \alpha_n) x_n.$$

Let $p \in F_T$. We may assume $\displaystyle\lim_{n\to\infty} \|x_n - p\| \neq 0$.

Then we have $\|y_n - p\| \le \|x_n - p\|$ and hence

$$\|x_{n+1} - p\| = \|\alpha_n(Ty - p) + (1 - \alpha_n)(x_n - p)\| \le$$

$$\le \|x_n - p\| \cdot \left[1 - 2\alpha_n(1 - \alpha_n)\delta_E\left(\frac{\|Ty_n - x_n\|}{\|x_n - p\|}\right)\right], \tag{23}$$

where δ_E is the modulus of convexity of E defined by

$$\delta_E(\varepsilon) = \inf\left\{1 - \left\|\frac{1}{2}(x + y)\right\| : \|x\| \le 1,\ \|y\| \le 1,\ \|x - y\| \ge \varepsilon\right\}$$

for $0 \le \varepsilon \le 2$.

Now, it results from (23) that

$$\sum_{n=0}^{\infty} \alpha_n(1 - \alpha_n)\, \delta_E\left(\frac{\|Ty_n - x_n\|}{\|x_n - p\|}\right)$$

converges. But, since $\displaystyle\sum_{n=0}^{\infty} \alpha_n(1 - \alpha_n)$ diverges, we must have

$$\lim_{n} \inf \delta_E\left(\frac{\|Ty_n - x_n\|}{\|x_n - p\|}\right) = 0,$$

which implies

$$\lim_{n} \inf \|Ty_n - x_n\| = 0, \tag{24}$$

since δ_E is strictly increasing and continuous, and

$$\lim_{n\to\infty} \|x_n - p\| > 0.$$

Since

$$\|Tx_n - x_n\| \le \|Tx_n - Ty_n\| + \|Ty_n - x_n\| \le \|x_n - y_n\| + \|Ty_n - x_n\| =$$

$$= \beta_n \, \|Tx_n - x_n\| + \|Ty_n - x_n\| ,$$

we get

$$\|Tx_n - x_n\| \le \frac{1}{1 - \beta_n} \cdot \|Ty_n - x_n\| ,$$

and therefore by (24) we deduce that

$$\liminf_n \|Tx_n - x_n\| = 0. \tag{25}$$

Next

$$\|Tx_{n+1} - x_{n+1}\| \le \alpha_n \|Tx_{n+1} - Ty_n\| + (1 - \alpha_n) \|Tx_{n+1} - x_n\| \le$$

$$\le \alpha_n \|x_{n+1} - y_n\| + (1 - \alpha_n) \cdot (\|Tx_{n+1} - x_{n+1}\| + \|x_{n+1} - x_n\|) \le$$

$$\le \alpha_n [\alpha_n \|Ty_n - y_n\| + (1 - \alpha_n) \|x_n - y_n\|] + (1 - \alpha_n) \cdot$$

$$\cdot (\|Tx_{n+1} - x_{n+1}\| + \alpha_n \|Ty_n - x_n\|)$$

from which we get

$$\|Tx_{n+1} - x_{n+1}\| \le \alpha_n \|Ty_n - x_n\| + (1 - \alpha_n) (\|Ty_n - x_n\| + \|x_n - y_n\|) \le$$

$$\le \alpha_n (\beta_n \|Ty_n - Tx_n\| + (1 - \beta_n) \|Ty_n - x_n\|) +$$

$$+ (1 - \alpha_n) (\|Ty_n - x_n\| + \|x_n - y_n\|) \le$$

$$\le (1 + \alpha_n \beta_n - \alpha_n) \|x_n - y_n\| + (1 - \alpha_n \beta_n) \|Ty_n - x_n\| \le$$

$$\le \beta_n (1 + \alpha_n \beta_n - \alpha_n) \|x_n - Tx_n\| + (1 - \alpha_n \beta_n) \cdot$$

$$\cdot (\|Ty_n - Tx_n\| + \|Tx_n - x_n\|) \le$$

$$\le [\beta_n (1 + \alpha_n \beta_n - \alpha_n) + (1 - \alpha_n \beta_n)(1 + \beta_n)] \|x_n - Tx_n\| =$$

$$= [1 + 2\beta_n (1 - \alpha_n)] \|x_n - Tx_n\| .$$

Since $\sum_{n=0}^{\infty} \beta_n (1 - \alpha_n)$ converges and $\{\|x_n - Tx_n\|\}$ is bounded, it follows by Lemma 1.3 that $\lim_{n\to\infty} \|Tx_n - x_n\|$ exists and, by (25), that it equals zero. $\square$

Lemma 5.3. *For a nonexpansive map* $T : C \to X$, *the points* $x, y \in C$ *and* $0 \le \lambda \le 1$, *there exists* $g : [0, \infty) \to [0, \infty)$ *a strictly increasing continuous function with* $g(0) = 0$ *such that*

$$g (\|T[\lambda x + (1 - \lambda)y] - [\lambda Tx + (1 - \lambda)Ty]\|) \le \|x - y\| - \|Tx - Ty\| .$$

Lemma 5.4. *Suppose in addition to the previous statements that* E *has a Frechet differentiable norm. Then for every* $p_1, p_2 \in F_T$ *and* $0 < \lambda < 1$

$$\lim_{n\to\infty} \|\lambda x_n + (1 - \lambda)p_1 - p_2\|$$

exists.

Proof. Let's denote $S_{n,m} = T_{n+m-1}T_{n+m-2}\cdots T_{n+1}T_n$, where T_n is defined by (22). As T and T_n are nonexpansive, $S_{n,m}$ is nonexpansive as well and $x_{n+m} = S_{n,m}x_n$. We also denote

$$a_n = a_n(\lambda) = \|\lambda x_n + (1-\lambda)p_1 - p_2\|$$

and

$$d_{n,m} = \|S_{n,m}[\lambda x_n + (1-\lambda)p_1] - [\lambda x_{n+m} + (1-\lambda)p_1]\|.$$

By Lemma 5.3 we get

$$g(d_{n,m}) \le \|x_n - p_1\| - \|S_{n,m}x_n - S_{n,m}p_1\| = \|x_n - p_1\| - \|x_{n+m} - p_1\|.$$

Since $\lim_{n\to\infty} \|x_n - p_1\|$ exists, by Lemma 5.1 we conclude that

$$\lim_{n,m\to\infty} d_{n,m} = 0. \tag{26}$$

As

$$a_{n+m} = \|\lambda x_{n+m} + (1-\lambda)p_1 - p_2\| \le$$

$$\le d_{n+m} + \|S_{n,m}[\lambda x_n + (1-\lambda)p_1 - p_2]\| \le d_{n,m} + a_n,$$

it follows by (26) that

$$\lim_{n}\sup a_n \le \lim_{n,m\to\infty} d_{n,m} + \lim_{n\to\infty} \inf a_n = \lim_{n} \inf a_n,$$

which shows that $\lim_{n\to\infty} a_n$ exists. $\qquad\square$

Now we can prove the main results of this section, concerning the weak, respectively the strong convergence of the Ishikawa iteration process in a uniformly convex Banach space, when the operator T is assumed to be only nonexpansive.

Theorem 5.4. *Let E be a uniformly convex Banach space which satisfies Opial's condition or whose norm is Frechet differentiable, K be a bounded closed convex subset of E and $T : K \to K$ a nonexpansive mapping.*

Then for any initial guess x_0 in K, the Ishikawa process $\{x_n\}$ defined by (21), (22), with $\{\alpha_n\}, \{\beta_n\} \subset [0,1]$ satisfying (i), (ii), and (iii), converges weakly to a fixed point of T.

Proof. By Browder's theorem (Theorem 4.7), we know that if E is uniformly convex, then T has a fixed point and $I - T$ is demiclosed at 0, i.e., for any sequence $\{y_n\}$ in K, the conditions $y_n \to y$ and $y_n - Ty_n \to 0$ imply $y = Ty$.

If we denote by $\omega_w(x_n)$ the weak ω-limit set of the sequence $\{x_n\}$, that is,

$$\omega_w(x_n) = \{u \in E : u = \text{weak-}\lim_{k\to\infty} x_{n_k}, \text{ for some } n_k \nearrow \infty\},$$

then, by a direct consequence of Lemma 5.2, we may conclude that

$$\omega_w(x_n) \subset F_T.$$

To show that $\{x_n\}$ converges weakly to a fixed point of T, it suffices to show that $\omega_w(x_n)$ consists of exactly one point. To this end, we consider the case when E satisfies Opial's condition (the second case is similar).

Let $p \neq q$ in $\omega_w(x_n)$. Then $p = \text{weak-}\lim_{k\to\infty} x_{n_k}$ and $q = \text{weak-}\lim_{j\to\infty} x_{m_j}$, for some subsequences $\{n_k\}$ and $\{m_j\}$ converging to ∞.

By Lemma 5.1 and Opial's condition of E, we have

$$\lim_{n\to\infty} \|x_n - p\| = \lim_{k\to\infty} \|x_{n_k} - p\| < \lim_{k\to\infty} \|x_{n_k} - q\| = \lim_{j\to\infty} \|x_{m_j} - q\| <$$

$$< \lim_{j\to\infty} \|x_{m_j} - p\| = \lim_{n\to\infty} \|x_n - p\|,$$

which is a contradiction.

Therefore, the conclusion of the theorem holds in the case in which E satisfies Opial's condition. $\qquad\square$

Remarks.

1) If we take $\beta_n = 0$, for all $n \geq 0$, from Theorem 5.4 we find a result of Reich [Re79a], regarding the convergence of Mann iterative process;

2) Another generalization of Reich's theorem has been obtained by Deng [Dng96] under more general assumptions on the ambient space: E is assumed to be a (not necessarily uniform convex) Banach space which satisfies Opial's condition, while the sequences $\{\alpha_n\}$, $\{\beta_n\}$ that define the Ishikawa iteration process are supposed to satisfy

$$(a)\ 0 \leq \alpha_n \leq \alpha < 1 \text{ and } \sum_{n=1}^{\infty} \alpha_n = \infty,$$

respectively

$$(b)\ 0 \leq \beta_n \leq 1 \text{ and } \sum_{n=1}^{\infty} \beta_n < \infty.$$

However, it is easy to check that conditions (a) and (b) of Deng are more restrictive than the conditions (i), (ii) and (iii) of Tan and Xu, H.K. [TX93a];

3) In a recent paper, Zeng [Ze02a] showed that Theorem 5.4 is still valid if we replace conditions (i) and (ii) by the following one:

(c) For any subsequence $\{n_k\}_{k=0}^{\infty}$ of $\{n\}_{n=0}^{\infty}$, the series

$$\sum_{k=0}^{\infty} \alpha_{n_k}(1 - \alpha_{n_k})$$

diverges.

If, additionally, $T(K)$ is contained in a compact subset of E, then the Ishikawa iterative process converges strongly, as shown by the next theorem.

Theorem 5.5. *Suppose all assumptions in Theorem 5.4 are satisfied. If there exists a compact subset C of E such that $T(K) \subset C$, then the Ishikawa iteration process converges strongly to a fixed point of T.*

Proof. By Lemma 5.2 and the precompactness of $T(K)$, we get that $\{x_n\}$ admits a strongly convergent subsequence $\{x_{n_k}\}$, whose limit we shall denote by p. Then, again by a consequence of Lemma 5.2, it results $p = Tp$.

Since, by Lemma 5.1, the sequence $\{\|x_n - p\|\}$ is decreasing, it results that p is actually the strong limit of the sequence $\{x_n\}$.   □

Remark. In relation to similar results obtained by Senter and Dotson [SeD74] in the case of the Mann iteration process, it can be shown that one can replace the precompactness condition of $T(K)$ by the so-called condition A, see Theorem 3 in Tan and Xu, H.K. [TX93a].

5.4 Quasi-Nonexpansive Type Operators

One of the most general contractive-type definitions for which Picard iteration yields a unique fixed point is that of quasi-contractive operators given by Ciric, see Example 2.10, 1). This class contains, among other classes of contractive operators, the class of quasi-nonexpansive operators, including in turn the Kannan and Zamfirescu operators.

As we have shown, the Picard iteration converges for a larger class than the one of quasi-contractive operators, see Theorem 2.10 in Section 2.6. It is also known that the Mann iteration converges for this class of operators (Theorem 7 in Rhoades [Rh74a]) considered in Hilbert spaces. We included in Section 4.5 the corresponding result for Zamfirescu operators in Theorem 4.10 (4.9), in the case of a (uniformly) Banach space setting.

It is the aim of this section to present a convergence theorem for the Ishikawa iteration, corresponding to a typical representative of the class of quasi-contractive operators, i.e., the class of Zamfirescu operators.

Recall that, in a normed space E, an operator $T : E \to E$ is said to be quasi-contractive if there exists a number α, $0 \leq \alpha < 1$ such that for all x, y in E

$$\|Tx - Ty\| \leq k \cdot M(x, y),$$

where

$$M(x, y) := \max\left\{\|x - y\|, \|x - Tx\|, \|y - Ty\|, \|x - Ty\|, \|y - Tx\|\right\}.$$

Recall also that T is said to be a Zamfirescu operator if there exist the numbers α, β and γ, $0 \leq \alpha < 1$, $0 \leq \beta, \gamma < 0.5$ such that for any $x, y \in E$ at least one the following conditions is true:

(z_1) $\|Tx - Ty\| \leq \alpha \|x - y\|$;

(z_2) $\|Tx - Ty\| \leq \beta \left[\|x - Tx\| + \|y - Ty\|\right];$
(z_3) $\|Tx - Ty\| \leq \gamma \left[\|x - Ty\| + \|y - Tx\|\right].$

The main result of this section is given by the next theorem.

Theorem 5.6. *Let E be a uniformly convex Banach space, K a closed convex subset of E and $T : K \to K$ a Zamfirescu operator. Let $\{\alpha_n\}$, $\{\beta_n\}$ be two sequences in $[0,1]$ with $\{\alpha_n\}$ satisfying the condition*

$$(i) \quad \sum_{n=0}^{\infty} \alpha_n(1 - \alpha_n) \ diverges.$$

Then, for any $x_0 \in K$, the Ishikawa iteration process $I(x_0, \alpha_n, \beta_n, T)$ converges strongly to the unique fixed point of T.

Proof. Let $\{x_n\}$ be the Ishikawa iteration $I(x_0, \alpha_n, \beta_n, T)$, i.e., the sequence defined by

$$x_{n+1} = (1 - \alpha_n)x_n + \alpha_n Ty_n, \quad y_n = (1 - \beta_n)x_n + \beta_n Tx_n, \ n \geq 0,$$

with $x_0 \in K$, arbitrary. By Theorem 2.4 we know that T has a unique fixed point in E. Call it p. For any $x_0 \in K$ we have

$$\|x_{n+1} - p\| \leq \alpha_n \|Ty_n - p\| + (1 - \alpha_n) \|x_n - p\|.$$

As any Zamfirescu operator is quasi-nonexpansive, we get

$$\|Ty_n - p\| = \|Ty_n - Tp\| \leq \|y_n - p\|.$$

By the definition of $\{y_n\}$ we have

$$\|y_n - p\| \leq \beta_n \|Ty_n - p\| + (1 - \beta_n) \|x_n - p\| \leq \|x_n - p\|,$$

and therefore $\|x_{n+1} - p\| \leq \|x_n - p\|$, which shows that $\{\|x_n - p\|\}$ is non-increasing. For the rest of the proof see that of Theorem 4.9. $\square$

Theorem 5.7. *Let K be a nonempty closed convex subset of a Banach space E and $T : K \to K$ a quasi-contraction. Suppose $\alpha_n > 0$, for all $n \geq 0$ and $\sum_{n=0}^{\infty} \alpha_n = \infty$. Let $\{x_n\}$ be the sequence defined by*

$$x_0 \in K$$

$$y_n \in \mathrm{co}\left(\{x_i\}_{i=k_n}^n \cup \{Tx_i\}_{i=k_n}^n\right), \ n \geq 0 \tag{27}$$

$$x_{n+1} = (1 - \alpha_n)x_n + \alpha_n Ty_n, \ n \geq 0, \tag{28}$$

where $\{k_n\}$ is a non-decreasing sequence of positive integers such that $k_n \leq n$ and $\lim_{n \to \infty} k_n = +\infty$.
Then $\{x_n\}$ converges strongly to the unique fixed point of T.

Remarks.

1) Rhoades [Rh94a] extended Theorem 5.7 to the more general class of generalized φ-contractions defined by

$$\|Tx - Ty\| \leq \varphi(M(x, y)),$$

where $\varphi : [0, \infty) \to [0, \infty)$ satisfies the following conditions:
   (a) $0 < \varphi(t) < t$ for each $t > 0$ and $\varphi(0) = 0$;
   (b) φ is increasing on $(0, \infty)$;
   (c)   the function $g(t) = t/(t - \varphi(t))$ is non-increasing on $(0, \infty)$;
   2) It is important to mention that Rhoades' result has been proved for the Ishikawa iteration scheme defined by Xu, i.e., by considering

$$y_n \in \text{co}\left(\{x_i\}_{i=0}^n \cup \{Tx_i\}_{i=0}^n\right), \ n \geq 0 \tag{27'}$$

instead of (27);
   3) Ciric [Cir97] himself, Mishra and Kalinde [MKa98] extended the previous results concerning the convergence of the Ishikawa iteration for the class of quasi-contractive operators, to the general case of convex metric spaces, which include all normed linear spaces.

The next theorem extends Theorem 5.6 to arbitrary Banach spaces by simultaneously weakening the assumptions on the sequence $\{\alpha_n\}$. Theorem 5.8 also extends Theorem 4.10 from Mann iteration to the Ishikawa iteration.

Theorem 5.8. *Let E be an arbitrary Banach space, K a closed convex subset of E, and $T : K \to K$ an operator satisfying condition $(z_1) - (z_2)$. Let $\{x_n\}_{n=0}^\infty$ be the Ishikawa iteration defined by $(28) - (29)$ and $x_0 \in K$, where $\{\alpha_n\}$ and $\{\beta_n\}$ are sequences of positive numbers in $[0, 1]$ with $\{\alpha_n\}$ satisfying*

$$(ii) \sum_{n=0}^{\infty} \alpha_n = \infty.$$

Then $\{x_n\}_{n=0}^\infty$ converges strongly to the fixed point of T.

Proof. We use similar arguments to those in proving Theorem 4.10. Let $\{x_n\}_{n=0}^\infty$ be the Ishikawa iteration defined by

$$x_{n+1} = (1 - \alpha_n)x_n + \alpha_n T y_n, \tag{28}$$

$$y_n = (1 - \beta_n)x_n + \beta_n T x_n, \ n \geq 0, \tag{29}$$

and $x_0 \in K$ arbitrary. Then

$$\|x_{n+1} - p\| = \left\|(1 - \alpha_n)x_n + \alpha_n T y_n - (1 - \alpha_n + \alpha_n)p\right\| =$$

$$= \left\|(1 - \alpha_n)(x_n - p) + \alpha_n(T y_n - p)\right\| \leq$$

$$\leq (1 - \alpha_n)\|x_n - p\| + \alpha_n\|T y_n - p\|. \tag{30}$$

With $x := p$ and $y := y_n$, from (8) in Chapter 4, we obtain

$$\|Ty_n - p\| \le \delta \cdot \|y_n - p\|, \tag{31}$$

where δ is given by (7) in the same Chapter 4. Further we have

$$\|y_n - p\| = \|(1 - \beta_n)x_n + \beta_n Tx_n - (1 - \beta_n + \beta_n)p\| =$$
$$= \|(1 - \beta_n)(x_n - p) + \beta_n(Tx_n - p)\| \le$$
$$\le (1 - \beta_n)\|x_n - p\| + \beta_n\|Tx_n - p\|. \tag{32}$$

Again by (8) in Chapter 4, this time with $x := p$; $y := x_n$, we find that

$$\|Tx_n - p\| \le \delta\|x_n - p\| \tag{33}$$

and hence, by (29) - (33) we obtain

$$\|x_{n+1} - p\| \le \left[1 - (1 - \delta)\alpha_n(1 + \delta\beta_n)\right] \cdot \|x_n - p\|,$$

which, by the obvious inequality

$$1 - (1 - \delta)\alpha_n(1 + \delta\beta_n) \le 1 - (1 - \delta)^2\alpha_n,$$

implies

$$\|x_{n+1} - p\| \le \left[1 - (1 - \delta)^2\alpha_n\right] \cdot \|x_n - p\|, \quad n = 0, 1, 2, \ldots. \tag{34}$$

Now, by (34) we inductively obtain

$$\|x_{n+1} - p\| \le \prod_{k=0}^{n} \left[1 - (1 - \delta)^2\alpha_k\right] \cdot \|x_0 - p\|, \quad n = 0, 1, 2, \ldots. \tag{35}$$

Using the fact that $0 \le \delta < 1$, $\alpha_k, \beta_n \in [0,1]$, and $\sum_{n=0}^{\infty} \alpha_n = \infty$, by (ii) it results that

$$\lim_{n\to\infty} \prod_{k=0}^{n} \left[1 - (1 - \delta)^2\alpha_k\right] = 0,$$

which by (35) implies

$$\lim_{n\to\infty} \|x_{n+1} - p\| = 0,$$

i.e., $\{x_n\}_{n=0}^{\infty}$ converges strongly to p.   □

Remark.
Condition (i) in Theorem 5.6 is slightly more restrictive than condition (iv) in Theorem 5.8, the latter known as a *necessary* condition for the convergence of Mann and Ishikawa iterations. Indeed, by virtue of (i) we cannot have $\alpha_n \equiv 0$ or $\alpha_n \equiv 1$ and hence

$$0 < \alpha_n(1 - \alpha_n) < \alpha_n, \quad n = 0, 1, 2, \ldots,$$

which shows that (i) always implies (ii).

But there exist values of $\{\alpha_n\}$ satisfying (ii), e.g., $\alpha_n \equiv 1$, such that (i) is not true.

Corollary 5.2. *Let E be an arbitrary Banach space, K a closed convex subset of E, and $T : K \to K$ a Kannan operator, i.e., an operator satisfying (8) in Chapter 2. Let $\{x_n\}_{n=0}^{\infty}$ be the Ishikawa iteration defined by $(28) - (29)$ and $x_0 \in K$, with $\{\alpha_n\}$, $\{\beta_n\} \subset [0, 1]$ satisfying (ii).*

Then $\{x_n\}_{n=0}^{\infty}$ converges strongly to the fixed point of T.

Corollary 5.3. *Let E be an arbitrary Banach space, K a closed convex subset of E, and $T : K \to K$ a Chatterjea operator, i.e., an operator satisfying (34) in Chapter 2. Then the Ishikawa iteration $\{x_n\}_{n=0}^{\infty}$ defined by $(28) - (29)$ and $x_0 \in K$, with $\{\alpha_n\}$, $\{\beta_n\} \subset [0, 1]$ satisfying (ii) converges strongly to the fixed point of T.*

Remark.

It is quite obvious that Theorem 4.10 is properly contained in Theorem 5.8, and it is obtained for $\beta_n \equiv 0$.

On the other hand, due to the fact that, except for (ii), no other conditions are required for $\{\alpha_n\}$, $\{\beta_n\}$, by Theorem 5.8 we may obtain, in particular, the convergence theorem regarding the convergence of Picard iteration in the class of Zamfirescu operators, see Chapter 2, for $\alpha_n \equiv 1$, $\beta_n \equiv 0$, as well as a convergence theorem for the Krasnoselskij iteration, for $\beta_n \equiv 0$ and $\alpha_n = \lambda \in [0, 1]$, see Chapter 3.

5.5 The Equivalence Between Mann and Ishikawa Iterations

As shown in Section 5.1, in order to approximate fixed points of Lipschitzian pseudo-contractive operators, we really need Ishikawa iteration. However, this iterative scheme, which is actually a two-step Mann iteration, is computationally more complicated than the former. Even if in the last two decades numerous papers were devoted to the study of Ishikawa or very complicated Ishikawa-type iterative methods, from a practical point of view, when two or more fixed point iterative schemes are known to be convergent in a certain class of mappings, it is natural to choose the simplest method amongst them.

This was shown partly in Chapter 4, where we illustrated by Example 4.3 a situation when Krasnoselskij iteration suffices to approximate fixed points. More discussions can be find in Chapter 9, where we compare some fixed point iterative methods with respect to their rate of convergence.

Very recently some new results were published, which show that, for certain classes of operators, Mann and Ishikawa iterations are actually equivalent. This also points to the conclusion that the use of Mann iteration would be recommended in those circumstances. It is the aim of this small section to present a sample result in this field, without proof (the original proof is extremely long).

Theorem 5.9. *Let X be a real Banach space, K a nonempty closed convex subset of X, and $T : K \to K$ be a Lipschitzian, strongly pseudocontractive map with $Fix\,(T) \neq \emptyset$. Let $\{x_n\}_{n=0}^{\infty}$ be the Ishikawa iteration defined by*

$$x_{n+1} = (1 - \alpha_n)x_n + \alpha_n T y_n, \tag{36}$$

$$y_n = (1 - \beta_n)x_n + \beta_n T x_n, \ n \geq 0,$$

and $x_0 \in K$, and $\{u_n\}_{n=0}^{\infty}$ be the Mann iteration defined by

$$u_{n+1} = (1 - \alpha_n)u_n + \alpha_n T u_n, \tag{37}$$

and $u_0 = x_0 \in K$, where $\{\alpha_n\}$ and $\{\beta_n\}$ are sequences of positive numbers in $[0, 1]$ satisfying

$$\lim_{n \to \infty} \alpha_n = \lim_{n \to \infty} \beta_n = 0 \ and \ \sum_{n=0}^{\infty} \alpha_n = \infty.$$

Then T possesses a unique fixed point x^* and the following assertions are equivalent:
(i) the Mann iteration (37) converges to x^*;
(ii) the Ishikawa iteration (36) converges to x^*.

Remark. Since T in Theorem 5.9 has a unique fixed point, it would be more natural to consider $u_0 \neq x_0$ as well as weaker conditions on the sequences $\{\alpha_n\}$ and $\{\beta_n\}$ that define the Ishikawa iteration, in light of the results we presented in Chapter 4, and also to construct the Mann iteration by using a sequence $\{\alpha'_n\}$ which is different from the one defining the Ishikawa iteration.

5.6 Bibliographical Comments

Example 5.1 at the beginning of Chapter 5 is due to Rhoades [Rho91].

§5.1.

The Ishikawa iterative process was first introduced by Ishikawa [Ish74] in 1974, in order to approximate fixed points of Lipschitzian pseudocontractive operators, because in the case T is only pseudocontractive, the Mann iteration does not converge generally to the fixed point of T, as it was pointed out by Hicks and Kubicek [HK77a], see Exercise 4.1.

The Ishikawa iteration is one of the answers that were given by different authors to this problem, until Chidume and Mutangadura [CMu01] constructed their example, see Exercise 4.2.

The content of this section is mainly taken from Ishikawa [Ish74], except for Remarks 1-4 which are taken from Rhoades [Rho91].

§5.2.

Theorem 5.2 and Corollary 5.1 are taken from Chidume [Ch98b]. Theorem 5.2 is a significant generalization of most of the related results in literature. Among these, we mention Theorem 2 of Deng [Dg93b], Theorem 4.2 of Tan and Xu, H.K. [TX93c], and Theorem 1 of Reich [Re79c].

The other results of this section (Definition 5.1 and Theorem 5.3) are taken from Gu, Feng [Gu01d]. Several results due to Chang [Ca97b], Chidume [Ch94b], [Chi95]; Deng and Ding [DDi95]; Ding [Din81], [Din88] and Tan and Xu, H.K. [TX93a] are generalized or extended by Theorem 5.3.

In q-uniformly smooth Banach spaces, Huang, Z. [HZ00b] weakened the Lipschitz assumption in Theorem 5.3 to the continuity of the operator T, by imposing, *in compensation,* that the range of T is bounded. However, in this case, the assumptions on the sequences $\{\alpha_n\}$, $\{\beta_n\}$ involve the smoothness order q. A result that extends Theorem 4.12 from Mann iteration to Ishikawa iteration in the case of Lipschitzian strictly pseudocontractive operators was obtained in Zeng, L. [Ze02b].

§5.3.

The property of a Banach space to satisfy Opial's condition was first considered in Opial [Op67b], see also Exercise 3.18.

All the results contained in this section are taken from Tan and Xu, H.K. [TX93a]. Thus, Lemma 5.1 is Lemma 2 there, Lemma 5.2 is Lemma 3, Lemma 5.4 is Lemma 4, while Theorem 5.4 is Theorem 1 in the same paper. Lemma 5.3 is given in Bruck [Bru74].

Theorem 5.5 is Theorem 2 in the same paper by Tan and Xu, H.K. [TX93a].

For details in the case when the norm of E is Frechet differentiable in the proof of Theorem 5.4, see Tan and Xu, H.K. [TX93a], pp. 306-307.

§5.4.

The main result of the section, i.e., Theorem 5.6, is taken from Rhoades [Rho76], Theorem 8, while Theorem 5.7 is taken from Xu, H.K. [TX93b], with the correction indicated by Ciric [Cir97].

For a comparison of different contractive conditions involved in fixed point theorems, see Rhoades [Rh77b]. The contractive condition in this section is involved in a fixed point theorem of Ciric [Cir74], regarding the convergence of Picard iteration, see also Chapter 2.

Other related results were obtained by Sastry, Babu and Rao [SBS01], [SBS02]. Theorem 5.7 which gives the convergence of Ishikawa iteration in the general case of quasi-contractive mappings was obtained (in an incomplete form) by Xu, H.K. [XuH92] and then completed by Ciric [Cir97]. We gave here its correct version.

Theorem 5.8 and Corollaries 5.2 and 5.3 are taken from Berinde [Be04c].

§5.5.

Theorem 5.9 is due to Rhoades and Soltuz [RS03c]. For other related results see also [RS03a], [RS03b], [RS04a]-[RS04e], [So03a], [So04a-So04b] and [CCK03].

Exercises and Miscellaneous Results

5.1. Prove that for any x, y, z in a Hilbert space H and for any real number λ, we have

$$\| \lambda x + (1 - \lambda) y - z \|^2 = \lambda \| x - z \|^2 + (1 - \lambda) \| y - z \|^2 - \lambda (1 - \lambda) \| x - y \|^2.$$

5.2. Let X be a real Banach space and J be a normalized duality mapping. Then for any given $x, y \in X$, the following inequality holds:

$$\|x + y\|^2 \leq \|x\|^2 + 2 \langle y, j(x + y) \rangle, \; \forall j(x + y) \in J(x + y).$$

5.3. Prove that X is a uniformly smooth Banach space (or, equivalently, X^* is a uniformly convex Banach space) if and and only if J is single-valued and uniformly continuous on any bounded subset of X.

5.4. Gu, Feng (2001)
Let X be a uniformly smooth real Banach space, let K be a nonempty closed convex subset of X and let $T : K \to K$ be a L-Lipschitzian Φ-strongly pseudo-contractive mapping, with $L \geq 1$. Let $\{\alpha_n\}$ and $\{\beta_n\}$ be two sequences of positive numbers in $[0, 1]$ satisfying $\lim_{n \to \infty} \alpha_n = \lim_{n \to \infty} \beta_n = 0$ and $\sum_{n=0}^{\infty} \alpha_n = \infty$. If $F(T) \neq \emptyset$, then for any given $x_0 \in K$, the Ishikawa iterative sequence $\{x_n\}_{n=0}^{\infty}$ defined by

$$x_{n+1} = (1 - \alpha_n)x_n + \alpha_n T y_n,$$

$$y_n = (1 - \beta_n)x_n + \beta_n T x_n, \; n \geq 0,$$

converges strongly to the unique fixed point of T in K. (T is said to be Φ-strongly pseudo-contractive if $U := I - T$ is Φ-strongly accretive, i.e., for any $x, y \in K$, there exists $j(x + y) \in J(x + y)$ and a strictly increasing function $\Phi : [0, \infty) \to [0, \infty)$ with $\Phi(0) = 0$ such that $\langle Tx - Ty, j(x + y) \rangle \geq \Phi(\|x - y\|) \|x - y\|$).

5.5. Prove Lemma 5.3, Theorem 5.3, Theorem 5.7 and Theorem 5.9.

6

Other Fixed Point Iteration Procedures

The aim of this chapter is to present some other iterative procedures, less frequently used to approximate fixed points: Mann and Ishikawa iterations with errors, modified Mann and Ishikawa iterations, Kirk's iteration etc.

6.1 Mann and Ishikawa Iterations with Errors

The idea of considering fixed point iteration procedures *with errors* comes from practical numerical computations. Although they are related to the stability problem of fixed point iterations, see Section 7.1 in the next Chapter, we however inserted this topic here as a distinct Section, due to the considerable amount of research done by several authors, that complements in some sense the stability problem of fixed point iteration procedures.

Definition 6.1. Let K be a subset of a linear normed space E and let $T : K \to X$ be a mapping. The sequence $\{x_n\}$ in E defined by

$$x_0 \in K \tag{1}$$

$$x_{n+1} = (1 - \alpha_n)x_n + \alpha_n T y_n + u_n, \tag{2}$$

$$y_n = (1 - \beta_n)x_n + \beta_n T x_n + v_n , \ n \geq 0, \tag{3}$$

where $\{\alpha_n\}$ and $\{\beta_n\}$ are two sequences in $[0,1]$ and $\{u_n\}$ and $\{v_n\}$ are two summable sequences in E, i.e.,

$$\sum_{n=0}^{\infty} \|u_n\| < \infty , \ \sum_{n=0}^{\infty} \|v_n\| < \infty, \tag{4}$$

is called *the Ishikawa iteration with errors*.

Remark. If we take $\beta_n = 0$ and $v_n \equiv 0_E$, from the Ishikawa iteration with errors we obtain the *Mann iteration with errors*.

We give without proof one of the first results of this type on the fixed point iteration procedures with errors.

Theorem 6.1. *Let K be a nonempty closed subset of a uniformly smooth Banach space E. Let $T : K \to X$ be Lipschitzian (with constant $L \geq 1$) and strictly pseudocontractive (with constant $t > 1$). Let $\{u_n\}$, $\{v_n\}$ be two summable sequences in E, and let $\{\alpha_n\}$, $\{\beta_n\}$ be two real sequences in $[0,1]$ satisfying*

$$(i) \quad \lim_{n \to \infty} \alpha_n = 0 \text{ and } \sum_{n=0}^{\infty} \alpha_n = \infty; \quad (ii) \limsup_{n \to \infty} \beta_n < k/L(L+1),$$

where $k = (t-1)/t$.

If the range $T(K)$ of T is bounded, then $\{x_n\} \subset K$ generated by (1)-(3) converges strongly to the unique fixed point of T.

Remarks.

1) For null sequences $\{u_n\}$, $\{v_n\}$, from (1)-(3) we find the usual Ishikawa iteration;

2) However, there is no explanation how we can take $u_n, v_n \in E$ in order to be sure that $x_n \in K$, for all $n \geq 0$, see Example 6.1;

3) It was argued that the notion of iterative process with errors given in Definition 6.1 is not fully satisfactory, because the occurrence of errors is random, while the conditions (4) imposed on the error terms imply, in particular, that they tend to zero as n tends to infinity, which is therefore unreasonable.

Example 6.1. Let $E = l_2$, $K = \{x \in E : \|x\| \leq 1\}$, and define $T : K \to E$ by $Tx = -4x$.

Then it is easy to see that T is Lipschitzian and strongly pseudocontractive with the unique fixed point $x^* = (0,0,0,...)$. Take $x_0 = (1,0,0,...)$ and set $\alpha_n = \beta_n = 1/(n+2)$. Then

$$y_0 = (1 - \beta_0)x_0 + \beta_0 T x_0 = -3/2x_0 \notin K.$$

Thus Ty_0 cannot be computed. Observe that neither the Mann nor the Ishikawa iteration is well defined in this case.

An other concept of iterative process with errors is given by the next definition.

Definition 6.2. Let K be a nonempty convex subset of a Banach space E and $T : K \to X$ a mapping. The sequence $\{x_n\}_{n=1}^{\infty}$ defined iteratively by

$$x_0 \in K, \tag{5}$$

$$x_{n+1} = a_n x_n + b_n T y_n + c_n u_n, \tag{6}$$

$$y_n = a'_n x_n + b'_n T x_n + c'_n v_n, \ n \geq 0, \tag{7}$$

where $\{u_n\}$, $\{v_n\}$ are bounded sequences in K and $\{a_n\}$, $\{b_n\}$, $\{c_n\}$, $\{a'_n\}$ $\{b'_n\}$ and $\{c'_n\}$ are sequences in $[0, 1]$ such that

$$a_n + b_n + c_n = a'_n + b'_n + c'_n = 1, \ n \geq 0, \tag{8}$$

is still called *Ishikawa iteration sequence with errors*.

Remark. If $b'_n = c'_n = 0$, $n \geq 0$, then the sequence $\{x_n\}$ will be called *Mann iteration with errors*. There are however serious objections to the definition of Xu, too. It was pointed out by that if the range of T is bounded, the Xu's definition reduces to that of Liu and moreover, from a practical point of view, the construction of Xu cannot be carried out.

The following theorem extends Ishikawa's original result to both the case of iterative processes with errors and to the slightly more general class of Lipschitzian hemicontractions (in the case of Hilbert spaces).

Theorem 6.2. *Let K be a compact convex subset of a real Hilbert space H and $T : K \to K$ a continuous hemicontractive map. Let $\{a_n\}$, $\{b_n\}$, $\{c_n\}$, $\{a'_n\}$, $\{b'_n\}$ and $\{c'_n\}$ be real sequences in $[0, 1]$ satisfying the following conditions:*

(i) $a_n + b_n + c_n = a'_n + b'_n + c'_n = 1, \ n \geq 0$;

(ii) $\lim\limits_{n \to \infty} b_n = \lim\limits_{n \to \infty} b'_n = 0$; *(iii)* $\sum\limits_{n=0}^{\infty} c_n < \infty$; $\sum\limits_{n=0}^{\infty} c'_n < \infty$;

(iv) $\sum \alpha_n \beta_n = \infty$; $\sum\limits_{n=0}^{\infty} \alpha_n \beta_n \delta_n < \infty$, *where* $\delta_n = \|T x_n - T y_n\|^2$;

(v) $0 \leq \alpha_n \leq \beta_n < 1, \ n \geq 0$, *where* $\alpha_n = b_n + c_n$; $\beta_n = b'_n + c'_n$.

Then the Ishikawa iteration with errors $\{x_n\}_{n=0}^{\infty}$ defined by (5)-(7) converges strongly to a fixed point of T.

Proof. The existence of a fixed point of T follows from Schauder's fixed point theorem (since T is continuous). Let $x^* \in F_T$ be a fixed point of T. By Lemma 1.8 we have

$$\|(1 - \lambda)x + \lambda y\|^2 = (1 - \lambda)\|x\|^2 + \lambda\|y\|^2 - \lambda(1 - \lambda)\|x - y\|^2, x, y \in H, \lambda \in [0, 1]$$

Since T is hemicontractive, we have

$$\|T x - T x^*\|^2 \leq \|x - x^*\|^2 + \|x - T x\|^2.$$

So, after straightforward calculations we find that

$$\|x_{n+1} - x^*\|^2 \leq \|x_n - x^*\|^2 - \alpha_n \beta_n (1 - 2\beta_n)\|x_n - T x_n\|^2 +$$

$$+\alpha_n \beta_n \|T x_n - T y_n\|^2 + M(c_n + c'_n), \tag{8'}$$

where $M > 0$ is a constant.

Since K is compact and T is continuous, the sequence $\{\|x_n - Tx_n\|\}$ is bounded. By assumptions $(ii)-(iv)$, the compactness of K and the continuity of T, we have that $\lim_{n\to\infty} \inf \|x_n - Tx_n\| = 0$.

Again by the compactness of K, this implies that there exists a subsequence $\{x_j\}$ of $\{x_n\}$ which converges to a fixed point of T, say x^*.

Let $\psi_n = \|x_n - x^*\|^2$, $\sigma_n = \alpha_n \beta_n \|Tx_n - Ty_n\|^2 + M(c_n + c'_n)$.

Then $\psi_n \geq 0$, $\sigma_n \geq 0$ $(n \geq 0)$ and $\sum_{n=0}^{\infty} \sigma_n < \infty$ by conditions (iii) and (iv). Thus, the inequality (8') yields $\psi_{n+1} \leq \psi_n + \sigma_n$, $\forall n \geq 0$, which, by Lemma 1.7, part (ii), leads to $\psi_n \to 0$ as $n \to \infty$, i.e., $x_n \to x^*$ as $n \to \infty$. $\square$

Remark.
The second part of assumption (iv) in Theorem 6.2 is rather difficult to check. Recently, some results based on simpler assumptions on the parameters that define the iterations were obtained.

Theorem 6.3. *Let K be a compact convex subset of a uniformly convex Banach space E satisfying Opial's condition and let $T : K \to K$ be a nonexpansive mapping with $F_T \neq \emptyset$. Assume that $\{a_n\}$, $\{b_n\}$, $\{c_n\}$, $\{a'_n\}$, $\{b'_n\}$ and $\{c'_n\}$ are real sequences in $[0,1]$ satisfying (i), (ii) and either*
   1) $a_n \in [a, 1]$, $b_n \in [a, b]$, $b'_n \in [0, b]$ *for some $a, b \in \mathbb{R}$ with $0 < a \leq b < 1$,*
or
   2) $a'_n, b_n \in [a, 1]$, $b'_n \in [a, b]$ *for some $a, b \in \mathbb{R}$ with $0 < a \leq b < 1$.*
   Then the Ishikawa iteration with errors $\{x_n\}$ defined by (5)-(7) converges weakly to a fixed point of T.

Remark. For two operators $S, T : K \to K$, the iterative process defined by $x_0 \in K$ (9)

$$x_{n+1} = a_n x_n + b_n S y_n + c_n u_n, \quad n \geq 0 \tag{10}$$

$$y_n = a'_n x_n + b'_n T x_n + c'_n v_n, \quad n \geq 0, \tag{11}$$

where $\{a_n\}$, $\{b_n\}$, $\{c_n\}$, $\{a'_n\}$, $\{b'_n\}$, $\{c'_n\}$ are real sequences in $[0, 1]$ satisfying (i) and (iii), and $\{u_n\}$, $\{v_n\}$ are bounded sequences in K, is an Ishikawa type common fixed point iteration that reduces to (5)-(7), if $S \equiv T$.

Theorem 6.4. *Let E be a uniformly convex Banach space. Let K be a closed convex subset of E and let $S, T : K \to K$ be nonexpansive operators with a common fixed point (i.e., $F_S \cap F_T \neq \emptyset$). Then for the sequence defined by (9)-(11) the following hold:*
   1) *If $a_n, a'_n \in [a, 1]$, $b_n \in [a, 1]$, $b'_n \in [0, b]$ for some $a, b \in \mathbb{R}$ with $0 < a \leq b < 1$, then $x_{n_j} \rightharpoonup p$, implies $p \in F_S$;*
   2) *If $a'_n, b_n \in [a, 1]$ and $b'_n \in [a, b]$ for some $a, b \in \mathbb{R}$ with $0 < a \leq b < 1$ then $x_{n_j} \rightharpoonup p$, implies $p \in F_T$;*
   3) *If $a_n, a'_n \in [a, 1]$ and $b_n, b'_n \in [a, b]$ for some $a, b \in \mathbb{R}$ with $0 < a \leq b < 1$ then $x_{n_j} \rightharpoonup p$, implies $p \in F_S \cap F_T$.*

6.2 Modified Mann and Ishikawa Iterations

The aim of this section is to show that, considering the n-th iterate T^n instead of T in the relations that define the Mann and Ishikawa iterations, we obtain new iterative processes that converge strongly to the fixed points of some classes of Lipschitzian and contractive type operators.

Definition 6.3. Let K be a nonempty subset of a normed linear space E and let $T : K \to K$ be a mapping.

1) T is said to be *asymptotically nonexpansive* if there exists a sequence $\{k_n\}_{n=1}^{\infty}$ in $[1, \infty)$ with $\lim_{n \to \infty} k_n = 1$ such that

$$\|T^n x - T^n y\| \leq k_n \|x - y\| , \text{ for all } x, y \in K \text{ and } n \geq 1;$$

2) T is said to be *uniformly L-Lipschitzian* with constant $L > 0$ if

$$\|T^n x - T^n y\| \leq L \|x - y\| , \text{ for all } x, y \in K \text{ and } n \geq 1;$$

3) T is said to be $k-$*strict asymptotically pseudocontractive* if there exist a sequence $\{k_n\}_{n=1}^{\infty}$ in $[1, \infty)$ with $\lim_{n \to \infty} k_n = 1$ and a constant k in $[0, 1)$ such that

$$\|T^n x - T^n y\|^2 \leq k_n^2 \|x - y\|^2 + k \|(x - T^n x) - (y - T^n y)\|^2 ,$$

for all $x, y \in K$ and $n \geq 1$;

4) T is said to be *asymptotically demicontractive* if $F_T \neq \emptyset$ and there exist a sequence $\{k_n\}_{n=1}^{\infty}$ in $[1, \infty)$ with $\lim_{n \to \infty} k_n = 1$ and a constant k in $[0, 1)$ such that for all $x \in K$, $p \in F_T$ and $n \geq 1$,

$$\|T^n x - p\|^2 \leq k_n^2 \|x - p\|^2 + k \|x - T^n x\|^2 . \tag{13}$$

Definition 6.4. Let K be a nonempty convex subset of a normed linear space E, $T : K \to K$ a mapping and $\{\alpha_n\}_{n=1}^{\infty}$ and $\{\beta_n\}_{n=1}^{\infty}$ two sequences in $[0, 1]$. The sequence $\{x_n\}_{n=0}^{\infty}$ defined by

$$\begin{cases} x_0 \in K \\ y_n = (1 - \beta_n)x_n + \beta_n T^n x_n, \\ x_{n+1} = (1 - \alpha_n)x_n + \alpha_n T^n y_n , \ n \geq 0 \end{cases} \tag{14}$$

will be called the *modified Ishikawa iterative process*.

Remarks.

1) If we take $\beta_n = 0$ for each $n \geq 0$ in (14), we find the *modified Mann iteration scheme*;

2) If T is asymptotically nonexpansive, then T is both uniformly $\sup_{n \geq 1}\{k_n\}$-Lipschitzian and $0-$strict asymptotically pseudocontractive;

3) Each k–strict asymptotically pseudocontractive mapping with a non-empty fixed point set is asymptotically demicontractive.

We will need the following auxiliary result.

Lemma 6.1. *Let K be a nonempty convex subset of a normed linear space E and let $T : K \to K$ be a uniformly L-Lipschitzian operator. If $r_n = \|x_n - T^n x_n\|$, $n \geq 0$, where $\{x_n\}_{n=0}^{\infty}$ is the modified Ishikawa iteration associated to T, then*

$$\|x_n - Tx_n\| \leq r_n + r_{n-1}L(1 + 3L + 2L^2), \ n \geq 1.$$

Theorem 6.5. *Let K be a nonempty bounded closed convex subset of a Hilbert space H and let $T : K \to K$ be a completely continuous, uniformly L-Lipschitzian and asymptotically demicontractive mapping. Suppose that the sequence $\{k_n\}$ appearing in (13) satisfies*

$$\sum_{n=0}^{\infty} (k_n - 1) < \infty. \tag{15}$$

Assume that $\{\alpha_n\}_{n=0}^{\infty}$ and $\{\beta_n\}_{n=0}^{\infty}$ are real sequences in $[0, 1]$ satisfying

$$0 < a \leq \alpha_n, \ \ n \geq 0; \tag{16}$$

$$0 < b \leq \beta_n \leq \min \left\{ 1 - k - c, \ \frac{\sqrt{1 + 4(1 - d)L^2} - 1}{2L^2} \right\}, \ \ n \geq 0; \tag{17}$$

$$\alpha_n - k\beta_n \leq 1 - k, \ \ n \geq 0, \tag{18}$$

where k is the constant appearing in (13), and a, b, c are constants with $c + d > 0$, $0 \leq c < 1 - k$ and $0 \leq d < 1$.

Then the modified Ishikawa iteration $\{x_n\}_{n=0}^{\infty}$ defined by (14) converges strongly to some fixed point of T in K.

Proof. Since T is asymptotically demicontractive, $F_T \neq \emptyset$. Let $p \in F_T$. By using (13), (14) and Lemma 1.8 with $z = 0$), we obtain for $n \geq 0$

$$\|x_{n+1} - p\|^2 = \|(1 - \alpha_n)(x_n - p) + \alpha_n(T^n y_n - p)\|^2 =$$

$$= (1 - \alpha_n) \|x_n - p\|^2 + \alpha_n \|T^n y_n - p\|^2 - \alpha_n(1 - \alpha_n) \|x_n - T^n y\|^2$$

$$\leq (1 - \alpha_n) \|x_n - p\|^2 + \alpha_n \left(k_n^2 \|y_n - p\|^2 + k \|y_n - T^n y_n\|^2 \right) -$$

$$- \alpha_n(1 - \alpha_n) \|x_n - T^n y_n\|^2, \tag{19}$$

$$\|y_n - p\|^2 = \|(1 - \beta_n)(x_n - p) + \beta_n(T^n x_n - p)\|^2 =$$

$$= (1 - \beta_n) \|x_n - p\|^2 + \beta_n \|T^n x_n - p\|^2 - \beta_n(1 - \beta_n) \|x_n - T^n x_n\|^2$$

$$\leq (1 - \beta_n) \|x_n - p\|^2 + \beta_n \left(k_n^2 \|x_n - p\|^2 + k \|x_n - T^n x_n\|^2 \right) -$$

$$-\beta_n(1 - \beta_n) \|x_n - T^n y_n\|^2 =$$

$$= (1 - \beta_n + \beta_n k_n^2) \|x_n - p\|^2 + \beta_n(k - 1 + \beta_n) \|x_n - T^n x_n\|^2 \qquad (20)$$

and

$$\|y_n - T^n y_n\|^2 = \|(1 - \beta_n)(x_n - T^n y_n) + \beta_n(T^n x_n - T^n y_n)\|^2 =$$

$$= (1 - \beta_n) \|x_n - T^n y_n\|^2 + \beta_n \|T^n x_n - T^n y_n\|^2 -$$

$$-\beta_n(1 - \beta_n) \|x_n - T^n x_n\|^2 \leq (1 - \beta_n) \|x_n - T^n y_n\|^2 +$$

$$+L^2 \beta_n \|x_n - y_n\|^2 - \beta_n(1 - \beta_n) \|x_n - T^n x_n\|^2 \leq$$

$$\leq (1 - \beta_n) \|x_n - T^n y_n\|^2 + [L^2 \beta_n^3 - \beta_n(1 - \beta_n)] \|x_n - T^n x_n\|^2 . \qquad (21)$$

Substituting (20) and (21) in (19) and canceling, we obtain that

$$\|x_{n+1} - p\|^2 \leq [1 - \alpha_n + \alpha_n k_n^2(1 - \beta_n + \beta_n k_n^2)] \|x_n - p\|^2 +$$

$$+\alpha_n[k_n^2 \beta_n(k - 1 + \beta_n) + k\beta_n(L^2 \beta_n^2 - 1 + \beta_n)] \|x_n - T^n x_n\|^2 +$$

$$+[-\alpha_n(1 - \alpha_n) + \alpha_n k(1 - \beta_n)] \|x_n - T^n y_n\|^2 =$$

$$= \left\{1 + \alpha_n[k_n^2(1 + \beta_n(k_n^2 - 1)) - 1]\right\} \|x_n - p\|^2 -$$

$$-\alpha_n \beta_n[(1 - k - \beta_n)k_n^2 + k(1 - \beta_n - L^2 \beta_n^2)] \|x_n - T^n x_n\|^2 -$$

$$-\alpha_n[1 - \alpha_n - k(1 - \beta_n)] \|x_n - T^n y_n\|^2 =$$

$$= [1 + \alpha_n(k_n^2 - 1)(1 + \beta_n k_n^2)] \|x_n - p\|^2 -$$

$$-\alpha_n \beta_n[(1 - k - \beta_n)k_n^2 + k(1 - \beta_n - L^2 \beta_n^2)] \|x_n - T^n x_n\|^2 -$$

$$-\alpha_n(1 - k - \alpha_n + k\beta_n) \|x_n - T^n y_n\|^2 , \qquad (22)$$

which is valid for all $n \geq 0$ and $p \in F_T$.

Since K is bounded, by (15)-(18) and (22) it follows that there exists $M > 0$ such that

$$\|x_{n+1} - p\|^2 \leq \|x_n - p\|^2 + M(k_n - 1) - ab(c + kd) \|x_n - T^n x_n\|^2 , \qquad (23)$$

for all $n \geq 0$ and $p \in F_T$.

Using again the boundedness of K, by (15) and (23) we obtain that

$$\sum_{n=0}^{\infty} \|x_n - T^n x_n\|^2 < \infty,$$

which implies $\lim\limits_{n \to \infty} \|x_n - T^n x_n\| = 0$. As T is uniformly L-Lipschitzian, by Lemma 6.1 we get

$$\lim_{n \to \infty} \|x_n - T x_n\| = 0. \qquad (24)$$

Now, since K is bounded and closed and T is completely continuous, it follows that $\{Tx_n\}_{n=0}^{\infty}$ has a subsequence $\{Tx_{n_i}\}_{i=0}^{\infty}$ such that $\lim\limits_{i \to \infty} Tx_{n_i} = q$, for some $q \in K$.

From (24) it results that $\lim\limits_{i \to \infty} x_{n_i} = q$, and as T is continuous, we get $q \in F_T$.

Using (23) with $p = q$, it results that

$$\|x_{n+1} - q\|^2 \le \|x_n - q\|^2 + M(k_n - 1) \tag{25}$$

for all $n \ge 0$, hence by virtue of (15), (25) and Lemma 1.7, part (ii), we obtain that $\|x_n - q\| \to 0$ as $n \to \infty$, i.e., $\lim\limits_{n \to \infty} x_n = q$.   □

Remark. In the particular case $\beta_n = 0$, for all $n \ge 0$, by Theorem 6.5 we obtain a convergence result for the modified Mann iterative process.

6.3 Ergodic and Other Fixed Point Iteration Procedures

In this section we want to survey other important iteration procedures that have been considered by several authors in order to approximate the fixed points of several classes of mappings.

Following the idea of Krasnoselskij iteration, which is in fact the Picard iteration corresponding to the mean operator

$$U_\lambda = (1 - \lambda)I + \lambda T = a_0 I + a_1 T,$$

with $a_0 + a_1 = 1$, we can extend it to a convex combination involving the first k iterates of T. For this iteration we have

Theorem 6.6. *Let X be a Banach space and $T : X \to X$ a c-contraction.*

Let $\{x_n\}_{n=0}^{\infty}$ be the sequence defined by

$$x_0 \in X$$

$$x_{n+1} = \alpha_0 x_n + \alpha_1 T x_n + \alpha_2 T^2 x_n + \ldots + \alpha_k T^k x_n , \quad n \ge 0,$$

where $k \ge 1$ is an integer and $\alpha_i \in [0,1]$, $i = 0, 1, \ldots, k$ such that $\alpha_1 > 0$ and $\sum\limits_{i=0}^{k} \alpha_i = 1$.

Then the sequence $\{x_n\}$ converges strongly to the unique fixed point of T.

Proof. We define $F : X \to X$ by

$$Fx = \alpha_0 x + \alpha_1 T x + \alpha_2 T^2 x + \ldots + \alpha_k T^k x, \quad \text{for all } x \text{ in } X. \tag{26}$$

Then we show that F is a c−contraction and hence, by the mapping contraction principle, we get the conclusion.   □

Remarks.

1) If we consider in (26) $\alpha_0 = \alpha_1 = \ldots = \alpha_k = \dfrac{1}{k+1}$, then F will be the Cesaro mean

$$C_n[T]x = \frac{1}{k+1} \cdot \sum_{i=0}^{k+1} T^i x, \quad \text{for } x \in X \text{ and } n \geq 1;$$

2) An early result, which opened the general ergodic theory of nonlinear operators, shows that the Cesaro mean converges weakly to a fixed point of a nonexpansive self-operator T of a closed bounded convex subset of a Hilbert space. This reads as follows

Theorem 6.7. *Let K be a bounded closed convex subset of a Hilbert space H and $T : K \to K$ a nonexpansive operator. Then for each $x \in K$ the Cesaro means $\{C_n[T]x\}_{n=0}^{\infty}$ converge weakly to a fixed point of T.*

It was further proved that if T is an odd map, than the convergence in Theorem 6.7 is strong, and extended this theorem to L^p spaces.

Due to the fact that in the nonlinear case the Cesaro means have usually only weak convergence for nonexpansive operators, some authors considered some nonlinear analogues of the ergodic theorems. We shall present here such an iteration.

Let E be a Banach space and $T : E \to E$ a nonexpansive operator. Consider a sequence $\alpha = \{\alpha_n\}$ in $[0, 1]$ and define inductively $\{A_n^\alpha x\}$ by

$$\begin{cases} A_0^\alpha x = x, \\ A_{n+1}^\alpha x = \alpha_{n+1} x + (1 - \alpha_{n+1}) T A_n^\alpha x. \end{cases} \tag{27}$$

Remarks.

1) If T is positively homogeneous (i.e., $T(\lambda x) = \lambda T x$, for any $\lambda \geq 0$ and any $x \in E$) and $\alpha_n = \dfrac{1}{n+1}$, then by (27) we find

$$A_n^\alpha x = \frac{1}{n+1} S_n x,$$

where

$$\begin{cases} S_0 x = x \\ S_{n+1} x = x + T(S_n x), \end{cases} \tag{28}$$

and so $\{A_n^\alpha x\}$ is a nonlinear generalization of the Cesaro means;

2) If T is linear, then by (27) we find the Cesaro means.

We present here a result for a special class of Banach spaces.

Theorem 6.8. *Let* $\{\alpha_n\}_{n=1}^{\infty}$ *be a sequence in* $[0,1]$ *such that*

(i) $\lim\limits_{n \to \infty} \alpha_n = 0$;

(ii) $\sum\limits_{n=1}^{\infty} \alpha_n = +\infty$; *(iii)* $\sum\limits_{n=1}^{\infty} |\alpha_{n+1} - \alpha_n| < +\infty$.

Let E *be a uniformly convex and uniformly smooth Banach space with a weakly sequentially continuous duality mapping* $J : E \to E^*$, *let* K *be a nonempty closed convex subset of* E *and let* $T : K \to K$ *be a mapping such that* $F_T \neq \emptyset$.

Then for any $x \in K$, *the sequence* $\{A_n^\alpha x\}_{n=0}^{\infty}$ *given by (27) converges strongly to* $p = Px$, *where* P *is a sunny nonexpansive retraction of* K *into* F_T.

(Recall that if P *is a sunny retraction of* K *into* F_T, *then*

$$\langle x - p, \ J(z - p) \rangle \ \leq \ 0, \quad \text{for any } z \in F_T.)$$

Remark. As we have already seen, there is a close connection between fixed point iterative processes and summability methods of sequences. In this context, we want to present an analogous result to Baillon's nonlinear ergodic theorem, by using the Abel means (or method of summation).

Theorem 6.9. *Let* H *be a real Hilbert space. Let* K *be a nonempty closed convex subset of* H *and* $T : K \to K$ *be a nonexpansive mapping. If* $F_T \neq \emptyset$, *then for each* $x \in K$, *the Abel means, i.e., the generalized sequence* $\{A_r[T]x\}_{0<r<1}$ *given by*

$$A_r[T]x = (1 - r) \sum_{n=0}^{\infty} r^n T^n x, \ \ 0 < r < 1,$$

converges weakly to a fixed point of T *as* $r \nearrow 1$.

A Mann-type fixed point iteration procedure, obtained by replacing Tx_n in the well-known recurrence

$$x_{n+1} = c_n x_n + (1 - \alpha_n)Tx_n$$

by a Dirichlet summability method $D_{s_n}^{(u)}[T]x_n$, is also known to converge weakly to a fixed point of T.

Definition 6.5. Let E be a Banach space and $\{u_n\}$ a bounded sequence in a convex subset K of E. Define

$$r_m(x) = \sup\{\|u_n - x\| : n \geq m\},$$

and denote by c_m the unique point in K with the property that

$$r_m(c_m) = \inf\{r_m(x) : x \in K\}.$$

Then $\lim\limits_{n \to \infty} c_n = c$, and c is called the *asymptotic center* of $\{u_n\}$.

The following two results are interesting by themselves.

Theorem 6.10. *Let K be a closed convex subset of a real Hilbert space and $T : K \to K$ be a nonexpansive map with a fixed point. Then for any x in K and any strongly regular matrix A, the $A-$transform of $\{T^n x\}$ converges weakly to a fixed point p of T, which is the asymptotic center of $\{T^n x\}$.*

We shall end this section by inserting one result regarding the Figueiredo fixed point iteration. Let H be a Hilbert space, K a nonempty bounded closed convex subset of H and $T : K \to K$ be a nonexpansive operator.

Theorem 6.11. *Let K contain 0 and $T : K \to K$ be nonexpansive. Then, for any $x_0 \in K$, the sequence $\{x_n\}_{n=0}^{\infty}$ defined by*

$$x_n = T_n^{n^2} x_{n-1}, \quad n = 1, 2, \ldots,$$

where $T_n x = n/(n+1)Tx$, converges strongly to a fixed point of T.

6.4 Perturbed Mann Iteration

It is possible to consider a perturbation of the Mann iteration procedure to approximate fixed points of several classes of mappings in Banach spaces more general than Hilbert spaces. The idea in constructing such kind of methods is to check that such a method provides an approximate fixed point sequence.

Definition 6.6. Let E be a normed linear space and $T : E \to E$ be a mapping. A sequence $\{x_n\} \subset E$ satisfying $\lim_{n \to \infty} \|x_n - Tx_n\| = 0$, is called an *approximate fixed point sequence* for T.

In the previous Chapters we met several approximate fixed point sequences. In connection to Exercise 3.17, we give one more example of approximate sequence.

Example 6.2. Let K be a nonempty subset of a Banach space E and let $T : K \to E$ be a nonexpansive mapping. For $x_0 \in K$, define the Mann sequence $\{x_n\}$ by

$$x_{n+1} := (1 - c_n)x_n + c_n Tx_n, \quad n = 0, 1, 2, \ldots \qquad (29)$$

where $\{c_n\} \subset [0, 1]$ is a sequence of real numbers satisfying $\sum_{n=0}^{\infty} c_n = \infty$.

(a) If $\{x_n\} \subset K$ for all positive integers and $\{x_n\}$ is bounded, then $\{x_n\}$ is an approximate fixed point sequence of T;

(b) If K is closed and T is completely continuous, then T has a fixed point and the sequence $\{x_n\}$ defined by (29) converges strongly to a fixed point of T.

As shown by the previous example and other convergence theorems presented in this book, an approximate fixed point sequence considered in connection with some compactness-type assumptions either on T or on its domain, could ensure the convergence of that sequence to a fixed point of T.

This explains why in some convergence theorems for certain classes of mappings more general than the class of nonexpansive mappings, the condition $\lim_{n \to \infty} \|x_n - Tx_n\| = 0$ is explicitly assumed as part of the hypothesis. The main aim of this section is to consider a perturbed Mann iteration that will provide approximate fixed point sequences for Lipschitzian pseudocontractive mappings in Banach spaces.

To this end we need two sequences of real numbers in $(0, 1]$, $\{\lambda_n\}$ and $\{\theta_n\}$, satisfying the following conditions: (i) $\lim_{n \to \infty} \theta_n = 0$; (ii) $\lambda_n (1 + \theta_n) \le 1, \sum \lambda_n \theta_n = \infty$, $\lim_{n \to \infty} \dfrac{\lambda_n}{\theta_n} = 0$; (iii) $\lim_{n \to \infty} (\dfrac{\theta_{n-1}}{\theta_n} - 1)/(\lambda_n \theta_n) = 0$.

Examples of sequences satisfying these conditions are:

$$\lambda_n = \frac{1}{(n+1)^a}, \ \theta_n = \frac{1}{(n+1)^b}, \ 0 < b < a \text{ and } a + b < 1.$$

Lemma 6.2 provides an approximate fixed point sequence for Lipschitzian pseudocontractive mappings in a real Banach space.

Lemma 6.2. *Let K be a nonempty closed convex subset of a real Banach space E. Let $T : K \to K$ be a Lipschitzian pseudocontractive mapping with Lipschitz constant $L \ge 0$ and $F_T \ne \emptyset$. Let $\{x_n\}$ be a sequence generated from arbitrary $x_1 \in K$ by*

$$x_{n+1} := (1 - \lambda_n)x_n + \lambda_n T x_n - \lambda_n \theta_n (x_n - x_1), \ n = 0, 1, 2, \ldots \quad (30)$$

Then $\lim_{n \to \infty} \|x_n - Tx_n\| = 0$.

Remark. The sequence $\{x_n\}$ given by (30) will be called in the following *a perturbed Mann iteration*. By using Lemma 6.2 and other auxiliary results one can prove each of the next four sample convergence theorems for perturbed Mann iteration (proofs which are left to the reader).

Theorem 6.12. *Let K be a nonempty closed convex subset of a real Banach space E. Let $T : K \to K$ be a Lipschitzian pseudocontractive mapping with Lipschitz constant $L \ge 0$ and $F_T \ne \emptyset$. Suppose T is completely continuous. Then the perturbed Mann iteration $\{x_n\}$ given by (30), with $\{\lambda_n\}$ and $\{\theta_n\}$, satisfying (i)-(iii), converges strongly to a fixed point of T.*

Theorem 6.13. *Let K be a nonempty closed convex and bounded subset of a real Banach space E. Let $T : K \to K$ be a Lipschitzian pseudocontractive mapping with Lipschitz constant $L \ge 0$. Suppose T is completely continuous. Then T has a fixed point in K and the perturbed Mann iteration $\{x_n\}$ given by (30), with $\{\lambda_n\}$ and $\{\theta_n\}$, satisfying (i)-(iii), converges strongly to a fixed point of T.*

Theorem 6.14. *Let K be a nonempty closed convex subset of a real Banach space E with uniformly Gateaux differentiable norm. Let $T : K \to K$ be a Lipschitzian pseudocontractive mapping with Lipschitz constant $L \geq 0$ and $F_T \neq \emptyset$. Suppose every closed convex and bounded subset of K has the fixed point property for nonexpansive self mappings. Then the perturbed Mann iteration $\{x_n\}$ given by (30), with $\{\lambda_n\}$ and $\{\theta_n\}$, satisfying (i)-(iii), converges strongly to a fixed point of T.*

Theorem 6.15. *Let K be a nonempty closed convex and bounded subset of a real Banach space E. Let $T : K \to K$ be a uniformly continuous pseudocontractive map. Let the perturbed Mann iteration $\{x_n\}$ be given by (30), with $\{\lambda_n\}$ and $\{\theta_n\}$, satisfying (i)-(iii). Suppose $\|Tx_{n+1} - Tx_n\| = o(\theta_n)$ and T is completely continuous. Then T has a fixed point and $\{x_n\}$ converges strongly to a fixed point of T.*

6.5 Viscosity Approximation Methods

In Chapter 3, in order to prove Theorem 3.1 (Browder-Gohde-Kirk fixed point theorem in Hilbert spaces), we used a particular averaged mapping $U_s : C \to C$, defined by (see also Exercise 3.2)

$$U_s(x) := (1 - s)v_0 + sTx, \ x \in C \tag{31}$$

where $v_0 \in C$ was fixed and $0 < s < 1$, and $T : C \to C$ was a certain mapping. It is known by the proof of Theorem 3.1 that, if T is nonexpansive, then U_s is a s-contraction, and hence U_s has a unique fixed point x_s, for any $s \in (0, 1)$ and, moreover, that $x_s \to p$, as $s \to 1$, where p is a fixed point of T.

As we have remarked in Chapter 3, even if the proof presented there for Theorem 3.1 is more constructive than that given to the corresponding version of Theorem 3.1 in uniformly Banach spaces (Theorem 1.2), however, the proof of Theorem 3.1 does not provide direct information on a certain method for computing the fixed points of T. The so called *viscosity methods* are just the ones appropriate for supplying this situation.

The current development of viscosity approximation methods is based on replacing the constant v_0 in (31) by a certain contraction f. In this way we obtain a method for selecting a particular fixed point of the nonexpansive mapping T. To introduce this class of methods, we first remind some known facts.

Let H be a Hilbert space, C be a closed convex subset of H and $f : C \to C$ a contraction with coefficient $\alpha \in (0, 1)$. Denote by $\mathcal{C}$ the collection of all contractions on C. Let now $T : C \to C$ be a nonexpansive mapping with $F_T \neq \emptyset$.

For any real number $t \in (0, 1)$ and a given contraction $f \in \mathcal{C}$, define the mapping $T_t^f : C \to C$ by

$$T_t^f x := (1 - t)f(x) + tTx, \ x \in C. \tag{31'}$$

It is easy to show that T_t^f is a contraction with coefficient $1 - (1 - \alpha)t$, where α is the contraction coefficient of f. Denote by $x_t := x_t^f$ the unique fixed point of T_t^f in C (by Theorem 1.1).

Definition 6.5. Let H be a Hilbert space, C a closed convex subset of H. The *metric projection* or *nearest point projection* of H onto C, denoted by P_C, is defined, for any $x \in H$, as the only point in C with the property

$$\|x - P_C x\| = \inf\{\|x - y\| : y \in C\}.$$

The following well known characterization of the metric projection P_C is useful in proving convergence theorems for viscosity approximation methods.

Lemma 6.3. *Let H be a Hilbert space and C a closed convex subset of H. Given $x \in H$ and $y \in C$, then $y = P_C x$ if and only if the following inequality holds*

$$\langle x - y, y - z \rangle \geq 0, \ \forall z \in C.$$

We start with an early result regarding viscosity approximation methods.

Theorem 6.16. *Let H be a Hilbert space, C and U_t given by (31) and $t \in (0, 1)$. Let u_t be the unique fixed point of U_t, i.e.,*

$$u_t = (1 - t)v_0 + tTu_t.$$

Then, as $t \to 1$, u_t converges strongly to a fixed point of T which is closest to v_0, that is, the nearest point projection of v_0 onto F_T.

Definition 6.6. Let E be a Banach space and C, K subsets of E. A mapping $P : C \to K$ is called *sunny* if

$$P[tx + (1 - t)Px] = Px, \text{ for } x \in C \text{ with } tx + (1 - t)Px \in C \text{ and } t \geq 0.$$

Remark. We note that if E is a Hilbert space and K is closed and convex, then the metric projection and the sunny nonexpansive retraction from C onto K coincide, that is, when T is a nonexpansive mapping on C, then the sunny nonexpansive retraction from C onto $Fix(T)$ is just the metric projection. This, however, is not valid for an arbitrary Banach space.

Lemma 6.4. *Let E be a smooth Banach space and let J be the duality mapping from E into E^*. Let C be a convex subset of E, let K be a subset of C and let P be a retraction from C onto K. Then the following are equivalent:*
  (i) *$\langle x - Px, J(Px - y) \rangle \geq 0$ for all $x \in C$ and $y \in K$;*
  (ii) *P is both sunny and nonexpansive.*

Remark. The previous lemma shows that there is at most one sunny nonexpansive retraction from C onto K. The next lemma transposes Lemma 6.3 from the Hilbert space setting to Banach spaces.

Lemma 6.5. *Let C be a closed convex subset of a smooth Banach space E. Let K be a subset of C and let P be the unique sunny nonexpansive retraction from C onto K. Let $f : C \to C$ be a mapping and let $z \in K$. Then the following are equivalent:*

(i) z is a fixed point of $P \circ f$;

(ii) z is a solution of the variational inequality $\langle f(z) - z, J(z - y) \rangle \geq 0$, for all $y \in K$.

Proof. By Lemma 6.5, we immediately deduce that (i) implies (ii). To prove the converse let us denote $y = P \circ f(z)$ to get

$$\langle f(z) - z, J(z - P \circ f(z)) \rangle \geq 0.$$

On the other hand, putting $x = f(z)$ and $y = z$ in (i) of Lemma 6.5, we also have

$$\langle f(z) - P \circ f(z), J(P \circ f(z) - z) \rangle \geq 0.$$

Now, by the previous two inequalities we obtain

$$\langle P \circ f(z) - z, J(z - P \circ f(z)) \rangle \geq 0,$$

which implies (i). □

We now state a result which extend Theorem 6.16 from Hilbert spaces to uniformly smooth Banach spaces. This result is important by itself and will be crucial in proving Theorem 6.18.

Theorem 6.17. *Let C be a bounded closed convex subset of a uniformly smooth Banach space E and let $T : C \to C$ be a nonexpansive mapping. Fix $u \in C$ and define a net $\{y_\alpha\}$ in C by $y_\alpha = (1 - \alpha)Ty_\alpha + \alpha u$ for $\alpha \in (0,1)$. Then $\{y_\alpha\}$ converges strongly to Pu as α tends to $+0$, where P is the unique sunny nonexpansive retraction from C onto $\mathrm{Fix}\,(T)$.*

We remark that in Theorem 6.17, the net $\{y_\alpha\}$ is well defined, by Theorem 1.1, see the arguments above.

The main result of this section is contained in the next theorem.

Theorem 6.18. *Let C be a bounded closed convex subset of a uniformly smooth Banach space E. Let $T : C \to C$ be a nonexpansive mapping, let P be the unique sunny nonexpansive retraction from C onto $\mathrm{Fix}\,(T)$ and let f be a contraction on C. Define a net $\{x_\alpha\}$ in C by*

$$x_\alpha = (1 - \alpha)Tx_\alpha + \alpha f(x_\alpha), \text{ for } \alpha \in (0,1).$$

Then as α tends to $+0$, $\{x_\alpha\}$ converges strongly to the unique point $z \in C$ satisfying $P \circ f(z) = z$.

Proof. Define a net $\{y_\alpha\}$ in C by $y_\alpha = (1-\alpha)Ty_\alpha + \alpha f(z)$, for $\alpha \in (0,1)$ and $z \in C$ satisfying $P \circ f(z) = z$. Then by Theorem 6.17, $\{y_\alpha\}$ converges strongly to $P \circ f(z) = z$. For every $\alpha \in (0,1)$, we have

$$\|x_\alpha - y_\alpha\| \le (1-\alpha)\|Tx_\alpha - Ty_\alpha\| + \alpha \|f(x_\alpha) - f(z)\|$$

$$\le (1-\alpha)\|x_\alpha - y_\alpha\| + \alpha r \|x_\alpha - z\|$$

which yields $\|x_\alpha - y_\alpha\| \le r\|x_\alpha - z\|$. Using the last inequality, we get

$$\|x_\alpha - z\| \le \|x_\alpha - y_\alpha\| + \|y_\alpha - z\| \le r\|x_\alpha - z\| + \|y_\alpha - z\|,$$

from which we deduce

$$\lim_{\alpha \to +0} \|x_\alpha - z\| \le \frac{1}{1-r} \lim_{\alpha \to +0} \|y_\alpha - z\| = 0,$$

which completes the proof. □

The result given by Theorem 6.17 can be also established for the Halpern iteration procedure.

Theorem 6.19. *Let E, C, T, P and u be as in Theorem 6.17. Define a sequence $\{y_n\}$ in C by $y_1 \in C$ and $y_{n+1} = (1-\alpha_n)Ty_n + \alpha_n u$ for $n \in \mathbb{N}$, where $\{\alpha_n\}$ is a real sequence in $(0,1)$ satisfying*

(C_1) $\lim\limits_{n\to\infty} \alpha_n = 0$; (C_2) $\sum\limits_{n=1}^{\infty} \alpha_n = \infty$ *and* (C_3) *Either* $\sum\limits_{n=1}^{\infty} |\alpha_{n+1} - \alpha_n| = \infty$
or $\lim\limits_{n\to\infty} \dfrac{\alpha_{n+1}}{\alpha_n} = 0$. *Then $\{y_n\}$ converges strongly to Pu.*

The previous theorem, established in a Hilbert space setting, can be similarly extended to uniformly smooth Banach spaces.

Theorem 6.20. *E, C, T, P, f and z be as in Theorem 6.18. Define a sequence $\{x_n\}$ in C by $x_1 \in C$ and $x_{n+1} = (1-\alpha_n)Tx_n + \alpha_n f(x_n)$ for $n \in \mathbb{N}$, where $\{\alpha_n\}$ is a real sequence in $(0,1)$ satisfying (C_1), (C_2) and (C_3) in Theorem 6.19.*

Then $\{x_n\}$ converges strongly to z.

Proof. Define a sequence $\{y_n\}$ in C by $y_n = (1-\alpha_n)Ty_n + \alpha_n f(z)$, for $n \in \mathbb{N}$ and $z \in C$ satisfying $P \circ f(z) = z$. Then by Theorem 6.19, $\{y_n\}$ converges strongly to $P \circ f(z) = z$. For every $n \in \mathbb{N}$, we have

$$\|x_{n+1} - y_{n+1}\| \le (1-\alpha_n)\|Tx_n - Ty_n\| + \alpha_n \|f(x_n) - f(z)\|$$

$$\le (1-\alpha_n)\|x_n - y_n\| + \alpha_n r \|x_n - z\|$$

$$\le (1-\alpha_n + \alpha_n r)\|x_n - y_n\| + \alpha_n r \|y_n - z\|$$

$$\le (1-\alpha_n + \alpha_n r)\|x_n - y_n\| + (\alpha_n - \alpha_n r)\frac{r\|y_n - z\|}{1-r}.$$

No, by Lemma 1.2, (ii), we obtain $\lim\limits_{n\to\infty} \|x_n - y_n\| = 0$,
which implies

$$\lim_{n\to\infty} \|x_n - z\| = 0,$$

as required. $\square$

6.6 Bibliographical Comments

§6.1.

The fixed point iterations with errors were introduced by Liu, Lishan [LL95a], [LL95b]. As shown by Osilike [Os98c], it appears that fixed point iterations with errors are deeply related to the problem of stability of fixed point iterations, see Chapter 7.

Definition 6.1 belongs to Liu, Lishan [LL95a], together with Theorem 6.1, which is Theorem 2 there. Theorem 6.1 improves and generalizes several results in the literature, and answers positively an open problem posed by Chidume [Ch90a].

A very similar result is given in Liu, Lishan [LL95b] for Lipschitzian local strictly pseudo-contractive operators on uniformly smooth Banach spaces.

Example 6.1 is taken from Chidume and Moore [ChM97]. The same authors (Chidume and Moore [ChM99]) argued that the notion of iterative process with errors given in Definition 6.1 is not fully satisfactory.

Definition 6.2 is due to Xu, Y.G. [XuY98]. Theorem 6.2 and Theorem 6.3 are taken from Kim, G.E., Kiuchi, H. and Takahashi [KKT04]. There are however serious objections to the definition of Xu, too, see Rhoades [Rho04].

From Theorem 6.2 we obtain the convergence of the Ishikawa iteration with errors for Lipschitzian hemicontractive operators in Hilbert spaces, see Corollary 1 in Chidume and Moore [ChM99].

These results could be extended to continuous hemicontractive operators with bounded range, defined on uniformly Banach space, see Huang, Z. [HZ98a] respectively, Huang, Z. [HZ00a], for the case of multivalued φ-hemicontractive mappings, or to completely continuous asymptotically nonexpansive operators in uniformly convex Banach spaces, see Huang, Z. [HZ99a].

The iterative processes of the form (9)-(11) were defined by Das and Debata [DaD85], and Xu, Y.G. [XuY98].

The last result in this section was considered in order to include also the problem of approximating the common fixed points of two operators. Theorem 6.4 is Theorem 3.3 in Kim, G.E.; Kiuchi, H. and Takahashi [KKT04].

§6.2.

Definitions 6.3 and 6.4 are taken from Jiang, Y.-L., Chun and Kim, Ki Hong [JCK00], where they are Definitions 1.1 and 1.2, respectively, but they contain concepts introduced by Goebel and Kirk, Liu, Q. [LiQ96] and Schu [Sc91b]. Lemma 6.1 is taken from Schu [Sc91d].

Theorem 6.5 restores Theorem 2.1 in Jiang, Y.-L., Chun and Kim, Ki Hong [JCK00]. Theorem 2.2 in the same paper gives a similar result by replacing the assumption "asymptotically demicontractive" in Theorem 6.5 by "$k-$strict asymptotically pseudocontractive", and extends related results due to Schu [Sc91d] and Liu, Q. [LiQ96].

For other results on the modified Ishikawa iteration with errors in the class of completely continuous asymptotically nonexpansive operators, see also the work of Huang, Z. [HZ99a].

§6.3.

Kirk's iteration was introduced in 1971. Theorem 6.6 is taken from Harder and Hicks [HH88a]. The iteration (27) was introduced by Wittmann [Wit92] in connection with Halpern's iteration scheme, see Halpern [Hal67]. Theorem 6.7 is an early result of Baillon [Bai75], which opened the general ergodic theory of nonlinear operators and shows that the Cesaro mean converges weakly to a fixed point of a nonexpansive self-operator T of a closed bounded convex subset of a Hilbert space.

Baillon [Ba76a] further proved that if T is an odd map, than the convergence in Theorem 6.7 is strong, and extended this theorem to L^p spaces.

Theorem 6.8 extends Theorem in Wittmann [Wit92] from Hilbert spaces to the case of uniformly convex and uniformly smooth Banach spaces with a weakly sequentially continuous duality mapping. This result is taken from Shimizu [Shi97].

Theorem 6.9 is due to Rode [Rod82]. The result mentioned in the Remark following Theorem 6.9 is due to Yoshimoto [Yos02].

Definition 6.5 can be found in Edelstein [Ede66]. Theorems 6.10, 6.11 are due to Bruck [Bk78a].

For other related results, see the excellent pioneering survey of Rhoades [Rho91].

Theorem 6.11 is adapted after Harder and Hicks [HH88b]. The iteration scheme appearing in Theorem 6.11 is attributed to Figueiredo in Istratescu [Ist81].

§6.4.

The results in Section 6.4 are taken from Chidume and Zegeye [ChZ04] and, respectively, Chidume [Chi02]. Example 6.2 is taken from the paper of Ishikawa [Ish76].

Lemma 6.2 is Theorem 3.1, while Theorem 6.16 is Theorem 3.3 in Chidume and Zegeye [ChZ04]. Theorem 6.14 is Theorem 5.14 in Chidume [Chi02], while Theorem 6.15 and Theorem 6.17 are, respectively, Corollary 5.15 and Theorem 5.20, in the same paper.

The proof of Theorem 6.16 given in Chidume and Zegeye [ChZ04] uses a result in Morales and Jung [MoJ00]. The existence of a path for Lipschitz pseudocontractive maps was first established by Morales [Mrl90]. Note also that Bruck [Bru74] studied the perturbed iteration (30) for approximating solutions of the equation $Au = 0$ in a Hilbert space, where A is an m-accretive operator.

Bruck considered the sequence $\{x_n\}$ defined by the initial guess x_1 and

$$x_{n+1} = x_n - \lambda_n(Ax_n + \theta_n(x_n - x_1)),$$

which is just the perturbed Mann iteration (30), if we take $A = I - T$.

Bruck required that $\{\lambda_n\}$ and $\{\theta_n\}$ are *acceptably paired sequences*, i.e., they satisfy appropriate conditions with respect to a strictly increasing sequence $\{n(i)\}_{i=1}^{\infty}$ of positive integers. A prototype of acceptably paired sequences is given by $\lambda_n = n^{-1}, \theta_n = (\log\log n)^{-1}, n(i) = i^i$. Reich [Re78e] also studied the recursion formula (30) for Lipschitz accretive operators on real uniformly convex Banach spaces with a duality mapping that is weakly sequentially continuous at zero and with $\{\lambda_n\}$ and $\{\theta_n\}$ satisfying conditions slightly stronger than (i)-(iii) in Section 6.4.

From a computational point of view, it is clear that the perturbed Mann iteration (30) is superior to the Ishikawa iteration method. So, Theorems 6.14-6.17 appear to be the most general convergence theorems for approximating fixed points of Lipschitzian pseudocontractive operators.

§6.5.

The first result regarding the strong convergence of the path $\{x_s\}$ defined as the unique fixed point of U_s given by (31), as $s \to 1$, to a fixed point for a nonexpansive self mapping T of a nonempty closed convex and bounded subset C of a Hilbert space, given by Theorem 6.16, was obtained by Browder [Br67b]. The corresponding result for the discrete version $\{x_n\}$,

$$x_{n+1} = (1 - \alpha_n)u + \alpha_n Tx_n, \; n \geq 0,$$

where $u \in C$ is fixed and $\{\alpha_n\}$ is a sequence of real numbers in $[0, 1]$, was obtained independently by Halpern [Hal67] in Hilbert spaces as well. Halpern also pointed out that the conditions $\lim_{n\to\infty} \alpha_n = 0$ and $\sum_{n=0}^{\infty} \alpha_n = \infty$ are necessary for the convergence of $\{x_n\}$ to a fixed point of T. It is not known if generally they are also sufficient.

Ten years later, Lions [Lns77] improved the result of Halpern, still in Hilbert spaces, by considering the following assumptions on the parameters sequence $\{\alpha_n\}$: (i) $\lim_{n\to\infty} \alpha_n = 0$; (ii) $\sum_{n=0}^{\infty} \alpha_n = \infty$; (iii) $\lim_{n\to\infty} \frac{\alpha_n - \alpha_{n-1}}{\alpha_n^2} = 0$. As, both Halpern's and Lions' conditions on the sequence $\{\alpha_n\}$ excluded the common value $\alpha_n = (1 + n)^{-1}$, Wittmann [Wit92] obtained the convergence of $\{x_n\}$, again in Hilbert spaces, under the conditions (i) and (ii) above and (iii') $\sum_{n=0}^{\infty} |\alpha_{n+1} - \alpha_n| < \infty$, see also Theorem 6.8.

The continuous version of the Halpern's algorithm was also extensively studied. Reich [Rei80] extended Browder's result to uniformly smooth Banach spaces, while in 1981 Kirk obtained the same result in arbitrary Banach spaces under the additional assumption that T has precompact range, see Chidume [Chi03]. Morales and Jung [MoJ00] established a more general result in a Banach space which has Gateaux differentiable norm.

The rest of this section is mainly adapted from Xu, H.K. [XuH04] and Suzuki, T. [Sz07b]: Theorem 6.18, Theorem 6.19 and Theorem 6.20 are, respectively, Theorem 4.1, Theorem 3.2 and Theorem 3.2 in Xu, H.K. [XuH04], while Theorem 6.17 is taken from Reich, S. [Rei80]. The (short) proofs of Theorem 6.18 and 6.20 are due to Suzuki, T. [Sz07b]. Lemma 6.4 is due to Goebel, K. and Reich, S. [GbR84], p. 48, while Lemma 6.5 is Proposition 1 in Suzuki, T. [Sz07b].

The convergence theorems of the type considered in Section 6.5 seem to have been first called of 'viscosity' type in Moudafi [Mou00]. Many other authors contributed to this topic, considering non-self mappings or more than one mapping: Marino and Trombetta [MaT92], Singh, S.P. and Watson, B. [SWa88], who extended the result of Browder and Halpern to nonexpansive non-self mappings satisfying Rothe's boundary condition, Bauschke [Bau96] and many others considered finitely many maps, while Schu [Sch89] combined Halpern's and Mann iteration to approximate Lipschitzian pseudocontractive mappings in Hilbert spaces, to quote only a few important moments in the development of this topic.

For other results on the topic see also Xu, H.K. and Yin [XYi95], Osilike [Os04b], Chidume, C.E., Li, J.L. and Udomene, A. [ChL04], O'Hara, J.G., Pillay, P. and Xu, H.K. [OPX03], Jung, J.S. and Kim, S.S. [JK98a], Li, G. and Kim, J.K. [LiK01], Chidume, C.E. [Chi03], [Chi04], Ahmed, M.A. and Zeyada, F.M. [AhZ02], Nakajo, K. and Takahashi, W. [NaT03], Suzuki, T. [Suz03], Takahashi, W. [Tak01], Takahashi, W. and Kim, G.E. [TK98b], Zegeye, H. and Prempeh, E. [ZPr02].

At the end of these comments, we want to mention the generalized projection method used in approximating fixed points in Banach spaces by Alber [Alb96], Alber and Guerre-Delabriere [AlG94], [AlG97], [AlG01], Alber, Guerre-Delabriere and Zelenko [AGZ98], Alber and Notik [AlN95], based on a generalization of the metric projection in Hilbert spaces - a technique that was intensively used by Browder and Petryshyn [BrP67].

Exercises and Miscellaneous Results

6.1. Prove: (a) Theorem 6.1; (b) Theorem 6.3; (c) Theorem 6.4; (d) Theorem 6.7; (e) Theorem 6.8; (f) Theorem 6.9; (g) Theorem 6.10; (h) Theorem 6.11; (i) Lemma 6.1; (j) Lemmas 6.2-6.6; (k) Theorem 6.18.

6.2. Reinermann (1969)
Let H be a Hilbert space, $K \subset H$ be nonempty closed bounded and convex. Let T be an asymptotically nonexpansive selfmap of K. If $\sum(k_n^2 - 1) < \infty$ and $\epsilon \leq \alpha_n \leq 1 - \epsilon$, for all $n \in \mathbb{N}$ and some $\epsilon > 0$, then the modified Mann iteration $\{x_n\}$ defined by $x_0 \in K$ and

$$x_{n+1} = (1 - \alpha_n)x_n + \alpha_n T^n x_n, n \geq 0,$$

is an approximate fixed point sequence of T, that is, $\lim_{n \to \infty} \|x_n - Tx_n\| = 0$.

6.3. Schu (1991)
Let H be a Hilbert space, $K \subset H$ be nonempty closed bounded and convex. Let $T : K \to K$ be an uniformly L-Lipschitzian and asymptotically pseudo-contractive with $\{k_n\} \subset [1, \infty)$. Assume $\sum(q_n^2 - 1) < \infty$, where $q_n = 2k_n - 1$, for all $n \geq 1$, $\alpha_n, \beta_n \in [0, 1]$, $\epsilon \leq \alpha_n \leq \beta_n \leq b$, for all integers $n \geq 1$ and some $\epsilon > 0$, with $b \in (0, L^{-1}[(1 + L^2)^{1/2} - 1])$. Then the modified Ishikawa iteration $\{x_n\}$ defined by $x_0 \in K$ and

$$x_{n+1} = (1 - \alpha_n)x_n + \alpha_n T^n y_n, \ y_n = (1 - \beta_n)x_n + \beta_n T^n x_n, \ n \geq 0,$$

is an approximate fixed point sequence of T.

(Recall that $T : K \to K$ is called asymptotically pseudocontractive with $\{k_n\} \subset [1, \infty)$ if, for all $x, y \in K$,

$$\langle T^n x - T^n y, x - y \rangle \leq k_n \|x - y\|^2)$$

6.4. Chidume and Zegeye (2003)
Let K be a nonempty closed bounded and convex subset of a real Banach space E. Let $T : K \to K$ be an uniformly L-Lipschitzian, uniformly asymptotically regular with sequence $\{\epsilon_n\}$ and asymptotically pseudocontractive with sequence $\{k_n\}$ such that for $\lambda_n, \theta_n \in (0, 1)$, $\forall n \geq 0$, the following conditions are satisfied:
(i) $\sum \lambda_n \theta_n = \infty$, $\lim_{n \to \infty} \frac{\lambda_n}{\theta_n} = 0$; $\lambda_n(1 + \theta_n) \leq 1$;
(ii) $\lim_{n \to \infty} \theta_n = 0$; $\lim_{n \to \infty} \lambda_n/\theta_n = 0$; $\lim_{n \to \infty} (\frac{\theta_{n-1}}{\theta_n} - 1)/(\lambda_n \theta_n) = 0$, $\lim_{n \to \infty} \frac{\epsilon_{n-1}}{\lambda_n \theta_n^2} = 0$;
(iii) $k_{n-1} - k_n = o(\lambda_n \theta_n^2)$;
(iv) $k_n - 1 = o(\theta_n)$.
Let a sequence $\{x_n\}$ be iteratively generated from $x_1 \in K$ by

$$x_{n+1} := (1 - \lambda_n)x_n + \lambda_n Tx_n - \lambda_n \theta_n(x_n - x_1), \ n = 0, 1, 2, \ldots$$

Then $\{x_n\}$ is an approximate fixed point sequence of T.

6.5. Bruck (1974)

Let $T : K \to K$ be demicontinuous and pseudocontractive. Then T has a fixed point in K and whenever $\{\lambda_n\}$ and $\{\theta_n\}$ are acceptably paired,

$$\lambda_n(1 + \theta_n) \leq 1, \ \text{ for all } n \geq 0,$$

$z \in K$ and $x_0 \in K$, the sequence $\{x_n\}$ defined by

$$x_{n+1} = (1 - \lambda_n)x_n + \lambda_n T x_n + \lambda_n \theta_n (z - x_n), \ \ n \geq 0$$

remains in K and converges strongly to the fixed point of T closest to z. (For the concept of sequences acceptably paired, see Definition 8.4.)

6.6. Schu (1989)

Let $T : K \to K$ be Lipschitzian (with constant $L \geq 0$) and pseudocontractive; let $\{\lambda_n\}$ and $\{\alpha_n\}$ be sequences in $(0, 1)$ with

$$\lim_{n \to \infty} \lambda_n = 1, \quad \lim_{n \to \infty} \alpha_n = 0$$

such that $(\{\alpha_n\}, \{\mu_n\})$ has property (A), $(1 - \mu_n)(1 - \lambda_n)^{-1}$ is bounded, and

$$\lim_{n \to \infty} (1 - \mu_n) / \alpha_n = 0,$$

where

$$k_n = (1 + \alpha_n^2(1 + L)^2)^{1/2} \text{ and } \mu_n = \lambda_n / k_n, \ \ n \geq 0.$$

Fix $z_0 \in K$ and define

$$z_{n+1} = \mu_{n+1}[(1 - \alpha_n)z_n + \alpha_n T z_n] + (1 - \mu_{n+1})\omega, \ \ n \geq 0.$$

Then $\{z_n\}$ converges strongly to the unique fixed point of T closest to ω.

(The previous fixed point iteration procedure is constructed in a similar manner to that of Ishikawa iteration, i.e., by composing two iterations: a Mann iteration and a Halpern type fixed point iteration procedure - which is in fact a Mann type iteration with a fixed term ω, see Section 6.5 in this chapter)

7

Stability of Fixed Point Iteration Procedures

Intuitively, a fixed point iteration procedure is numerically stable if, "small" modifications in the initial data or in the data that are involved in the computation process, will produce a "small" influence on the computed value of the fixed point.

It is the aim of this chapter to survey the most significant contributions to this area. To this end, we shall define a fixed point iteration procedure by a general relation of the form

$$x_{n+1} = f(T, x_n), \quad n = 0, 1, 2, ..., \tag{1}$$

where $T : X \to X$ is an operator and $x_0 \in X$, by tacitly considering that $f(T, x_n)$ in the right-hand side of (1) does contain all parameters that define the given fixed point iteration procedure.

For example, in the case of Mann iteration procedure $M(x_0, \alpha_n, T)$, $f(T, x_n)$ appearing in (1), given by the formula $f(T, x_n) = (1-\alpha_n)x_n + \alpha_n T x_n$ implicitly includes $\{\alpha_n\}$.

7.1 Stability and Almost Stability of Fixed Point Iteration Procedures

Let (X, d) be a metric space, $T : X \to X$ an operator with $F_T \neq \emptyset$ and $\{x_n\}_{n=0}^{\infty}$ a sequence obtained by a certain fixed point iteration procedure that ensure its convergence to a fixed point p of T.

In concrete applications, when calculating $\{x_n\}_{n=0}^{\infty}$, we usually follow the steps:

1. We choose the initial approximation $x_0 \in X$;

2. We compute $x_1 = f(T, x_0)$ but, due to various errors that occur during the computations (rounding errors, numerical approximations of functions,

derivatives or integrals etc.), we do not get the exact value of x_1, but a different one, say y_1, which is however close enough to x_1, i.e., $y_1 \approx x_1$.

3. Consequently, when computing $x_2 = f(T, x_1)$ we will actually compute x_2 as

$$x_2 = f(T, y_1),$$

and so, instead of the theoretical value x_2, we will obtain in fact another value, say y_2, again close enough to x_2, i.e., $y_2 \approx x_2, ...$, and so on.

In this way, instead of the theoretical sequence $\{x_n\}_{n=0}^{\infty}$, defined by the given iterative method, we will practically obtain an *approximate sequence* $\{y_n\}_{n=0}^{\infty}$. We shall consider the given fixed point iteration method to be numerically **stable** if and only if, for y_n close enough (in some sense) to x_n at each stage, the approximate sequence $\{y_n\}_{n=0}^{\infty}$ still converges to the fixed point of T.

Following basically this idea, the next concept of stability was introduced.

Definition 7.1. Let (X, d) be a metric space and $T : X \to X$ a mapping, $x_0 \in X$ and let us assume that the iteration procedure (1), that is, the sequence $\{x_n\}_{n=1}^{\infty}$ produced by (1), converges to a fixed point p of T.

Let $\{y_n\}_{n=0}^{\infty}$ be an arbitrary sequence in X and set

$$\varepsilon_n = d(y_{n+1}, f(T, y_n)), \quad \text{for} \quad n = 0, 1, 2, ... \tag{2}$$

We shall say that the fixed point iteration procedure (1) is T-*stable* or *stable with respect to* T if and only if

$$\lim_{n \to \infty} \varepsilon_n = 0 \Leftrightarrow \lim_{n \to \infty} y_n = p. \tag{3}$$

Remarks.

1) It is known that the Picard iteration is T-stable with respect to any α-contraction T and also with respect to any Zamfirescu mapping T, both these results being established in the framework of a metric space setting;

2) It has also been shown that in a normed linear space setting certain Mann iterations are T-stable with respect to any Zamfirescu mapping.

In the same setting, a similar result was proved for Kirk's iteration procedure, in the class of $c-$contractions $(0 \leq c < 1)$;

3) One of the most general contractive definition for which corresponding stability results have been obtained in the case of Kirk, Mann and Ishikawa iteration procedures in arbitrary Banach spaces appears to be the following class of mappings: for (X, d) a metric space, $T : X \to X$ is supposed to satisfy the condition

$$d(Tx, Ty) \leq a\, d(x, y) + L\, d(x, Tx) \tag{4}$$

for some $a \in [0, 1)$, $L \geq 0$ and for all $x, y \in D \subset X$.

Notice that any a-contractive and any Zamfirescu operator satisfy (4). Actually, condition (15) in Section 2.3 is exactly condition (4) above, with $a := \delta$ and $L = 2\delta$, where

$$\delta = \max\left\{\alpha, \frac{\beta}{1-\beta}, \frac{\gamma}{1-\gamma}\right\},$$

with α, β, γ the constants that are involved in Zamfirescu's contractive conditions $(z_1), (z_2)$ and (z_3), respectively.

However, if a mapping T satisfies only (4), it need not have a fixed point in general. But (as we have seen in Chapter 2, in the case of Zamfirescu mappings, Kannan mappings or weak contractions) if T has a fixed point and satisfies (4), then the fixed point is unique.

Consequently, we shall present in the following some general stability results for mappings satisfying (4).

Theorem 7.1. *Let (X, d) be a metric space and $T : X \to X$ a mapping satisfying (4). Suppose T has a fixed point x^*. Let $x_0 \in X$ and $x_{n+1} = Tx_n$, $n \geq 0$.*

Then $\{x_n\}$ converges strongly to x^* and is stable with respect to T (i.e., for $\{\varepsilon_n\}$ given by (2), the equivalence (3) holds).

Proof. Using triangle rule and (4) we get

$$d(y_{n+1}, x^*) \leq d(y_{n+1}, Ty_n) + d(Ty_n, x^*) \leq a\, d(y_n, x^*) + \varepsilon_n. \qquad (5)$$

Suppose $\lim_{n\to\infty} \varepsilon_n = 0$. Then, since $a \in [0, 1)$, it follows by Lemma 1.6 that $\lim_{n\to\infty} y_n = x^*$. Moreover, since by (4),

$$d(x_{n+1}, p) \leq ad(x_n, p),$$

it follows that $\lim_{n\to\infty} x_n = x^*$. Conversely, if $\lim_{n\to\infty} y_n = x^*$, then

$$\varepsilon_n = d(y_{n+1}, Ty_n) \leq d(y_{n+1}, x^*) + a\, d(y_n, x^*) \to 0,$$

as $n \to \infty$. $\qquad\qquad\square$

Theorem 7.2. *Let E be a normed linear space and $T : E \to E$ a mapping satisfying (4) (with $d(u, v) = \|a - v\|$). Suppose T has a fixed point x^*. Let x_0 be arbitrary in E and define*

$$z_n = (1 - \beta_n)x_n + \beta_n Tx_n, \quad n \geq 0$$

and

$$x_{n+1} = (1 - \alpha_n)x_n + \alpha_n Tz_n, \quad n \geq 0,$$

where $\{\alpha_n\}$ and $\{\beta_n\}$ are sequences in $[0, 1]$ such that $0 < \alpha \leq \alpha_n$, for some α. Let $\{y_n\}$ be any given sequence in E and define

$$s_n = (1 - \beta_n)y_n + \beta_n Ty_n, \quad n \geq 0$$

$$\varepsilon_n = \|y_{n+1} - (1 - \alpha_n)y_n - \alpha_n Ts_n\|, \quad n \geq 0.$$

Then $\{x_n\}$ converges strongly to x^* and is stable with respect to T.

Proof. We have the following estimate

$$\|y_{n+1} - x^*\| \le \|y_{n+1} - (1 - \alpha_n)y_n - \alpha_n T s_n\| +$$

$$+ \|(1 - \alpha_n)(y_n - x^*) + \alpha_n(T s_n - x^*)\| \le$$

$$\le (1 - \alpha_n) \|y_n - x^*\| + \alpha_n \|T s_n - x^*\| + \varepsilon_n \le$$

$$\le (1 - \alpha_n) \|y_n - x^*\| + \alpha_n a \left[(1 - \beta_n) \|y_n - x^*\| + \beta_n a \|y_n - x^*\|\right] + \varepsilon_n =$$

$$= [(1 - \alpha_n) + \alpha_n a(1 - \beta_n(1 - a))] \|y_n - x^*\| + \varepsilon_n \le$$

$$\le [1 - \alpha_n(1 - a)] \|y_n - x^*\| + \varepsilon_n \le [1 - \alpha(1 - a)] \|y_n - x^*\| + \varepsilon_n.$$

Now, suppose $\lim_{n \to \infty} \varepsilon_n = 0$. Since $a < 1$ and $\alpha > 0$, it results by Lemma 1.6, part (i), that $\lim_{n \to \infty} y_n = x^*$. Since $\|x_{n+1} - (1 - \alpha_n)x_n - \alpha_n T z_n\| = 0$, it also results

$$\lim_{n \to \infty} x_n = x^*.$$

For the converse, assume $\lim_{n \to \infty} y_n = x^*$ holds. Then it follows easily that

$$\varepsilon_n = \|y_{n+1} - (1 - \alpha_n)y_n - \alpha T s_n\| \le \|y_{n+1} - x^*\| + \|y_n - x^*\| \to 0$$

as $n \to \infty$, that completes the proof. □

Remarks

1) A result similar to Theorems 7.1 and 7.2 can be proved in a normed linear setting for Kirk's iteration procedure and for a a self-operator T satisfying (4);

2) There are several examples of fixed point iterations which are not stable with respect to certain operators;

3) It is well known that neither Picard iteration, nor Mann or Kirk's iterations are T-stable with respect to a nonexpansive self-operator of a closed convex bounded set in a Hilbert space, but the next theorem shows that Figueiredo's iteration is T−stable with respect to nonexpansive mappings.

Theorem 7.3. *Let K be a closed, bounded and convex subset of a Hilbert space H containing 0. If $T : K \to K$ is a nonexpansive mapping, then for any $x_0 \in K$ the sequence $\{x_n\}_{n=0}^{\infty}$, defined by*

$$x_n = T_n^{n^2} x_{n-1}, \qquad n = 1, 2, \ldots$$

and $T_n x = n/(n+1)Tx$, is T−stable.

Definition 7.2. *Suppose E is a real Banach space and T is a selfmap of E, with $F_T \ne \phi$. Let $x_0 \in E$ and let $\{x_n\}_{n=0}^{\infty}$ be an iteration procedure given by*

$$x_{n+1} = f(T, x_n), \quad n = 0, 1, 2, \ldots \tag{6}$$

that converges strongly to a fixed point $x^* \in F_T$.

Suppose $\{y_n\}_{n=0}^{\infty}$ is a sequence in E and $\{\varepsilon_n\}_{n=0}^{\infty}$ is a sequence of positive real numbers given by

$$\varepsilon_n = \|y_{n+1} - f(T, y_n)\|. \tag{7}$$

If $\sum_{n=0}^{\infty} \varepsilon_n < \infty$ implies $\lim_{n \to \infty} y_n = x^*$, then the iteration procedure defined by (6) is said to be *almost T-stable* or *almost stable with respect to T.*

Remark. Clearly, any T-stable iteration procedure is almost T-stable, but an almost T-stable procedure may fail to be T-stable.

The next theorem shows that, under certain assumptions, the Ishikawa iteration procedure is almost T-stable with respect to a Lipschitz φ-strongly pseudocontractive operator.

Theorem 7.4. *Suppose E is a real Banach space and $T : E \to E$ is a Lipschitzian (with constant L) φ-strongly pseudocontractive operator. Suppose $F_T \neq \emptyset$ and $\{\alpha_n\}_{n=0}^{\infty}$ and $\{\beta_n\}_{n=0}^{\infty}$ are real sequences in $[0, 1]$ satisfying the conditions*

$$(i) \sum_{n=0}^{\infty} \alpha_n = \infty; \quad (ii) \sum_{n=0n}^{\infty} \alpha_n \beta_n < \infty; \quad (iii) \sum_{n=0n}^{\infty} \alpha_n^2 < \infty.$$

Let $\{x_n\}_{n=0}^{\infty}$ be the Ishikawa iteration, given by $x_0 \in E$ and

$$z_n = (1 - \beta_n)y_n + \beta_n T y_n, \quad n \geq 0$$

$$x_{n+1} = (1 - \alpha_n)x_n + \alpha_n T z_n, \quad n \geq 0.$$

Suppose $\{y_n\}_{n=0}^{\infty}$ is a sequence in E and define $\{\varepsilon_n\}_{n=0}^{\infty}$ by

$$\varepsilon_n = \|y_{n+1} - (1 - \alpha_n)y_n - \alpha_n T s_n\|, \ n \geq 0.$$

$$s_n = (1 - \beta_n)y_n + \beta_n T y_n, \quad n \geq 0.$$

Then
1. *The sequence $\{x_n\}$ converges strongly to the fixed point p of T;*
2. *We have the error estimate*

$$\|y_{n+1} - p\| \leq [1 - \alpha_n r(p_n, p)] \|y_n - p\| +$$

$$+ \left[L^3 + 4L^2 + 3(L + 1)\right] \alpha_n^2 \|y_n - p\| + L(1 + L)\alpha_n \beta_n \|y_n - p\| + \varepsilon_n,$$

where $p_n = (1 - \alpha_n)y_n + \alpha_n T s_n$ and

$$r(p_n, p) = \frac{\varphi(\|p_n - p\|)}{1 + \varphi(\|p_n - p\|) + \|p_n - p\|};$$

3. $\sum_{n=0n}^{\infty} \varepsilon_n < \infty \Rightarrow \lim_{n \to \infty} y_n = p; \ 4. \ \lim_{n \to \infty} y_n = p \Rightarrow \lim_{n \to \infty} \varepsilon_n = 0.$

Proof. Since T is φ-strongly pseudocontractive, it results that for all $x, y \in E$ there exist $j(x - y) \in J(x - y)$ and a strictly increasing function $\varphi : [0, \infty) \to [0, \infty)$ with $\varphi(0) = 0$ such that

$$\langle Tx - Ty, \, j(x - y) \rangle \leq \|x - y\|^2 - \varphi(\|x - y\|) \, \|x - y\|.$$

This shows that if T has a fixed point, then the fixed point is unique. The rest of the proof is standard and we omit it. $\square$

Remarks.
1) If we set $\beta_n = 0$ for all $n \geq 0$ in Theorem 7.4, then we obtain a result which shows that the Mann iteration is almost T-stable;
2) The class of φ-strongly pseudocontractive operators with nonempty fixed point set is a proper subset of the class of φ-hemicontractive operators.

However, Theorem 7.4 can be easily extended to the class of φ-hemicontractive operators.

7.2 Weak Stability of Fixed Point Iteration Procedures

In this section we want to show that the concept of (almost) stability introduced in the previous section is slightly not very precise. As we stressed at the beginning of this Chapter, it is not natural that the sequence $\{y_n\}_{n=0}^{\infty}$ involved in the definition of (almost) stability be *arbitrary* taken. From a numerical point of view $\{y_n\}_{n=0}^{\infty}$ must be, in a certain sense, an **approximate sequence** of $\{x_n\}$.

By adopting a concept of such kind of approximate sequences, it is possible to introduce a weaker and more natural concept of stability, called *weak stability*. So, any stable iteration will be also weakly stable, but the reverse is not generally true.

Definition 7.3. Let (X, d) be a metric space and $\{x_n\}_{n=1}^{\infty} \subset X$ be a given sequence. We shall say that $\{y_n\}_{n=0}^{\infty} \in X$ is an *approximate sequence* of $\{x_n\}$ if, for any $k \in \mathbb{N}$, there exists $\eta = \eta(k)$ such that

$$d(x_n, y_n) \leq \eta, \quad \text{for all } n \geq k.$$

Remark. We can have approximate sequences of both convergent and divergent sequences. The following result will be useful in the sequel.

Lemma 7.1. *The sequence $\{y_n\}$ is an approximate sequence of $\{x_n\}$ if and only if there exists a decreasing sequence of positive numbers $\{\varepsilon_n\}$ converging to some $\eta \geq 0$ such that*

$$d(x_n, y_n) \leq \varepsilon_n, \quad \text{for any } n \geq k \text{ (fixed)}.$$

Proof. *Sufficiency.* We take $\eta(k) = \varepsilon_k, k = 0, 1, 2, ...$
Necessity. For $k = 1$ we find $\eta_1 > 0$ such that

$$d(x_n, y_n) \leq \eta_1, \ n = 1, 2, ...$$

Put $\varepsilon_1 = \eta_1$. For $k = 2$ we find $\eta_2 > 0$ such that

$$d(x_n, y_n) \leq \eta_2, \ n = 2, 3, ...$$

Put $\varepsilon_2 = \min\{\eta_1, \eta_2\}, ...$
We obtain in this way a decreasing sequence of positive numbers $\{\epsilon_n\}$ (which is convergent to some $\eta \geq 0$). □

Definition 7.4. Let (X, d) be a metric space and $T : X \to X$ be a map. Let $\{x_n\}$ be an iteration procedure defined by $x_0 \in X$ and

$$x_{n+1} = f(T, x_n), \ n \geq 0. \tag{8}$$

Suppose $\{x_n\}$ converges to a fixed point p of T. If for any approximate sequence $\{y_n\} \subset X$ of $\{x_n\}$

$$\lim_{n \to \infty} d(y_{n+1}, f(T, y_n)) = 0$$

implies

$$\lim_{n \to \infty} y_n = p,$$

then we shall say that (8) is *weakly T-stable* or *weakly stable with respect to T*. **Remarks.**

1) It is obvious that any stable iteration procedure is also weakly stable, but the reverse is generally not true;

2) All examples given by various authors that have studied the stability of fixed point iteration procedures - examples intended to illustrate non stable fixed point iteration procedures - do not consider approximate sequences of $\{x_n\}$. We present in detail some of the aforementioned examples, in order to show how important and natural is to restrict the stability concept to approximate sequences $\{y_n\}$ of $\{x_n\}$.

Example 7.1.
Let $\mathbb{R}$ denote the reals with the usual metric. Define $T : \mathbb{R} \to \mathbb{R}$ by $Tx = \frac{1}{2}x$. As T is an $\frac{1}{2}$-contraction, it follows by Theorem 7.2 that the Ishikawa iteration $\{x_n\}_{n=1}^{\infty}$ is T-stable, hence almost T-stable and weakly T-stable, too.

However, it has been claimed (and "proved" !) that the Ishikawa iteration is not T-stable. To show this, it was used the sequence $\{y_n\}_{n=1}^{\infty}$ given by

$$y_n = \frac{n}{1+n}, \ n \geq 0.$$

But this is obviously nonsense, because $x_n \to 0$ (the unique fixed point of T), while $y_n \to 1$ as $n \to \infty$, although, by construction, $\{y_n\}_{n=1}^{\infty}$ would have to be an approximate sequence of $\{x_n\}$.

Example 7.2.
Let $T : [0,1] \to [0,1]$ be given by

$$Tx = \frac{1}{2} \text{ if } 0 \le x \le \frac{1}{2} \text{ and } Tx = 0 \text{ if } \frac{1}{2} < x \le 1,$$

where $[0,1]$ is endowed with the usual metric. We have $F_T = \left\{\frac{1}{2}\right\}$.

It was shown by that the Picard iteration is not T-stable, by taking $\{y_n\}$ as an a priori divergent sequence.

We will show that the Picard iteration is also not weakly T-stable. This will imply, in particular, that it is indeed not T-stable.

Let $x_0 \in [0,1]$ and $x_{n+1} = Tx_n$, for $n = 0, 1, \ldots$
If $0 \le x_0 \le \frac{1}{2}$, then $x_1 = Tx_0 = \frac{1}{2}$ and if $\frac{1}{2} < x_0 \le 1$, then $x_1 = Tx_0 = 0$.

In either case, $x_n = \frac{1}{2}$ for $n \ge 2$ and thus $\lim_{n\to\infty} x_n = \frac{1}{2} = T\left(\frac{1}{2}\right)$.

Let $\{y_n\}$ be an approximate sequence of $\{x_n\}$. By Lemma 7.1 it results that there exists a decreasing sequence of positive numbers $\{\eta_n\}$ converging to some $\eta \ge 0$ such that

$$|x_n - y_n| \le \eta_n, \text{ for } n \ge k(\text{fixed}).$$

In particular, we can take $y_n = x_n + (-1)^n \cdot \eta_n$, $n \ge k$ which shows that

$$y_n = \frac{1}{2} + (-1)^n \eta_n, \text{ for each } n \ge 2.$$

Then

$$Ty_n = \begin{cases} \frac{1}{2}, & \text{if } n \text{ is odd} \\ 0, & \text{if } n \text{ is even} \end{cases}$$

and hence

$$|y_{n+1} - Ty_n| = \begin{cases} \left|y_{n+1} - \frac{1}{2}\right|, & \text{if } n \text{ is odd} \\ y_{n+1}, & \text{if } n \text{ is even} \end{cases} = \begin{cases} \left|y_{2p+2} - \frac{1}{2}\right|, & n = 2p+1 \\ y_{2p+1}, & n = 2p. \end{cases}$$

By $\lim_{n\to\infty} |y_{n+1} - Ty_n| = 0$ it results that

$$\lim_{p\to\infty} y_{2p+2} = \frac{1}{2} \text{ and } \lim_{p\to\infty} y_{2p+1} = 0$$

which shows that $\{y_n\}$ is not convergent in the whole.
Consequently, the Picard iteration is not weakly T-stable.

Example 7.3.

Let $T : [0, 1] \to [0, 1]$ be given by

$$Tx = 0, \text{ if } 0 \le x \le \frac{1}{2} \text{ and } Tx = \frac{1}{2}, \text{ if } \frac{1}{2} < x \le 1,$$

where $[0, 1]$ is again endowed with the usual metric.

Let $x_0 \in [0, 1]$ and $x_{n+1} = Tx_n$, for $n = 0, 1, 2, \ldots$

If $0 \le x_0 \le \frac{1}{2}$, then $x_1 = Tx_0 = 0$, while if $\frac{1}{2} < x_0 \le 1$, we have $x_1 = Tx_0 = \frac{1}{2}$. Therefore $x_n = 0$, for $n = 2, 3, \ldots$ and thus

$$\lim_{n \to \infty} x_n = 0 = T(0).$$

Let $\{y_n\} \subset [0, 1]$ be an approximate sequence of $\{x_n\}$. It results by Lemma 7.1 that there exists a decreasing sequence of positive numbers $\{\eta_n\}$ converging to some $\eta \ge 0$ such that

$$|x_n - y_n| \le \eta_n, \quad n \ge 0.$$

This gives $x_n - \eta_n \le y_n \le x_n + \eta_n$ and since $x_n = 0, n \ge 0,$ we get $0 \le y_n \le \eta_n, \quad n \ge 2$. We can choose $\{\eta_n\}$ such that $\eta_n \le \frac{1}{2}$, for all $n \ge 2$.

Hence $Ty_n = 0, \ n \ge 2$ and by $\lim_{n \to \infty} |y_{n+1} - Ty_n| = 0$ we get $\lim_{n \to \infty} y_n = 0 = T(0)$.

This shows that the Picard iteration is weakly T-stable. But, as known, the Picard iteration is not T-stable.

Remarks.

1) For other examples, see Harder and Hicks [HH88b]. Note that the Picard iteration is also not weakly T-stable for the operators T in Examples 1 and 2 in Harder and Hicks [HH88b], but is weakly T-stable for T in Example 5;

2) It is now very natural to suggest a comparison of the concepts of almost stability and that of weak stability. In fact, we can introduce a concept of almost weak stability.

An open problem.

It is easy to see that any weakly T-stable iteration is almost T-stable and hence the almost weak stability will be the weakest concept of stability for fixed point procedures.

It remains the task to identify, amongst the classes of operators for which a certain iteration is not T-stable or is not almost T-stable, the ones for which the iteration is weakly T-stable.

7.3 Data Dependence of Fixed Points

Let (X, d) be a metric space and $T : X \to X$ an operator such that $F_T \neq \emptyset$ and there exists a certain fixed point iteration procedure that converges to some fixed point $p \in F_T$.

Due to various reasons, when computing p we actually use a certain *approximate operator* U of T, that is an operator $U : X \to X$, such that for a suitable $\eta > 0$ we have

$$d(Tx, Ux) \leq \eta, \quad \text{for each} \quad x \in X.$$

Assume U has a fixed point q that can be computed by a certain method. Then the following question naturally arises:

Does q approximate p and, if yes, how can we estimate $d(p, q)$?

The first part of this section is intended to present some positive answers to the previous question, in the case of Picard iteration procedure.

Let $\varphi : \mathbb{R}_+ \to \mathbb{R}_+$ be a strict comparison function and denote

$$t_\eta = \sup\{t \in \mathbb{R}_+ \ : \ t - \varphi(t) \leq \eta\}, \quad \eta > 0. \tag{9}$$

Example 7.4. If $\varphi(t) = at$, $a \in (0, 1)$, then $t_\eta = \dfrac{\eta}{1 - a}$ and if $\varphi(t) = \dfrac{t}{1 + t}$, $t > 0$, then $t_\eta = \dfrac{1}{2}\left(\eta + \sqrt{\eta^2 + 4\eta}\right)$.

Remark. For t_η given by (9) we have $\lim\limits_{\eta \to 0} t_\eta = 0$.

Theorem 7.5. *Let (X, d) be a complete metric space and $T, U : X \to X$ be two mappings satisfying*
(i) T is a strict φ-contraction; (ii) $q \in F_U$;
(iii) there exists $\eta > 0$ such that

$$d(Tx, \ Ux) \leq \eta, \quad \text{for all} \ \ x \in X. \tag{10}$$

Then

$$d(p, q) \leq t_\eta,$$

where p is the unique fixed point of T, i.e., $\{p\} = F_T$.

Proof. By (i) and Theorem 2.7 we know that T is a Picard operator, i.e., $F_T = \{p\}$ and the Picard iteration $\{T^n x_0\}$ converges to p, for any $x_0 \in X$.

Using (i), (ii) and (iii) we have that

$$d(p, q) = d(Tp, \ Uq) \leq d(Tp, Tq) + d(Tq, Uq) \leq$$

$$\leq \varphi(d(p, q)) + \eta$$

and hence

$$d(p,q) - \varphi(d(p,q)) \leq \eta$$

which, by (9), gives

$$d(p,q) \leq t_\eta,$$

i.e., exactly the desired conclusion. □

Remark. Theorem 7.5 shows that if U is an approximate operator of T, then

$$d(p,q) \to 0 \quad \text{as} \quad \eta \to 0.$$

If T is a (c)-φ-contraction (i.e. φ is a (c)-comparison function), then we can give a more detailed estimate.

Theorem 7.6. *Let (X,d) be a complete metric space and $T : X \to X$ be a φ-contraction with φ a subadditive (c)-comparison function. Let $U : X \to X$ be an approximate operator of T, i.e., (10) holds, and $\{x_n\}_{n=0}^\infty$, $\{y_n\}_{n=0}^\infty$ be the Picard iterations associated to T, respectively to U, starting from $x_0 \in X$.*
If $q \in F_U$ and $F_T = \{p\}$ then

$$\text{1)} \quad d(y_n, p) \leq s(\eta) + s(d(x_n, x_{n+1})), \; n > 1; \tag{11}$$

$$\text{2)} \quad d(p,q) \leq s(\eta),$$

where $s(t)$ denotes the sum of the comparison series $\sum_{k=0}^\infty \varphi^k(t)$.

Proof. By Theorem 2.8 we know that $F_T = \{p\}$ and that $x_n \to p$ as $n \to \infty$, for any $x_0 \in X$.
As $y_1 = Ux_0$, $y_2 = Uy_1$, ..., $y_n = Uy_{n-1}$, $n > 1$ we have that

$$d(y_n, p) \leq d(y_n, x_n) + d(x_n, p) \tag{12}$$

and

$$d(y_n, x_n) = d(Uy_{n-1}, Tx_{n-1}) \leq$$

$$\leq d(Uy_{n-1}, Ty_{n-1}) + d(Ty_{n-1}, Tx_{n-1}) \leq \eta + \varphi(y_{n-1}, x_{n-1}).$$

By the subadditivity of φ and the previous inequality, a simple induction yields

$$d(y_n, x_n) \leq \eta + \varphi(\eta) + \ldots + \varphi^n(\eta), \quad n \geq 1.$$

Using now the estimate in Theorem 2.8 and taking into account that the sequence $\{S_n(\eta)\}$ of partial sums of the comparison series is nondecreasing, that is

$$S_n(\eta) \leq s(\eta), \quad \text{for each } n \in \mathbb{N}^*,$$

from (12) we get exactly

$$d(y_n, p) \leq s(\eta) + s(d(x_n, x_{n+1})),$$

where

$$s(\eta) = \sum_{k=0}^{\infty} \varphi^k(\eta), \quad \eta \geq 0.$$

To prove part 2) of the theorem, take $x_0 = q$, where $q \in F_U$. Then

$$y_n = q, \quad \text{for each } n \geq 1$$

and letting $n \to \infty$ in (11), we get

$$d(p, q) \leq s(\eta),$$

since s is continuous at zero and $d(x_n, x_{n+1}) \to 0$ as $n \to \infty$. $\square$

Remarks.
1) Similar results can be obtained for other classes of contractive type mappings;
2) We can derive an a priori estimate instead of the a posteriori estimate (11) that involves the displacement $d(x_n, x_{n+1})$.

Indeed, we know by the proof of Theorem 2.8 that

$$d(x_n, x_{n+1}) \leq \varphi^n(d(x_0, x_1))$$

and hence (11) becomes

$$d(y_n, p) \leq s(\eta) + s(\varphi^n(d(x_0, Tx_0))), \quad n \geq 1; \tag{13}$$

3) Using the fact that s is continuous at zero, the two estimates previously proved show that

$$\lim_{\eta \to 0} d(p, q) = 0,$$

i.e., for $\eta > 0$ small enough, the fixed point q of U does approximate p, the unique fixed point of T.

The continuous dependence of the fixed point on a parameter may be formulated in the following general context.

Let (X, d) be a metric space, (Y, τ) a topological space and $T : X \times Y \to X$ a family of operators depending on the parameter $\lambda \in Y$.

Assume that $T_\lambda := T(\,\cdot\,, \lambda)$, $\lambda \in Y$, has a unique fixed point x_λ^*, for any $\lambda \in Y$.

If we consider the operator $U : Y \to X$, given by

$$U(\lambda) = x_\lambda^*, \quad \forall \lambda \in Y,$$

then we are interested to find sufficient conditions on T that guarantee the continuity of U.

A typical result for this problem is given by the next theorem. However, all these results are established for the Picard iteration. To our best knowledge, the continuous dependence of the fixed points has not been studied so far for other fixed point iteration procedures.

Theorem 7.7. *Let (X,d) be a complete metric space and (Y,τ) a topological space. Let $T : X \times Y \to X$ be a continuous mapping for which there exists a strict comparison function φ such that*

$$d(T_\lambda x_1, T_\lambda x_2) \leq \varphi(d(x_1, x_2)),$$

for all $x_1, x_2 \in X$ and $\lambda \in Y$ (where $T_\lambda x := T(x, \lambda)$). Let x_λ^* be the unique fixed point of T_λ. Then the mapping $U : Y \to X$, given by

$$U(\lambda) = x_\lambda^*, \ \lambda \in Y,$$

is continuous.

Proof. Let $\lambda_1, \lambda_2 \in Y$. Then

$$d(x_{\lambda_1}^*, x_{\lambda_2}^*) = d(T(x_{\lambda_1}^*, \lambda_1), T(x_{\lambda_2}^*, \lambda_2) \leq$$

$$\leq d(T(x_{\lambda_1}^*, \lambda_1), T(x_{\lambda_2}^*, \lambda_1)) + d(T(x_{\lambda_2}^*, \lambda_1), T(x_{\lambda_2}^*, \lambda_2)) \leq$$

$$\leq \varphi(d(x_{\lambda_1}^*, x_{\lambda_2}^*)) + d(T_{\lambda_1} x_{\lambda_2}^*, T_{\lambda_2} x_{\lambda_2}^*).$$

Hence

$$d(x_{\lambda_1}^*, x_{\lambda_2}^*) - \varphi(d(x_{\lambda_1}^*, x_{\lambda_2}^*)) \leq d(T_{\lambda_1} x_{\lambda_2}^*, T_{\lambda_2} x_{\lambda_2}^*).$$

Since T is continuous and φ is a strict comparison function, for $\lambda_2 \to \lambda_1$ we get

$$d(T_{\lambda_1} x_{\lambda_2}^*, T_{\lambda_2} x_{\lambda_2}^*) \to 0,$$

which leads to

$$d(x_{\lambda_1}^*, x_{\lambda_2}^*) \to 0,$$

and this means that $d(U(\lambda_1), U(\lambda_2)) \to 0$ as $\lambda_2 \to \lambda_1$. ☐

We end this section by presenting a very general result regarding multivalued mappings in metric spaces.

Let (X, d) be a metric space. We denote

$$\mathcal{P}(X) = \{A \subset X : A \neq \emptyset\}, \ \mathcal{P}_{b\,cl}(X) = \{A \in \mathcal{P}(X) : A \text{ is closed and bounded}\}$$

and define the functional

$$D : \mathcal{P}(X) \times \mathcal{P}(X) \to \mathbb{R}_+, \ D(A, B) = \inf\{d(a,b)|a \in A, b \in B\}.$$

We also consider the following generalized functionals:

$$\rho : \mathcal{P}(X) \times \mathcal{P}(X) \to \mathbb{R}_+ \cup \{+\infty\}, \quad \rho(A, B) = \sup\{D(a, B)|a \in A\},$$

$$H_d : \mathcal{P}(X) \times \mathcal{P}(X) \to \mathbb{R}_+ \cup \{+\infty\}, \quad H_d(A, B) = \max\{\rho(A, B), \rho(B, A)\}.$$

It is well known that H_d is a metric on $\mathcal{P}_{b\,cl}(X)$, commonly called *Hausdorff-Pompeiu metric*, and that, if (X, d) is complete, then $(\mathcal{P}_{b\,cl}(X), H_d)$ is a complete metric space, too.

The next two Lemmas can easily be proved and will be needed in the following.

Lemma 7.2. *Let (X, d) be a metric space, $A, B \in \mathcal{P}(X)$ and $q \in \mathbb{R}, q > 1$ be given. Then for every $a \in A$, there exists $b \in B$ such that*

$$d(a, b) \leq qH_d(A, B).$$

Lemma 7.3. *Let (X, d) be a metric space, $A, B \in \mathcal{P}(X)$. Suppose that there exists $\eta \in \mathbb{R}, \eta > 0$, such that the following two conditions are satisfied:*
(i) for each $a \in A$, there exists $b \in B$ such that $d(a, b) \leq \eta$;
(ii) for each $b \in B$, there exists $a \in A$ such that $d(a, b) \leq \eta$;
Then $H_d(A, B) \leq \eta$.

Definition 7.5. Let $T : X \to \mathcal{P}(X)$ be a multivalued operator. An element $x^* \in X$ is a *fixed point* of T if and only if $x^* \in T(x^*)$. Denote, as in the single-valued case, by F_T or $Fix\,(T)$ the set of all fixed points of T.

Definition 7.6. Let (X, d) be a metric space and $T : X \to \mathcal{P}(X)$ be a multivalued operator. T is said to be a *(multivalued) weakly Picard operator* if and only if for each $x \in X$ and any $y \in T(x)$, there exists a sequence $\{x_n\}_{n \geq 0}$ such that:
(i) $x_0 = x, x_1 = y$;
(ii) $x_{n+1} \in T(x_n)$ for all $n = 0, 1, 2, \ldots$;
(iii) the sequence $\{x_n\}_{n \geq 0}$ is convergent and its limit is a fixed point of T.

A sequence $\{x_n\}_{n \geq 0}$ satisfying $(i) - (ii)$ in the previous definition is called *sequence of successive approximations of a multivalued operator* defined by the multivalued operator T and starting values (x, y).

Definition 7.7. Let (X, d) be a metric space and $T : X \to \mathcal{P}(X)$ be a multivalued weakly Picard operator of graph $Graph\,(T)$. Define the multivalued mapping $T^\infty : Graph\,(T) \to \mathcal{P}(F_T)$ by

$$T^\infty(x, y) := \{z \in F_T | \text{there exists a sequence of successive approximations of}$$

$$T \text{ starting from } (x, y) \text{ that converges to } z\}.$$

Definition 7.8. Let (X, d) be a metric space and $T : X \to \mathcal{P}(X)$ be a multivalued weakly Picard operator. T is said to be a *c-weakly Picard operator* if and only if there exists a single-valued selection t^∞ of T^∞ such that

$$d(x, t^\infty(x, y)) \leq cd(x, y), \quad \text{for all } (x, y) \in Graph\,(T).$$

Example 7.5. Let (X, d) be a complete metric space and $T : X \to \mathcal{P}(X)$ be a multivalued operator.

1) If T is a multivalued a-contraction, i.e., a mapping for which there exists a constant a, $0 < a < 1$, such that

$$H_d(T(x), T(y)) \le ad(x, y), \quad \text{for all } x, y \in X,$$

then T is a c-weakly multivalued Picard operator with $c = (1-a)^{-1}$;

2) If T is a multivalued operator for which there exist $\alpha, \beta, \gamma \in \mathbb{R}_+$, with $\alpha + \beta + \gamma < 1$ such that

$$H_d(T(x), T(y)) \le \alpha d(x, y) + \beta D(x, T(x)) + \gamma D(y, T(y)), \quad \text{for all } x, y \in X,$$

then T is a c-weakly multivalued Picard operator indexsubjectPicard operator! c-weakly multivaluedwith

$$c = (1-\gamma)(1-\alpha-\beta-\gamma)^{-1};$$

3) If T is a multivalued operator which satisfies the following two conditions:
(i) there exist $\alpha, \beta \in \mathbb{R}_+, \alpha + \beta < 1$ such that

$$H_d(T(x), T(y)) \le \alpha d(x, y) + \beta D(y, T(y)), \quad \text{for every } x \in X \text{ and every } y \in T(x);$$

(ii) T is a closed multivalued operator,
then T is a c-weakly multivalued Picard operator indexsubjectPicard operator! c-weakly multivaluedwith

$$c = (1-\beta)(1-\alpha-\beta)^{-1}.$$

The next theorem gives a very general result on the data dependence of fixed points for multivalued mappings.

Theorem 7.8. *Let (X, d) be a complete metric space and $T_1, T_2 : X \to \mathcal{P}(X)$ be two multivalued operators. Suppose that*
(i) T_i is a c_i-multivalued weakly Picard operator, $i \in \{1, 2\}$;
(ii) there exists $\eta > 0$ such that for all $x \in X$,

$$H_d(T_1(x), T_2(x)) \le \eta.$$

Then

$$H_d(Fix(T_1), Fix(T_2)) \le \eta \max\{c_1, c_2\}.$$

Proof. Let t_i be a selection of T_i, $i \in \{1, 2\}$. Then

$$H_d(Fix(T_1), Fix(T_2)) \le \max \left\{ \sup_{x \in Fix(T_2)} d(x, t_1(x))), \sup_{x \in Fix(T_1)} d(x, t_2(x))) \right\}.$$

Let $q > 1$. Then, by Lemma 7.2, we can choose t_i, for $i \in \{1, 2\}$, such that

$$d(x, t_1^\infty(x, t_1(x))) \le c_1 q H_d(Fix(T_2), Fix(T_1)), \quad \text{for all } x \in Fix(T_2)$$

and

$$d(x, t_2^\infty(x, t_2(x))) \le c_2 q H_d(Fix(T_1), Fix(T_2)), \quad \text{for all } x \in Fix(T_1).$$

Thus, by Lemma 7.3, we have

$$H_d(Fix(T_1), Fix(T_2)) \le q\eta \max\{c_1, c_2\},$$

and letting $q \searrow 1$, the conclusion follows. $\qquad\square$

In particular, by the previous theorem we may obtain a stability result for two multivalued contractions. A special version of it is the following

Corollary 7.1. *Let (X, d) be a complete metric space and $T_1, T_2 : X \to \mathcal{P}(X)$ be two multivalued contractions with contraction coefficient k, $k < 1$. Then*

$$H_d\left(Fix\left(T_1\right), Fix\left(T_2\right)\right) \leq (1 - k)^{-1} \sup_{x \in X} H_d\left(T_1(x), T_2(x)\right).$$

7.4 Sequences of Applications and Fixed Points

Let (X, d) be a metric space and $T : X \to X$ a given operator such that $F_T = \{p\}$.

A possible method to approximate the fixed point p of T would be the following one: construct a sequence of operators $\{T_n\}$ which approximate (uniformly) the operator T, i.e.,

$$T_n \to T \ (T_n \rightrightarrows T) \ \text{ as } n \to \infty,$$

such that for each n the set $F_{T_n} \neq \emptyset$ can be easily computed and, moreover, for any $x_n^* \in F_{T_n}$, we have

$$x_n^* \to p \ \text{ as } n \to \infty.$$

Theorem 7.9. *Let (X, d) be a complete metric space and $\{T_n\}$ a sequence of operators, $T_n : X \to X$, such that $F_{T_n} = \{x_n^*\}$, for each $n = 1, 2, \dots$.*
If the sequence $\{T_n\}$ converges uniformly to an a-contraction $T : X \to X$ with $F_T = \{x^*\}$, then

$$x_n^* \to x^* \ \text{ as } n \to \infty.$$

Proof. Let $\varepsilon > 0$ and choose a natural number N such that $n \geq N$ implies

$$d(T_n x, T x) < \varepsilon(1 - a), \ \text{ for all } x \in X,$$

where a is the contraction coefficient. Then, for $n \geq N$ we have

$$d(x_n^*, x^*) = d(T_n x_n^*, T x^*) \leq d(T_n x_n^*, T x_n^*) + d(T x_n^*, T x^*) < \varepsilon(1-a) + a d(x_n^*, x^*),$$

which yields

$$d(x_n^*, x^*) < \varepsilon, \ \text{ for all } n \geq N.$$

This proves that $\{x_n^*\}_{n=0}^{\infty}$ converges to x^* as $n \to \infty$. □

Remark. The uniform convergence of $\{T_n\}_{n=0}^{\infty}$ can be weakened to the pointwise convergence , if the operators T_n possess certain additional contractive properties, as in the next theorems.

Theorem 7.10. *Let* (X, d) *be a complete metric space and let us consider* $T_n, T : X \to X$ $(n \in \mathbb{N})$ *be operators such that*
(i) T_n *is a strict* φ-*contraction for all* $n \geq 0$;
(ii) $\{T_n\}_{n=0}^{\infty}$ *converges pointwisely to* T.
Then T *is a strict* φ-*contraction and*

$$x_n^* \to x^* \ as \ n \to \infty,$$

where $F_{T_n} = \{x_n^*\}$ *and* $F_T = \{x^*\}$.

Proof. We have

$$d(Tx, Ty) \leq d(Tx, T_n x) + d(T_n x, T_n y) + d(T_n y, Ty)$$

and by (ii) there exists a strict comparison function $\varphi : \mathbb{R}_+ \to \mathbb{R}_+$ such that

$$d(T_n x, T_n y) \leq \varphi(d(x, y)), \ \ \forall \ x, y \in X,$$

for each $n \in \mathbb{N}^*$. So

$$d(Tx, Ty) \leq d(T_n x, Tx) + \varphi(d(x, y)) + d(T_n y, Ty), \ \ \forall \ x, y \in X$$

and letting $n \to \infty$ we get by (ii) that

$$d(Tx, Ty) \leq \varphi(d(x, y)), \ \ \forall \ x, y \in X,$$

i.e., T is a strict φ-contraction with the same comparison function that appears in (i).

By Theorem 2.7 we have $F_{T_n} = \{x_n^*\}$, $n \geq 0$ and $F_T = \{x^*\}$. In order to prove that $x_n^* \to x^*$, we need the following estimate

$$d(x_n^*, x^*) \leq d(T_n x_n^*, Tx^*) \leq d(T_n x_n^*, T_n x^*) + d(T_n x^*, Tx^*) \leq$$

$$\leq \varphi(d(x_n^*, x^*)) + d(T_n x^*, Tx^*),$$

which gives

$$d(x_n^*, x^*) - \varphi(d(x_n^*, x^*)) \leq d(T_n x^*, Tx^*), \ \ n \geq 0. \tag{14}$$

Since φ is a strict comparison function and $d(T_n x^*, Tx^*) \to 0$ as $n \to \infty$, from (14) we get (see Remark following Example 7.4)

$$\lim_{n \to \infty} d(x_n, x^*) = 0,$$

i.e., $x_n^* \to x^*$ as $n \to \infty$. $\qquad\qquad\qquad\qquad\qquad\qquad\qquad\qquad\qquad\qquad$ □

Theorem 7.11. *Let (X, d) be a complete metric space and consider T_n, $T : X \to X$ $(n \in \mathbb{N})$ such that*
  (i)   *T is a strict φ-contraction;*
  (ii) *$\{T_n\}_{n=0}^{\infty}$ converges uniformly to T;*
  (iii) *$x_n^* \in F_{T_n} \neq \emptyset$, $n \geq 0$.*
Then $\{x_n\}_{n=0}^{\infty}$ converges to x^*, the unique fixed point of T.

Proof. Similarly to Theorem 7.10 we get

$$d(x_n^*, x^*) - \varphi(d(x_n^*, x^*)) \leq d(T_n x^*, T x^*), \quad n \geq 0$$

and using (ii), the conclusion follows. □

Remarks.
  1) If in Theorem 7.10 the operators T_n are strict φ_n-contractions, where $\{\varphi_n\}_{n=0}^{\infty}$ is a sequence of strict comparison functions, then the conclusion of Theorem 7.10 is generally not true.
  2) In locally compact metric spaces we have the following result.

Theorem 7.12. *Let (X, d) be a locally compact metric space and let T_n, $T : X \to X$ be such that*
  (i)   *T_n is a strict φ_n-contraction, for all $n \in \mathbb{N}$;*
  (ii) *T is a strict φ-contraction;*
  (iii) *$\{T_n\}_{n=0}^{\infty}$ converges pointwisely to T.*
If we denote $F_{T_n} = \{x_n^*\}_{n=0}^{\infty}$, $n \geq 0$ and $F_T = \{x^*\}$, then

$$\lim_{n \to \infty} x_n^* = x^*.$$

Remarks.
  1) For $\varphi_n(t) = a_n t$, $0 < a_n < 1$, $n \geq 0$ and $\varphi_n(t) = at$, $0 < a < 1$, from Theorem 7.12 we find an early result in this respect, i.e., Theorem 2 in Nadler [Nad69];
  2) Nadler [Nad69] also indicated a construction - which can be done in any infinite dimensional Banach space - of a sequence of contractions that converges pointwisely to a contraction without the sequence of their fixed points converging and so obtained the following characterization of finite dimensional Banach spaces by means of a typical property of sequences of contractions.

Theorem 7.13. *A separable or reflexive Banach space E is finite dimensional if and only if whenever a sequence of contraction mappings of E into E converges pointwisely to a contraction mapping T, the sequence of their fixed points converges to the unique fixed point of T.*

7.5 Bibliographical Comments

§7.1.

The concept of stability of a fixed point iteration procedure seems to be due to Ostrowski, as mentioned by Rhoades [Rho07], but has been systematically studied by Harder [Har87] in her Ph.D. thesis and published in the papers Harder and Hicks [HH88a], [HH88b]. The stability of the Picard iteration with respect to α-contractions and Zamfirescu mappings is given in Harder and Hicks [HH88b], Theorem 1 and Theorem 2, respectively.

Condition (4) appears in Osilike [Os95c]; Theorem 7.1 is Theorem 4, while Theorem 7.2 is Theorem 5 in the same paper. Theorem 7.3 is taken from Harder and Hicks [HH88b], while Theorem 7.4 is the main result in Osilike [Os98c], Theorem 1. For a stability result involving Kirk iteration, see Osilike and Udomene [OsU99], Theorem 6.

Other related results to those in this section may be found in Osilike [Os95b], [Os95c], [Os96b], [Os96d], [Os97a], [Os97b], [Os98c], [Os99b], [Os00c], Osilike and Udomene [OsU99], Kim, J.K., Liu, Z., Nam, Y.M. and Chun, S.A [KLN04], Liu, Z., Zhao, Y.L. and Lee, B.S. [LZL02], Agarwal, R.P., Cho, Y.J., Li, J. and Huang, N.-J. [ACL02], Zhou, H., Chang, S.S. and Cho, Y.J. [ZCC01], Liu, Z., Kang, S.M. and Cho, Y.J. [LKC04], Fang, Y.-P., Kim, J.K. and Huang, N.-J. [FKH02], Zhou, H. [ZH99a], Rhoades [Rho90], [Rh93a].

§7.2.

The content of this section is taken from Berinde [Be02b], [Be02d]. For other related results, see also Berinde [Be03d]. Examples 7.2. and 7.3 are Examples 3 and 4 in Harder and Hicks [HH88b]. The fact that the class of φ-strongly pseudocontractive operators with nonempty fixed point sets is a proper subset of the class of φ-hemicontractive operators, was shown by an example in Chidume and Osilike [ChO94].

§7.3.

The first part of this section is taken from Rus [Rus01], Chapter 7: Theorem 7.5 is Theorem 7.1.1 there. Theorem 7.6, together with the remarks following its proof, is taken from Berinde [Be97a], Chapter III, Theorem 3.1.2, while Theorem 7.7 is Theorem 7.1.2 in Rus [Rus01]. The last part of this section, devoted to data dependence of fixed points for multivalued mappings is adapted from Rus, Petrusel, A. and Sintamarian [RPS03]. Theorem 7.8 is actually Theorem 2.1 in that paper, while Corollary 7.1 is taken from Lim [Lim85]. For other related results, see Berinde [Be97a] and Petrusel, A., Rus, I.A. [PeR01].

§7.4.

Theorem 7.9 is due to Nadler [Nad69], Theorem 1. Theorem 7.10 is Theorem 7.2.1, Theorem 7.11 is Theorem 7.2.2, while Theorem 7.12 is Theorem 7.2.3, all in Rus [Rus01]. For other related results, see Rus [Ru04b].

Exercises and Miscellaneous Results

7.1. If c is a real number such that $0 < |c| < 1$ and $\{b_k\}_{k=0}^\infty$ is a sequence of real numbers such that $\lim_{k \to \infty} b_k = 0$, then $\lim_{n \to \infty} \left(\sum_{k=0}^n c^{n-k} b_k \right) = 0$.

7.2. Harder and Hicks (1988)

Let (X, d) be a complete metric space and $T : X \to X$ be a mapping for which there exist the real numbers α, β and γ satisfying $0 \le \alpha < 1$, $0 \le \beta < 0.5$ and $0 \le \gamma < 0.5$, such that, for each $x, y \in X$, at least one of the following is true:

(z_1) $d(Tx, Ty) \le \alpha\, d(x, y)$;

(z_2) $d(Tx, Ty) \le \beta[d(x, Tx) + d(y, Ty)]$;

(z_3) $d(Tx, Ty) \le \gamma[d(x, Ty) + d(y, Tx)]$.

Let p be the fixed point of T (see Theorem 2.4), $x_0 \in X$ and $\{x_n\}$ be the Picard iteration associated to T. Let also $\{y_n\}$ be a sequence in X and set $\epsilon_n = d(y_{n+1}, Ty_n)$, $n = 0, 1, 2, \ldots$. Then

$$d(p, y_{n+1}) \le d(p, x_{n+1}) + \sum_{k=0}^n 2\delta^{n+1-k} d(x_k, x_{k+1}) + \delta^{n+1} d(x_0, y_0) + \sum_{k=0}^n \delta^{n-k} \epsilon_k$$

where

$$\delta = \max \left\{ \alpha, \frac{\beta}{1 - \beta}, \frac{\gamma}{1 - \gamma} \right\}.$$

and $\lim_{n \to \infty} y_n = 0$ if and only if $\lim_{n \to \infty} \epsilon_n = 0$.

7.3. Lim (1985)

Let (X, d) be a complete metric space and $T, T_n : X \to \mathcal{P}_{b\,cl}(X)$ be multivalued k-contractions with contraction coefficient k, $k < 1$. If

$$H_d\left(T(x), T_n(x)\right) \to 0 \text{ as } n \to \infty, \text{ uniformly for all } x \in X,$$

then

$$H_d\left(Fix\,(T), Fix\,(T_n)\right) \to 0 \text{ as } n \to \infty.$$

7.4. Berinde (2004)

Let (X, d) be a metric space and $T : X \to X$ a mapping satisfying

$$d(Tx, Ty) \le a d(x, y) + L d(x, Tx), \ \forall x, y \in X.$$

Suppose T has a fixed point p. Let $x_0 \in X$ and $x_{n+1} = Tx_n$, $n \ge 0$. Then $\{x_n\}$ converges strongly to p and is *summable almost stable* with respect to T, i.e., for $\{\varepsilon_n\}$ given by $\varepsilon_n = d(y_{n+1}, f(T, y_n))$, $n = 0, 1, 2, \ldots$, the following implication holds

$$\sum_{n=0}^\infty \varepsilon_n < \infty \quad \Rightarrow \quad \sum_{n=0}^\infty d(y_n, p) < \infty.$$

7.5. Let (X, d) be a metric space, $T : X \to X$ a mapping and the following contractive conditions:

(a) There exist $a \in [0, 1)$ and $L \geq 0$ such that

$$d(Tx, Ty) \leq ad(x, y) + Ld(x, Tx), \quad \text{for all} x, y \in X; \qquad (15)$$

(b) There exist $h \in [0, 1)$ such that

$$d(Tx, Ty) \leq h \cdot \max\left\{d(x, y), d(x, Tx), d(y, Ty), d(x, Ty), d(y, Tx)\right\},$$
$$\text{for all } x, y \in X. \quad (16)$$

(c) There exist $h \in [0, 1)$ such that

$$d(Tx, Ty) \leq h \cdot \max\left\{d(x, y), d(x, Ty), d(y, Tx)\right\}, \quad \text{for all} \quad x, y \in X. \quad (17)$$

(d) There exist $h \in [0, 1)$ such that

$$d(Tx, Ty) \leq h \cdot \max\left\{d(x, y), \frac{1}{2}[d(x, Tx) + d(y, Ty)], d(x, Ty), d(y, Tx)\right\},$$
$$\text{for all } x, y \in X. \quad (18)$$

1) Show that (18) implies (16), that (17) implies (18) and hence that (17) implies (16);

2) Using $T : [0, 1] \to [0, 1]$ with the usual norm and $T(x) = 1/2$, if $0 \leq x < 1$ and $T(1) = 0$, show that conditions (15) and (18) are independent and that the class of mappings satisfying (18) is a proper subclass of (15);

3) Use an appropriate example to show that the class of Zamfirescu mappings, that is, those satisfying $(z_1) - (z_3)$ in Exercise 7.3, is independent of that of quasi-contractive mappings, that is, those satisfying (16).

7.6. Rhoades (1990)
Let (X, d) be a complete metric space and $T : X \to X$ a mapping satisfying (17). Let p be the fixed point of T. Let $x_0 \in X$ and $x_{n+1} = Tx_n$, $n \geq 0$ be the Picard iteration. Let $\{y_n\} \subset X$ and define $\{\varepsilon_n\}$ by

$$\varepsilon_n = d(y_{n+1}, Ty_n), \quad \text{for} \quad n = 0, 1, 2, \ldots$$

Show that $\{x_n\}$ converges to p and

$$\lim_{n \to \infty} \varepsilon_n = 0 \Leftrightarrow \lim_{n \to \infty} y_n = p,$$

that is, the Picard iteration is T-stable if T satisfies (17).

7.7. Rhoades (1990)
Let X be a normed linear linear space and $T : X \to X$ a mapping satisfying (17). Let p be the fixed point of T. Let $x_0 \in X$ and define the Mann iteration $\{x_n\}$ by $x_{n+1} = (1 - \alpha_n)x_n + \alpha_n Tx_n$, $n \geq 0$, where $\{\alpha_n\} \subset [0, 1]$ is a sequence

of real numbers satisfying the following conditions: (i) $\alpha_0 = 1$; (ii) $\sum \alpha_n = \infty$ and (iii) $\sum_{j=0}^{n} \prod_{i=j+1}^{n} (1 - \alpha_i + h\alpha_i)$ converges. Let $\{y_n\} \subset X$ and define $\{\varepsilon_n\}$ by

$$\varepsilon_n = \|y_{n+1} - (1 - \alpha_n)y_n - \alpha_n T y_n\|, \quad \text{for} \quad n = 0, 1, 2, \dots$$

Show that $\{x_n\}$ converges to p and

$$\lim_{n \to \infty} \varepsilon_n = 0 \Leftrightarrow \lim_{n \to \infty} y_n = p$$

that is, the Mann iteration is T-stable if T satisfies (17).

7.8. Rhoades (1993)

Let (X, d) be a complete metric space and $T : X \to X$ be a mapping satisfying (18). Let p be the fixed point of T. Let $x_0 \in X$ and $x_{n+1} = Tx_n$, $n \geq 0$ be the Picard iteration. Let $\{y_n\} \subset X$ and define $\{\varepsilon_n\}$ by

$$\varepsilon_n = d(y_{n+1}, Ty_n), \quad \text{for} \quad n = 0, 1, 2, \dots$$

Show that $\{x_n\}$ converges to p and

$$\lim_{n \to \infty} \varepsilon_n = 0 \Leftrightarrow \lim_{n \to \infty} y_n = p,$$

that is, the Picard iteration is T-stable if T satisfies (18).

7.9. Rhoades (1993)

Let X be a normed linear linear space and $T : X \to X$ a mapping satisfying (18). Let p be the fixed point of T. Let $x_0 \in X$ and define the Mann iteration $\{x_n\}$ by $x_{n+1} = (1 - \alpha_n)x_n + \alpha_n T x_n$, $n \geq 0$, where $\{\alpha_n\} \subset [0, 1]$ is a sequence of real numbers satisfying the following conditions: (i) $\alpha_0 = 1$; (ii) $\sum \alpha_n = \infty$ and (iii) $\sum_{j=0}^{n} \prod_{i=j+1}^{n} (1 - \alpha_i + h\alpha_i)$ converges. Let $\{y_n\} \subset X$ and define $\{\varepsilon_n\}$ by

$$\varepsilon_n = \|y_{n+1} - (1 - \alpha_n)y_n - \alpha_n T y_n\|, \quad \text{for} \quad n = 0, 1, 2, \dots$$

Show that $\{x_n\}$ converges to p and

$$\lim_{n \to \infty} \varepsilon_n = 0 \Leftrightarrow \lim_{n \to \infty} y_n = p,$$

that is, the Mann iteration is T-stable if T satisfies (18).

7.10. Osilike (1995)

Let X be a normed linear linear space and $T : X \to X$ a mapping satisfying (18) with $F_T \neq \emptyset$. Show that the Ishikawa iteration $\{x_n\}$ given by $x_0 \in X$ and

$$x_{n+1} = (1 - \alpha_n)x_n + \alpha_n T \left[(1 - \beta_n)x_n + \beta_n T x_n\right],$$

with $\alpha_n, \beta_n \in [0, 1]$ satisfying (i)-(iii) in Exercise 7.9, is stable with respect to T. What happens if T satisfies the more general condition (15), instead of condition (18)?

8

Iterative Solution of Nonlinear Operator Equations

Let E be a normed linear space, $F : E \to E$ an operator and let $f \in E$ be given. In order to solve the equation

$$Fx = f \tag{1}$$

we often follow the pattern: a) define an operator $T : E \to E$ in a certain manner (for example by $Tx = f + (I - T)x$, where I is the identity operator), and b) rewrite (1) equivalently as a fixed point problem

$$x = Tx. \tag{2}$$

Now, to this new problem we can apply a fixed point theorem as those presented in Chapters 2-6, in order to obtain a certain sequence $\{x_n\}$ that converges in some sense to the (unique) fixed point x^* of (2), that is to the (unique) solution x^* of (1).

At least two reasons motivate this approach.

First, the solvability of equation (1) is ensured if F possesses Lipschitzian or/and accretive properties. These properties arise naturally in practice: an early fundamental result of Browder [Br67a] states that the initial value problem

$$\frac{du}{dt} + Tu = 0; \quad u(0) = u_0 \tag{3}$$

is solvable if T is locally Lipschitzian and accretive. Secondly, there exists an intimate connection between the class of accretive / monotone type operators and the class of (pseudo) contractive operators, relationship expressed by the following statement: T is (strongly) pseudocontractive if and only if $U = I - T$ is (strongly) accretive. Therefore:

(a) to find a solution of (1) and

(b) to find a fixed point of (2) are, in most of the cases, twin problems and so the results obtained in approximating fixed points can be applied to solve nonlinear equations of the form (1), and vice versa.

It is the aim of this chapter to survey some of the most interesting results that have been obtained in direct relation to the iterative processes presented in the previous chapters of the book.

As the applications of Picard iteration are consistently covered in several monographs published so far, we will restrict our presentation in this chapter to Mann and Ishikawa iterations.

Actually, by means of some theorems presented in this chapter, one can obtain, as particular cases, the corresponding results for Krasnoselskij iteration or even for Picard iteration.

8.1 Nonlinear Equations in Arbitrary Banach Spaces

Theorem 8.1. *Suppose E is a real Banach space and $F : E \to E$ is a Lipschitzian strongly accretive operator. Let $\{\alpha_n\}_{n=0}^{\infty}$ and $\{\beta_n\}_{n=0}^{\infty}$ be real sequences satisfying*

(i) $0 \le \alpha_n, \beta_n < 1, \quad n \ge 0$;

(ii) $\lim\limits_{n \to \infty} \alpha_n = 0; \quad \lim\limits_{n \to \infty} \beta_n = 0$;

(iii) $\sum\limits_{n=0}^{\infty} \alpha_n = \infty$.

Then the sequence $\{x_n\}_{n=0}^{\infty}$ generated starting from any $x_0 \in E$ by

$$y_n = (1 - \beta_n)x_n + \beta_n(f + (I - F)x_n), \ n \ge 0$$
$$x_{n+1} = (1 - \alpha_n)x_n + \alpha_n(f + (I - F)y_n), \ n \ge 0,$$

converges strongly to the solution of equation $Fx = f$.

Proof. The existence of a solution of $Tx = f$ follows from Browder [Br67a], while the uniqueness follows from the strong accretivity condition on F:

$$\langle Fx - Fy, \ j(x - y) \rangle \ge k \|x - y\|^2, \quad (k > 0). \tag{4}$$

Let x^* denote the unique solution of (1). If we define $T : E \to E$ by

$$Tx = f + (I - F)x,$$

then x^* is a fixed point of T and T is Lipschitzian with constant $L_1 = 1 + L$, where L is the Lipschitz constant of F. Furthermore, from (4) we get

$$\langle (I - T)x - (I - T)y, \ j(x - y) \rangle \ge k \|x - y\|^2, \quad \forall x, y \in E,$$

which shows that T is strongly pseudo-contractive.

The rest of the proof consists now of standard arguments for a fixed point convergence theorem involving a Lipschitz strong pseudocontractive operator. $\qquad \square$

Remark. If we take $\beta_n = 0$ in Theorem 8.1, then we obtain a convergence result for the Mann iteration.

Corollary 8.1. *Suppose E and F are as in Theorem 8.1. Let $\{\alpha_n\}_{n=0}^{\infty}$ be a real sequence satisfying the following conditions:*

$$(i) \; 0 \leq \alpha_n \leq 1, \; n \geq 0; \quad (ii) \; \lim_{n\to\infty} \alpha_n = 0; \quad (iii) \; \sum_{n=0}^{\infty} \alpha_n = \infty.$$

Then the sequence $\{x_n\}_{n=0}^{\infty}$ given by

$$x_{n+1} = (1 - \alpha_n)x_n + \alpha_n(f + (I - F))x_n, \quad n \geq 0$$

converges strongly to the (unique) solution of the equation $Fx = f$, $f \in E$.

Remark. In certain practical circumstances, the operator F has the special form $Fx := x + Fx$. A typical convergence result for this situation is the next theorem.

Theorem 8.2. *Suppose E is a real Banach space and $F : E \to E$ is a Lipschitzian accretive operator.*

Let $\{\alpha_n\}_{n=0}^{\infty}$ and $\{\beta_n\}_{n=0}^{\infty}$ be real sequences satisfying (i)-(iii) in Theorem 8.1. Then the sequence $\{x_n\}_{n=0}^{\infty}$ generated from an arbitrary $x_0 \in E$ by

$$y_n = (1 - \beta_n)x_n + \beta_n(f - Tx_n), \; n \geq 0$$
$$x_{n+1} = (1 - \alpha_n)x_n + \alpha_n(f - Ty_n), \; n \geq 0,$$

converges strongly to the unique solution of the equation

$$x + Fx = f, \; f \in E. \tag{5}$$

Proof. The existence of the solution of equation (5) follows similarly from Browder [Br67a], while its uniqueness follows from the accretivity condition of F :

$$\langle Fx - Fy, \, j(x - y)\rangle \geq 0, \, \forall \, x, y \in E. \tag{6}$$

Let x^* denote the unique solution of (5). Define now $T : E \to E$ by

$$Tx = f - Fx.$$

Then x^* is a fixed point of T and T is Lipschitzian (with the same constant as F). Further, by (6) we have

$$\langle (I - T)x - (I - T)y, \, j(x - y)\rangle \geq \|x - y\|^2, \text{ for all } x, y \in E, \tag{7}$$

which shows that T is strongly pseudocontractive, with constant $k = 1$. Then we follow the standard arguments in proving a fixed point convergence theorem. $\qquad\square$

Remark.

The class of strongly accretive operators is a proper subclass of the class of φ-strongly accretive operators. The next Theorem 8.3 will present a very general result concerning the solution of nonlinear equations in the class of φ-strongly accretive and Lipschitzian operators.

Theorem 8.3. *Suppose E is a real Banach space and $T : E \to E$ is a Lipschitzian φ-strongly accretive operator. Suppose the equation $Tx = f$ has a solution and suppose $\{\alpha_n\}_{n=0}^{\infty}$ and $\{\beta_n\}_{n=0}^{\infty}$ are real sequences satisfying the following conditions:*

(i) $0 \le \alpha_n, \beta_n \le 1$;

(ii) $\sum\limits_{n=0}^{\infty} \alpha_n = \infty$; (iii) $\sum\limits_{n=0}^{\infty} \alpha_n^2 < \infty$; (iv) $\sum\limits_{n=0}^{\infty} \alpha_n \beta_n < \infty$. Then the Ishikawa iteration generated from an arbitrary $x_0 \in E$ by

$$y_n = (1 - \beta_n)x_n + \beta_n(f + (I - T)x_n), \quad n \ge 0, \tag{8}$$

$$x_{n+1} = (1 - \alpha_n)x_n + \alpha_n(f + (I - T)y_n), \quad n \ge 0 \tag{9}$$

converges strongly to the solution of the equation $Tx = f$.

Proof. It follows by the φ-accretivity property,

$$\langle Tx - Ty, j(x - y) \rangle \ge \phi(\|x - y\|)\|x - y\|, \quad x, y \in E, \tag{10}$$

that if $Tx = f$ has a solution, then this is unique. Let x^* denote this solution and let L be the Lipschitz constant of T. Define $S : E \to E$ by

$$Sx := f + (I - T)x.$$

Then x^* is a fixed point of S and S is Lipschitzian with constant $L_* = 1 + L$.

By (10) we have for all $x, y \in E$

$$\langle (I - S)x - (I - S)y, \, j(x - y) \rangle =$$

$$\langle Tx - Ty, \, j(x - y) \rangle \ge \varphi(\|x - y\|)\|x - y\| \ge$$

$$\ge \frac{\varphi(\|x - y\|)}{1 + \varphi(\|x - y\|) + \|x - y\|} \cdot \|x - y\|^2.$$

Denote

$$\sigma(x, y) = \frac{\varphi(\|x - y\|)}{1 + \varphi(\|x - y\|) + \|x - y\|} \in [0, 1), \quad \forall \, x, y \in E,$$

and thus we get

$$\langle (I - S)x - \sigma(x, y)x - ((I - S)y - \sigma(x, y)y), \, j(x - y) \rangle \ge 0,$$

and applying Lemma of Kato, see Exercise 4.12, it results that

$$\|x - y\| \leq \|x - y + r[(I - S)x - \sigma(x,y)x - ((I - S)y - \sigma(x,y)\,y\,)]\| ,\quad (11)$$

which is valid for all $x, y \in E$ and $r > 0$. By (9) we obtain

$$x_n = x_{n+1} + \alpha_n x_n - \alpha_n S y_n = (1 + \alpha_n)x_{n+1}+$$

$$+\alpha_n[(I - S)x_{n+1} - \sigma(x_{n+1}, x^*)\,x_{n+1}] - (1 - \sigma(x_{n+1}, x^*))\,\alpha_n x_n+$$

$$+(2 - \sigma(x_{n+1}, x^*))\,\alpha_n^2(x_n - S y_n) + \alpha_n(S x_{n+1} - S y_n).$$

But

$$x^* = (1 + \alpha_n)x^* + \alpha_n[(I - S)x^* - \sigma(x_{n+1}, x^*)\,x^*] - (1 - \sigma(x_{n+1}, x^*))\,\alpha_n x^*,$$

and so

$$x_n - x^* = (1 + \alpha_n)(x_{n+1} - x^*) + \alpha_n[(I - S)x_{n+1} - \sigma(x_{n+1}, x^*)\,x_{n+1}-$$

$$-((I - S)x^* - \sigma(x_{n+1}, x^*)\,x^*)] - (1 - \sigma(x_{n+1}, x^*))\,\alpha_n(x_n - x^*)+$$

$$+(2 - \sigma(x_{n+1}, x^*))\,\alpha_n^2(x_n - S y_n) + \alpha_n(S x_{n+1} - S y_n).$$

Hence, using (11), we get

$$\|x_n - x^*\| \geq (1 + \alpha_n)\left\| x_{n+1} - x^* + \frac{\alpha_n}{1 + \alpha_n}[(I - S)x_{n+1}-\right.$$

$$- \sigma(x_{n+1}, x^*)x_{n+1} - ((I - S)x^* - \sigma(x_{n+1}, x^*)\,x^*)]\| -$$

$$-(1 - \sigma(x_{n+1}, x^*))\,\alpha_n \|x_n - x^*\| -$$

$$-(2 - \sigma(x_{n+1}, x^*))\,\alpha_n^2 \|x_n - S y_n\| - \alpha_n \|S x_{n+1} - S y_n\| \geq$$

$$\geq (1 + \alpha_n)\|x_{n+1} - x^*\| - (1 - \sigma(x_{n+1}, x^*))\alpha_n \|x_n - x^*\| -$$

$$-(2 - \sigma(x_{n+1}, x^*))\,\alpha_n^2 \|x_n - S y_n\| - \alpha_n \|S x_{n+1} - S y_n\| ,$$

so that

$$\|x_{n+1} - x^*\| \leq \frac{1 + (1 - \sigma(x_{n+1}, x^*))\alpha_n}{1 + \alpha_n} \cdot \|x_n - x^*\| +$$

$$+2\alpha_n^2 \|x_n - S y_n\| + \alpha_n \|S x_{n+1} - S y_n\| . \quad (12)$$

On the other hand

$$\|y_n - x^*\| = \|(1 - \beta_n)(x_n - x^*) + \beta_n(S x_n - x^*)\| \leq$$

$$\leq (1 + \beta_n(L_* - 1))\|x_n - x^*\| \leq L_* \|x_n - x^*\| ,$$

$$\|x_n - S y_n\| \leq \|x_n - x^*\| + L_* \|y_n - x^*\| \leq (1 + L_*^2)\|x_n - x^*\| \quad (13)$$

and

$$\|S x_{n+1} - S y_n\| \leq L_* \|(1 - \alpha_n)(x_n - y_n) + \alpha_n(S y_n - y_n)\| \leq$$

$$\leq L_*(1 - \alpha_n)\beta_n(1 + L_*)\|x_n - x^*\| + \alpha_n(1 + L_*)L_*^2 \|x_n - x^*\| \leq$$

$$\leq [L_*(1+L_*)\beta_n + (1+L_*)L_*^2\alpha_n] \, \|x_n - x^*\| . \qquad (14)$$

Now, using (13) and (14) in (12) we obtain

$$\|x_{n+1} - x^*\| \leq \frac{1 + (1 - \sigma(x_{n+1}, x^*))\alpha_n}{1 + \alpha_n} \, \|x_n - x^*\| +$$

$$+[L_*(1+L_*)\alpha_n\beta_n + (L_*^3 + 3L_*^2 + 2)\alpha_n^2] \|x_n - x^*\| \leq$$

$$\leq [1 + (1 - \sigma(x_{n+1}, x^*))\alpha_n](1 - \alpha_n + \alpha_n^2) \|x_n - x^*\| +$$

$$+[L_*(1+L_*)\alpha_n\beta_n + (L_*^3 + 3L_*^2 + 2)\alpha_n^2] \|x_n - x^*\| =$$

$$\leq [1 - \alpha_n\sigma(x_{n+1}, x^*)] \|x_n - x^*\| +$$

$$+[L_*(1+L_*)\alpha_n\beta_n + (L_*^3 + 3L_*^2 + 3)\alpha_n^2] \|x_n - x^*\| . \qquad (15)$$

Set

$$a_n := \|x_n - x^*\| , \quad \delta_n := L_*(1+L_*)\alpha_n\beta_n + (L_*^3 + 3L_*^2 + 3)\alpha_n^2$$

and then inequality (15) can be written in the form

$$a_{n+1} \leq [1 + \delta_n]a_n - \alpha_n\frac{\phi(a_{n+1})}{1 + \phi(a_{n+1}) + a_{n+1}} \cdot a_n.$$

Since by (ii)-(iii) we have $\sum\limits_{n=0}^{\infty} \alpha_n = \infty$ and $\sum\limits_{n=0}^{\infty} \delta_n = \infty$, by Lemma 1.4 it results that $\lim\limits_{n\to\infty} a_n = 0$, i.e., $\lim\limits_{n\to\infty} x_n = x^*$.   □

We shall present now a more general result which extends Theorem 8.3 to the case of the Ishikawa iteration method with errors. To this end we need the following lemma.

Lemma 8.1. *Let X be a real Banach space and let $T : X \to X$ be a continuous and φ-strongly pseudocontractive operator. Then T has a unique fixed point.*

Theorem 8.4. *Suppose that E is a real Banach space, $T : E \to E$ is a uniformly continuous and ϕ-strongly accretive operator, and the range of either $I - T$ or T is bounded. For $f \in E$, define $S : E \to E$ by*

$$Sx = f + x - Tx, \quad \text{for all } x \in E.$$

Define the sequence $\{x_n\}_{n=0}^{\infty}$ by $x_0, u_0, v_0 \in E$, and

$$y_n = a'_n x_n + b'_n S x_n + c'_n v_n, \quad n \geq 0$$

$$x_{n+1} = a_n x_n + b_n S y_n + c_n u_n, \quad n \geq 0,$$

where $\{u_n\}_{n=0}^{\infty}$ and $\{v_n\}_{n=0}^{\infty}$ are arbitrary bounded sequences in X, and $\{a_n\}_{n=0}^{\infty}$, $\{b_n\}_{n=0}^{\infty}$, $\{c_n\}_{n=0}^{\infty}$, $\{a'_n\}_{n=0}^{\infty}$, $\{b'_n\}_{n=0}^{\infty}$ and $\{c'_n\}_{n=0}^{\infty}$ are real sequences in $[0,1]$ satisfying the following conditions:

  (a) $a_n + b_n + c_n = a'_n + b'_n + c'_n = 1, \quad 0 < a_n < 1, \, n \geq 0;$

(b) $\lim_{n\to\infty} b_n = \lim_{n\to\infty} b'_n = \lim_{n\to\infty} c'_n = \lim_{n\to\infty} \dfrac{c_n}{b_n + c_n} = 0;$

(c) $\sum_{n=0}^{\infty} b_n = +\infty.$

Then $\{x_n\}$ converges strongly to the unique solution of the equation $Tx = f$.

Proof. The equation $Tx = f$ is equivalent to the fixed point problem $x = Sx$, with $Sx = f + x - Tx$. Since T is ϕ–strongly accretive, it results that S is ϕ-strongly pseudocontractive. Moreover, as T is uniformly continuous, we obtain that S is continuous.

Now, applying Lemma 8.1, it results that the equation $Tx = f$ has a unique solution, for any $f \in X$. The rest of the proof is similar to that of Theorem 8.3. $\square$

Remarks.

1) A prototype for the numerical sequences that are involved in Theorem 8.4 is given by

$$a_n = 1 - \frac{1}{4\sqrt{n+1}} - \frac{1}{4(n+1)}; \quad b_n = \frac{1}{4\sqrt{n+1}}; \quad c_n = \frac{1}{4(n+1)};$$

$$a'_n = \frac{n+1}{n+3}, \quad b'_n = c'_n = \frac{1}{n+3}, \quad \text{for all } n \geq 0.$$

They depend neither on the geometric structure of the ambient Banach space, nor on the properties of the operator T;

2) The sequence $\{x_n\}$ defined in Theorem 8.4 is the Ishikawa iteration with errors associated to S in the sense of Xu, see Definition 6.2.

Note that, if we denote $a_n = b_n + c_n$, from Definition 6.2 it follows

$$x_{n+1} = a_n x_n + b_n T x_n + c_n = (1 - \alpha_n)x_n + \alpha_n T x_n + c_n(u_n - T x_n)$$

so that, if the rage of T is bounded, Like in Theorem 8.4, then $v_n = u_n - T x_n$ is a bounded sequence and Xu's definition reduces to that of Liu, i.e., Definition 6.1, since, using the definition of Xu, it always assumed that $\sum c_n < \infty$. Moreover, the construction of Xu cannot be carried out in practice. Indeed, in order to determine the values of a_n, b_n and c_n in Definition 6.2, it is necessary to know the value of u_n for each n. But, if u_n is an unknown arbitrary bounded sequence, its values are not known;

3) Taking $a'_n = 1$, $b'_n = c'_n = 0$, for all $n \geq 0$, by Theorem 8.4 we obtain a result regarding the convergence of the Mann iteration with errors to the unique solution of $Tx = f$;

4) The next example illustrates some of the assumptions involved in the previous theorems.

Example 8.1. Let $\mathbb{R}$ denote the reals with the usual norm and define $T : \mathbb{R} \to \mathbb{R}$ by $Tx = x - \frac{1}{2}\cos x$. Then T is Lipschitzian and strongly accretive, the range of $I - T$ is bounded, but the range of T is not bounded.

A similar result to that in Theorem 8.4 can be formulated for the equations of the form $x + Tx = f$.

Theorem 8.5. *Let E be a real Banach space, and $T : E \to E$ be a uniformly continuous and ϕ-strongly accretive operator, such that the range of either $I + T$ or T is bounded. For any fixed $f \in E$, define $S : E \to E$ by $Sx = f - Tx$, for all $x \in E$. Define the sequences $\{x_n\}_{n=0}^{\infty}$, $\{u_n\}_{n=0}^{\infty}$, $\{v_n\}_{n=0}^{\infty}$, $\{a_n\}_{n=0}^{\infty}$, $\{b_n\}_{n=0}^{\infty}$, $\{c_n\}_{n=0}^{\infty}$, $\{a_n'\}_{n=0}^{\infty}$, $\{b_n'\}_{n=0}^{\infty}$, and $\{c_n'\}_{n=0}^{\infty}$ as in Theorem 8.4.*

Then $\{x_n\}_{n=0}^{\infty}$ converges strongly to the unique solution of the equation

$$x + Tx = f.$$

Proof. Set $A = I + T$. Then $A : X \to X$ is uniformly continuous and ϕ-strongly accretive, and the range of either $I - A$ or A is bounded. Then $x + Tx = f$ is equivalent to the fixed point problem $x = Sx$, with

$$Sx = f - Tx = f - (A - I)x = f + x - Ax, \ \forall \, x \in A.$$

Apply Theorem 8.4 to obtain the conclusion. $\qquad\qquad\qquad\qquad\qquad\square$

8.2 Nonlinear Equations in Smooth Banach Spaces

The aim of this Section is to show how some assumptions on the operator T or/and on the parameters that define a certain iteration procedure can be weakened, by transferring them into restrictions on the geometry of the underlying Banach space. We shall restrict the presentation to two sample results. To extend the area of applications, the second convergence theorem will be given for multivalued mappings.

Theorem 8.6. *Let E be a real uniformly smooth Banach space and let $T : E \to E$ be a Lipschitzian (with constant $L > 0$) ϕ-strongly accretive mapping. For any given $f \in E$, define the mapping $S : E \to E$ by $Sx = f - Tx + x$, for each $x \in E$.*

Let $\{\alpha_n\}_{n=0}^{\infty}$ and $\{\beta_n\}_{n=0}^{\infty}$ be two sequences of real numbers in $[0, 1]$ satisfying

(i) $\lim\limits_{n \to \infty} \alpha_n = \lim\limits_{n \to \infty} \beta_n = 0;$ *(ii)* $\sum\limits_{n=0}^{\infty} \alpha_n = \infty.$

Then the sequence $\{x_n\}_{n=0}^{\infty}$ defined by $x_0 \in E$ and

$$y_n = (1 - \beta_n)x_n + \beta_n Sx_n, \ n \geq 0,$$
$$x_{n+1} = (1 - \alpha_n)x_n + \alpha_n Sy_n, \ n \geq 0$$

converges strongly to the unique solution of the equation $Tx = f$.

Proof. Since T is Lipschitzian and φ-strongly accretive, it results that S is continuous and ϕ-strongly pseudocontractive. Then by Lemma 8.1 it follows that S has a unique fixed point, i.e., the equation $Tx = f$ has a unique solution. The rest of the proof is standard.   □

Remarks.
1) If $\beta_n = 0$ for all $n \geq 0$, Theorem 8.6 gives a convergence result for the Mann iterative process for solving the equation $Tx = f$;
2) Theorem 8.6 does not require the unnecessary condition that $S(T)$, the set of solutions of S, is nonempty.

In order to ensure the appropriate framework for presenting the next results in this section, we need to consider some additional notions to those introduced in Chapter 1.

Let E be a real normed linear space with the dual E^*.

Definition 8.1. For $q > 1$, the mapping $J_q : E \to 2^{E^*}$, defined by

$$J_q(x) = \left\{ x^* \in E^* : \langle x, x^* \rangle = \|x\|^2 \, , \, \|x^*\|^2 = \|x\|^{q-1} \right\},$$

is called the *generalized duality mapping* ($\langle \cdot, \cdot \rangle$ denotes in this context the generalized duality pairing).

Remarks.
1) For $q = 2$ we obtain the normalized duality mapping $J = J_2$ that has been used in several convergence theorems presented in this book;
2) It is well known, see Exercise 8.11, that if E is smooth then J_q is single-valued and

$$J_q(x) = \|x\|^{q-2} J(x) \, , \quad x \neq 0.$$

This will enable us to denote the single-valued generalized duality map by j_q.

Definition 8.2. A multivalued mapping $A : E \to 2^E$ is said to be *accretive* if, for all $x, y \in D(A)$, there exists $j(x - y) \in J(x - y)$ such that

$$\langle u - v, j(x - y) \rangle \geq 0, \quad \text{for each } u \in Ax \text{ and } v \in Ay.$$

The map A is called $m-accretive$ if it is accretive and $R(I + rA) = E$, for all $r > 0$ ($R(T)$ denotes the range of T).

The map A is called *strongly accretive* if for all $x, y \in D(A)$, there exist $j(x - y) \in J(x - y)$ and $k > 0$ such that for all $u \in Ax$ and $v \in Ay$:

$$\langle u - v, j(x - y) \rangle \geq k \, \|x - y\|^2 .$$

A map T with domain $D(T)$ in E and range $R(T)$ in 2^E is called *pseudocontractive* if, for each $x, y \in D(T)$, there exists $j(x - y) \in J(x - y)$ such that

$$\langle u - v, \, j(x - y) \rangle \leq \|x - y\|^2 \, , \quad \text{for each } u \in Tx \text{ and } v \in Ty,$$

and it is called *strongly pseudocontractive* if, for each $x, y \in D(T)$, there exists $j(x - y) \in J(x - y)$ and a constant $k \in (0, 1)$ such that

$$\langle u - v,\; j(x - y) \rangle \le k \, \|x - y\|^2 , \quad \text{for each } u \in Tx \text{ and } v \in Ty.$$

Remarks.

1) If E is a Hilbert space, an accretive mapping is also called *monotone*.

2) A mapping A is (strongly) accretive if and only if $T = I - A$ is (strongly) pseudocontractive. As in the case of single-valued operators, a zero of A is a fixed point of $T := I - A$ and vice versa.

3) In a real q-uniformly smooth Banach space (typical examples of such spaces are the Lebesgue L_p, the sequences l_p and the Sobolev W_p^m spaces, for $1 < p < \infty$), see Exercise 8.12, the following inequality holds

$$\|x + y\|^q \le \|x\|^q + q \, \langle y, j_q(x) \rangle + c_q \, \|y\|^q , \tag{16}$$

for all $x, y \in E$ and some real constant $c_q > 0$.

4) Note also that the uniformly smooth spaces have norms that are uniformly Gateaux differentiable (for some related concepts, see Chapter 6).

Definition 8.3. A mapping $A : E \to 2^E$ is said to satisfy the *linear growth condition* if $\|Ax\| \le c \, (1 + \| x \|)$, for all $x \in D(A)$ and for some $c > 0$.

Definition 8.4. Two sequences $\{\lambda_n\}$ and $\{\theta_n\}$ of positive real numbers are called *acceptably paired* if $\{\theta_n\}$ is non-increasing and there exists a strictly increasing sequence $\{n(i)\}_{i=1}^{\infty}$ of positive integers such that

$$(i) \;\; \liminf_{i \to \infty} \theta_{n(i)} \sum_{j=n(i)}^{n(i+1)-1} \lambda_j > 0; \quad (ii) \;\; \lim_{i \to \infty} [\theta_{n(i)} - \theta_{n(i+1)}] \sum_{j=n(i)}^{n(i+1)-1} \lambda_j = 0;$$

$$(iii) \;\; \limsup_{i \to \infty} \theta_{n(i)} \sum_{j=n(i)}^{n(i+1)-1} \lambda_j < \infty.$$

Remarks.

1) In the previous definition it is not necessary that $\lim_{n \to \infty} \theta_n = 0$;

2) An example of acceptably paired sequences is given by

$$\lambda_n = 1/n, \;\; \theta_n = (\log \log \; n)^{-1}, \;\; n \ge 1, \;\; n(i) = i^i.$$

Theorem 8.7. *Let E be a reflexive Banach space with a uniformly Gateaux differentiable norm, and such that every weakly compact convex subset of E has the fixed point property for nonexpansive mappings. Let $A : E \to 2^E$ be a m-accretive mapping. If $A^{-1}(0) \ne \emptyset$, then, for each $x \in E$, the strong limit*

$$\lim_{t \to \infty} J_t(x), \;\; \text{where } J_t = (I - tA)^{-1}, \; t > 0,$$

exists and belongs to $A^{-1}(0)$ and, if $A^{-1}(0) = \emptyset$, then for each $x \in E$ we have

$$\lim_{t \to \infty} \| J_t(x) \| = \infty.$$

Now we can prove the main result of this section.

Theorem 8.8. *Let E be a real q-uniformly smooth Banach space and $A : D(A) = E \to 2^E$ a $m-$accretive mapping which satisfies the linear growth condition. Suppose that $\{\lambda_n\}$ and $\{\theta_n\}$ are acceptably paired, with $\sum \lambda_n^q < \infty$ and $\lim_{n \to \infty} \theta_n = 0$. Let x_1 and z be arbitrary in E. Define the sequence $\{x_n\}$ by*

$$x_{n+1} = x_n - \lambda_n (u_n + \theta_n (x_n - z)), \quad u_n \in Ax_n, \tag{17}$$

for all $n \geq 0$. If $A^{-1}(0) \neq \emptyset$ then $\{x_n\}$ converges strongly to $x^* \in A^{-1}(0)$, and if $A^{-1}(0) = \emptyset$, then $\|x\| \to \infty$ as $n \to \infty$.

Proof. Note that if A is $m-$accretive, then $\theta^{-1}A$ is also accretive, for $\theta > 0$. Thus for each i and any $z \in E$, there exists a unique $y_i \in E$ such that

$$z \in y_i + \theta_i^{-1} A y_i$$

and hence

$$J_{1/\theta_i}(z) := (I - (1/\theta_i)A)^{-1}(z) = y_i.$$

In the sequel y_i will be defined as above, while $x^* \in A^{-1}(0)$ will denote the limit of y_i defined by

$$\lim_{i \to \infty} y_i = \lim_{1/\theta_i \to \infty} J_{1/\theta_i}(z) = \lim_{t \to \infty} J_t(z) = x^*,$$

guaranteed by Reich's theorem, see Exercise 8.7.

Let $n \geq i \geq 2$. Then, by (17), for $u_{n-1} \in Ax_{n-1}$ we have that

$$x_n - y_i = x_{n-1} - y_i - \lambda_{n-1}(u_{n-1} + \theta_{n-1}(x_{n-1} - z)),$$

and hence, by (16),

$$\|x_n - y_i\|^q = \|x_{n-1} - y_i - \lambda_{n-1}(u_{n-1} + \theta_{n-1}(x_{n-1} - z))\|^q \leq$$

$$\leq \|x_{n-1} - y_i\|^q - q\lambda_{n-1} \langle u_{n-1} + \theta_{n-1}(x_{n-1} - z), j_q(x_{n-1} - y_i) \rangle +$$

$$+ c_q \lambda_{n-1}^q \|u_{n-1} + \theta_{n-1}(x_{n-1} - z)\|^q \leq$$

$$\leq \|x_{n-1} - y_i\|^q - q\lambda_{n-1} \langle u_{n-1} + \theta_i(x_{n-1} - z), j_q(x_{n-1} - y_i) \rangle -$$

$$- q\lambda_{n-1}(\theta_{n-1} - \theta_i) \langle x_{n-1} - z, j_q(x_{n-1} - y_i) \rangle +$$

$$+ c_q \lambda_{n-1}^q \|u_{n-1} + \theta_{n-1}(x_{n-1} - z)\|^q. \tag{18}$$

Since A is accretive and $-\theta_i(y_i - z) \in Ay_i$, $u_{n-1} \in Ax_{n-1}$, we get

$$\langle u_{n-1} + \theta_i(y_i - z), j_q(x_{n-1} - y_i) \rangle \geq 0,$$

which gives

$$\langle u_{n-1} + \theta_i(x_{n-1} - z),\ j_q(x_{n-1} - y_i)\rangle = \langle u_{n-1} + \theta_i(y_i - z),\ j_q(x_{n-1} - y_i)\rangle +$$

$$+\theta_i \langle x_{n-1} - y_i,\ j_q(x_{n-1} - y_i)\rangle \geq \theta_i \|x_{n-1} - y_i\|^q.$$

For $p, q > 1$ such that $\dfrac{1}{p} + \dfrac{1}{q} = 1$ we have

$$|\langle x_{n-1} - z,\ j_q(x_{n-1} - y_i)\rangle| \leq \|x_{n-1} - z\|\,\|x_{n-1} - y_i\|^{q-1} \leq$$

$$\leq \frac{1}{q}\|x_{n-1} - z\|^q + \frac{1}{p}\|x_{n-1} - y_i\|^{p(q-1)} \leq$$

$$\leq \frac{1}{q}\,(\,\|x_{n-1} - y_i\| + \|y_i\| + \|z\|)^q + \frac{1}{p}\|x_{n-1} - y_i\|^q \leq$$

$$\leq \frac{1}{q}d_1\,(\,\|x_{n-1} - y_i\|^q + \|y_i\|^q + \|z\|^q\,) + \frac{1}{p}\|x_{n-1} - y_i\|^q,$$

for some $d_1 > 0$. Now using the linear growth condition we have that

$$\begin{aligned}\|u_{n-1} + \theta_{n-1}(x_{n-1} - z)\|^q &\leq (\,\|u_{n-1}\| + \|x_{n-1}\| + \|z\|)^q \leq\\ &\leq d'\,(1 + 2\,\|x_{n-1}\| + \|z\|)^q \leq\\ &\leq d'\,(1 + 2\,\|x_{n-1} - y_i\| + 2\,\|y_i\| + \|z\|)^q \leq\\ &\leq d_2\,(1 + \|x_{n-1} - y_i\|^q + \|y_i\|^q + \|z\|^q),\end{aligned}$$

for some $d', d_2 > 0$.

These last estimates together with (18) yield

$$\|x_n - y_i\|^q \leq \|x_{n-1} - y_i\|^q - q\lambda_{n-1}\theta_i\|x_{n-1} - y_i\|^q +$$

$$+q\lambda_{n-1}(\theta_i - \theta_{n-1})\frac{1}{2}d_1\|x_{n-1} - y_i\|^q + \frac{1}{q}d_1\|y_i\|^q + \frac{1}{q}d_1\|z\|^q +$$

$$+\frac{1}{p}\|x_{n-1} - y_i\|^q + c_q\lambda_{n-1}^q d_2\,[1 + \|x_{n-1} - y_i\|^q + \|y_i\|^q + \|z\|^q] =$$

$$= \|x_{n-1} - y_i\|^q - \left[q\lambda_{n-1}\theta_i - d_1\lambda_{n-1}(\theta_i - \theta_{n-1}) - \frac{q}{p}\lambda_{n-1}(\theta_i - \theta_{n-1}) - \right.$$

$$\left. -c_q\lambda_{n-1}^q d_2\right]\|x_{n-1} - y_i\|^q + d_1\lambda_{n-1}(\theta_i - \theta_{n-1})\,(\|y_i\|^q + \|z\|^q) +$$

$$+c_q\lambda_{n-1}^q d_2\,(\|y_i\|^q + \|z\|^q + 1) \leq \|x_{n-1} - y_i\|^q -$$

$$-q\lambda_{n-1}\theta_i - d_1\lambda_{n-1}(\theta_i - \theta_{n-1}) - \frac{q}{p}\lambda_{n-1}(\theta_i - \theta_{n-1}) - c_q\lambda_{n-1}^q d_2$$

$$\|x_{n-1} - y_i\|^q + (d_1\lambda_{n-1}(\theta_i - \theta_{i-1}) + c_q\lambda_{n-1}^q d_2)\cdot(\|y_i\|^q + \|z\|^q + 1) \leq$$

$$\leq \|x_{n-1} - y_i\|^q - (q\lambda_{n-1}\theta_i - d_3\lambda_{n-1}(\theta_i - \theta_{n-1}) - c_q\lambda_{n-1}^q d_2)\,\|x_{n-1} - y_i\|^q +$$

$$+(d_1\lambda_{n-1}(\theta_i - \theta_{n-1}) + c_q\lambda_{n-1}^2 d_2)\cdot(\|y_i\|^q + \|z\|^q + 1) \le$$

$$\le (1 - b_{n-1,i})\|x_{n-1} - y_i\|^q + a_{n-1,i}(\|y_i\|^q + \|z\|^q + 1), \qquad (19)$$

where $d_3 = \max\left\{d_1, \dfrac{q}{p}\right\}$,

$$b_{n-1,i} = q\lambda_{n-1}\theta_i - d_3\lambda_{n-1}(\theta_i - \theta_{n-1}) - c_q\lambda_{n-1}^q d_2 \quad \text{and}$$

$$a_{n-1,i} = d_1\lambda_{n-1}(\theta_i - \theta_{n-1}) + c_q\lambda_{n-1}^q d_2.$$

Let now take $i = n(i)$ and $n = n(i+1)$ and iterate (19) from $n(i)$ on, to get that

$$\|x_{n(i+1)} - y_{n(i)}\|^2 \le \exp\left(-\sum_{j=n(i)}^{n(i+1)-1} b_{j,n(i)}\right)\|x_{n(i)} - y_{n(i)}\|^2 +$$

$$+ \sum_{j=n(i)}^{n(i+1)-1} a_{j,n(i)}\left(\|y_{n(i)}\|^q + \|z\|^q + 1\right). \qquad (20)$$

Using conditions (i)-(iii) in the definition of acceptably paired sequences, on the one hand, and the fact that $\sum\limits_{n=1}^{\infty}\lambda_n^q < \infty$, on the other hand, it results that there exists $\delta \in (0,1)$ such that

$$\exp\left(-\sum_{j=n(i)}^{n(i+1)-1} b_{j,n(i)}\right) \le \delta$$

and that

$$e_{n(i)} = \left(\sum_{j=n(i)}^{n(i+1)-1} a_{j,n(i)}\right) \to 0 \quad \text{as } i \to \infty.$$

Therefore, (20) yields

$$\|x_{n(i+1)} - y_{n(i)}\|^q \le \delta\|x_{n(i)} - y_{n(i)}\|^q + \varepsilon_{n(i)}\left(\|y_{n(i)}\|^q + \|z\|^q + 1\right),$$

and hence

$$\|x_{n(i+1)} - y_{n(i)}\| \le \delta^{1/q}\|x_{n(i)} - y_{n(i)}\| + \varepsilon_{n(i)}^{1/q}\left(\|y_{n(i)}\| + \|z\| + 1\right). \qquad (21)$$

In a similar manner we obtain

$$\|x_n - y_{n(i)}\| \le D^{1/q}\|x_{n(i)} - y_{n(i)}\| + \varepsilon_{n(i)}^{1/q}\left(\|y_{n(i)}\| + \|z\| + 1\right) \qquad (22)$$

for some $D < \infty$.

Using now the accretivity property of A, it results that

$$\left\|y_{n(i)} - y_{n(i+1)}\right\| \le \left\|y_{n(i)} - y_{n(i+1)} + \frac{1}{\theta_{n(i+1)}}(Ay_{n(i)} - Ay_{n(i+1)})\right\| \le$$

$$\le \frac{\theta_{n(i)} - \theta_{n(i+1)}}{\theta_{n(i+1)}}\left(\left\|y_{n(i)}\right\| + \|z\|\right) = \left(\frac{\theta_{n(i)}}{\theta_{n(i+1)}} - 1\right)\left(\left\|y_{n(i)}\right\| + \|z\|\right). \quad (23)$$

Again from $(i) - (iii)$ in the definition of acceptably paired sequences we get that

$$\lim_{i \to \infty}\left(\frac{\theta_{n(i)}}{\theta_{n(i+1)}} - 1\right) = 0.$$

Hence by (21) and (23) we deduce

$$\left\|x_{n(i)+1} - y_{n(i+1)}\right\| \le \left\|x_{n(i+1)} - y_{n(i)}\right\| + \left\|y_{n(i)} - y_{n(i+1)}\right\| \le$$

$$\le \delta^{1/2}\left\|x_{n(i)} - y_{n(i)}\right\| + \alpha_{n(i)}\left(\left\|y_{n(i)}\right\| + \|z\| + 1\right), \quad (24)$$

where $\alpha_{n(i)} = \varepsilon_{n(i)}^{1/q} + \theta_{n(i)}/\theta_{n(i+1)} - 1 \to 0$ as $i \to \infty$. Moreover, by (23) we obtain

$$\frac{1 - \alpha_{n(i)}}{\left\|y_{n(i+1)}\right\| + \|z\| + 1} \le \frac{1}{\left\|y_{n(i)}\right\| + \|z\| + 1}$$

which together with (24) yields

$$(1 - \alpha_{n(i)}) \cdot \frac{\left\|x_{n(i+1)} - y_{n(i+1)}\right\|}{\left\|y_{n(i+1)}\right\| + \|z\| + 1} \le \delta^{1/q} \cdot \frac{\left\|x_{n(i)} - y_{n(i)}\right\|}{\left\|y_{n(i)}\right\| + \|z\| + 1} + \alpha_{n(i)}.$$

Since $\alpha_{n(i)} \to 0$ as $i \to \infty$ and $\delta^{1/q} < 1$, we get that

$$\lim_{i \to \infty}\left(\left\|x_{n(i)} - y_{n(i)}\right\| / \left(\left\|y_{n(i)}\right\| + \|z\| + 1\right)\right) = 0,$$

and hence, by (21), it results that

$$\lim_{i \to \infty} \max_{n(i) \le n \le n(i+1)} \frac{\left\|x_n - y_{n(i)}\right\|}{\left\|y_{n(i)}\right\| + \|z\| + 1} = 0.$$

This shows that $\left\|x_n - y_{n(i)}\right\| \to 0$ as $n, i \to \infty$. Since the weakly compact subsets of E have the fixed point property for nonexpansive mappings, and the uniformly smooth Banach spaces have uniformly Gateaux differentiable norms, by Theorem 8.7 we get the conclusion. □

Remarks
1) The explicit scheme (17) can be written as an implicit scheme

$$x_{n+1} + \lambda_n(u_{n+1} + \theta_n(x_{n+1} - z)) = x_n + e_n,$$

with the error term $e_n = \lambda_n(u_{n+1} - u_n + \theta_n(x_{n+1} - x_n))$, for $u_n \in Ax_n$.
    It is possible to obtain a convergence result for the implicit scheme if $\sum \|e_n\| < \infty$, see Theorem 3.6 in Chidume and Zegeye [ChZ02];
2) Theorem 8.8 extends several results in literature.

8.3 Nonlinear m-Accretive Operator Equations in Reflexive Banach Spaces

We end this chapter with a result that complements the results presented in the previous sections, for the case of reflexive Banach spaces. An estimation of the rate of convergence for a Mann type iteration is also obtained in this case. This will naturally link the material in Chapter 8 to the next one.

Theorem 8.9. *Let E be a real reflexive Banach space, and $T : D(T) \subset E \to E$ be an m-accretive and locally Lipschitzian operator (with constant L). Suppose $D(T)$ is open and denote by $x^* \in D(T)$ the unique solution of the equation $x + Tx = f$, $f \in E$. Suppose $\{\alpha_n\}_{n=0}^{\infty}$ is a real sequence satisfying the following conditions*

(i) $0 \le \alpha_n \le 1/2(L^2 + 2L + 2)$, $n \ge 0$; *(ii)* $\sum_{n=0}^{\infty} \alpha_n = \infty$.

Then there exists a closed convex neighborhood V of x^* contained in $D(T)$ and, for any $x_0 \in V$, a sequence $\{x_n\}_{n=0}^{\infty} \subset V$ such that by setting

$$p_n = (1 - \alpha_n)x_n + \alpha_n(f - Tx), \quad n \ge 0$$

the sequence $\{p_n\}$ satisfies the condition

$$\|p_n - x_{n+1}\| = \inf\{\|p_n - x\| \mid x \in B\}, \quad \forall\, n \ge 0$$

and converges strongly to x^*. Moreover, if

$$\alpha_n = 1/2(L^2 + 2L + 2), \quad \text{for all } n \ge 0,$$

then

$$\|p_n - x^*\| \le \rho^n \|p_0 - x^*\|,$$

where $\rho = (1 - 1/4(L^2 + 2L + 2)) \in (0,1)$.

Proof. Since T is m-accretive, then for any $f \in E$, the equation

$$x + Tx = f \tag{25}$$

has a unique solution, $x^* \in D(T)$.

Define $S : D(T) \to E$ by $Sx = f - Tx$, for all $x \in D(T)$. Then x^* is a fixed point of T and S is locally Lipschitzian (with constant L). Furthermore, $(-S)$ is accretive and hence for all $r > 0$ and $x, y \in D(T)$ we have

$$\|x - y\| \le \|x - y - r(Sx - Sy)\|. \tag{26}$$

We may assume $L \ge 1$ (if $L < 1$, then S is a locally L-contraction and the conclusion follows by the results already established).

Let $B(y, r) = \{x \in E \,/\, \|x - y\| \le r\}$ be the closed ball.

Since $D(T)$ is open, there exists $r_1 > 0$ such that $B(x^*, r_1) \subset D(T)$. As S is locally Lipschitzian, there exists $r_2 > 0$ such that S is Lipschitzian on $B(x^*, r_2)$.

Let $r = \min\{r_1, r_2\}$. Then $B(x^*, r) \subset D(T)$ and S is Lipschitzian on $B(x^*, r)$. Let $V = B(x^*, r/2L)$. For any $x_0 \in V$, we have

$$\|Sx_0 - x^*\| \leq r/2 < r$$

and so $Sx_0 \in B(x^*, r)$. This shows that

$$p_0 = (1 - \alpha_0)x_0 + \alpha_0 Sx_0 \in B(x^*, r).$$

Since E is reflexive, there exists $x_1 \in V$ such that

$$\|p_0 - x_1\| = \inf\{\|p_0 - x\| : x \in V\}.$$

Thus

$$p_1 = (1 - \alpha_1)x_1 + \alpha_1 Sx_1 \in B(x^*, r).$$

By continuing this process we obtain the sequences $\{p_n\}$ in $B(x^*, r\}$ and $\{x_n\}$ in V satisfying the conditions

$$p_n = (1 - \alpha_n)x_n + \alpha_n Sx_n, \quad n \geq 0, \tag{27}$$

$$\|p_n - x_{n+1}\| = \inf\{\|p_n - x\| : x \in V\}, \quad n \geq 0.$$

Thus

$$\|x_n - x^*\| \leq \|p_{n-1} - x^*\|, \quad n \geq 1.$$

We prove now that $\lim_{n \to \infty} p_n = x^*$. Indeed, from (27) we have

$$x_n = p_n + \alpha_n x_n - \alpha_n Sx_n = (1 + \alpha_n)p_n - \alpha_n Sp_n +$$

$$+\alpha_n^2(x_n - Sx_n) + \alpha_n(Sp_n - Sx_n). \tag{28}$$

Using the fact that $x^* = Sx^*$, i.e.,

$$x^* = (1 + \alpha_n)x^* - \alpha_n Sx^*,$$

by (28) and (26) we obtain

$$\|x_n - x^*\| = \|(1 + \alpha_n)(p_n - x^*) - \alpha_n(Sp_n - Sx^*) + \alpha_n^2(x_n - Sx_n) +$$

$$+\alpha_n(Sp_n - Sx_n)\| \geq (1 + \alpha_n)\left\|p_n - x^* - \frac{\alpha_n}{1 + \alpha_n}(Sp_n - Sx_n)\right\| -$$

$$-\alpha_n^2 \|x_n - Sx_n\| - \alpha_n \|Sp_n - Sx_n\| \geq$$

$$\geq (1 + \alpha_n)\|p_n - x^*\| - \alpha_n^2 \|x_n - Sx_n\| - \alpha_n \|Sp_n - Sx_n\|.$$

Therefore

$$\|p_n - x^*\| \leq \frac{1}{1 + \alpha_n} \|x_n - x^*\| + \alpha_n^2 \|x_n - Sx_n\| + \alpha_n \|Sp_n - Sx_n\| \leq$$

$$\leq (1 - \alpha_n + \alpha_n^2) \|x_n - x^*\| + (1 + L)\alpha_n^2 \|x_n - x^*\| +$$
$$+ L(1 + L)\alpha_n^2 \|x_n - x^*\| . \tag{29}$$

So

$$\|p_n - x^*\| \leq \left[1 - \frac{1}{2}\alpha_n\right] \|p_{n-1} - x^*\| \leq$$

$$\leq \exp\left(-\frac{1}{2}\sum_{j=0}^{n} \alpha_j\right) \|p_0 - x^*\| \to 0 \quad \text{as } n \to \infty.$$

If we set in (29)

$$\alpha_n = 1/2(L^2 + 2L + 2), \quad n \geq 0,$$

then we obtain

$$\|p_n - x^*\| \leq \rho \|p_{n-1} - x^*\| \leq \rho^n \|p_0 - x^*\| , \tag{30}$$

that completes the proof.   □

Remarks.
1) Note, however, that the iteration $\{p_n\}$ for which the convergence order estimation (30) is obtained, is actually a Krasnoselskij iteration, with

$$\lambda = 1/(L^2 + 2L + 2), \quad n \geq 0;$$

2) The proof of Theorem 8.9 can be adapted to prove a similar result for an Ishikawa type iteration procedure stated in the following without proof.

Theorem 8.10. *Suppose $E, T, D(T), S$ and x^* are like in Theorem 8.9. Suppose $\{\alpha_n\}_{n=0}^{\infty}$ and $\{\beta_n\}_{n=0}^{\infty}$ are real sequences satisfying the conditions*
(i) $0 \leq \alpha_n \leq 1/2(L^2 + 2L + 2)$, $n \geq 0$;
(ii) $0 \leq \beta_n \leq 1/4(L^2 + 2L + 2)$, $n \geq 0$;
(iii) $\sum_{n=0}^{\infty} \alpha_n = \infty$.
Then there exists a closed neighborhood V of x^* contained in $D(T)$ and, for any given $x_0 \in V$, a sequence $\{x_n\}_{n=0}^{\infty}$ of elements of V such that by setting

$$y_n = (1 - \beta_n)x_n + \beta_n Sx_n, \quad n \geq 0,$$
$$p_n = (1 - \alpha_n)x_n + \alpha_n Sy_n, \quad n \geq 0,$$

the sequence $\{p_n\}$ satisfies the condition

$$\|p_n - x_{n+1}\| = \inf\{\|p_n - x\| : x \in V\}, \quad n \geq 0$$

and converges strongly to x^*, the unique solution of $x + Tx = f$, $f \in E$.

Moreover, if $\alpha_n = 1/2(L^2 + 2L + 2)$ *and* $\beta_n = 1/4(L^2 + 2L + 2)$, $n \geq 0$, *then*

$$\|p_n - x^*\| \leq \rho^n \|p_0 - x^*\| , \quad n \geq 0,$$

where

$$\rho = (1 - 1/8(L^2 + 2L + 2)) \in (0, 1).$$

8.4 Bibliographical Comments

For a relationship between φ-monotone operators and φ-contractive operators, see for example Berinde [Be93a], while for various applications of Picard iteration in solving nonlinear operator equations, see for instance Rus [Ru79c], Dugundji and Granas [DuG82], Berinde [Be97a])

§8.1.

Theorem 8.1 is Corollary 6 in Chidume and Osilike [ChO98], Corollary 8.1 is Corollary 7, while Theorem 8.2 is Corollary 9, both in the same paper. Theorem 8.3 is Theorem 1 in Osilike [Os99a].

Lemma 8.2 is proved in Liu, Z. and Kang, S.M. [LK01c]. Theorems 8.4 and 8.5 are taken from the same work. The example in Remark 1 following the proof of Theorem 8.4 is also taken from Liu, Z. and Kang, S.M. [LK01c]. For the Remark 2) following Theorem 8.4, see Rhoades [Rho04].

Example 8.1 is taken from Chidume and Osilike [ChO99], while Exercise 8.4 is taken from Corollary 3.2 in Barbu [Bar76].

As shown by Examples 3.1 and 3.2 in Liu, Z. and Kang, S.M. [LK01c], the assumptions (a), (b) and (c) in Theorem 8.4 are different from those of Chidume [Ch98a] and Xu, Y.G. [XuY98]. Nevanlinna [Nev79] indicated a technique for constructing acceptably paired sequences.

Similar results, but for Ishikawa iteration with errors in the non-convex form, were obtained in Yin, Liu, Z. and Lee, B.S. [YLL00].

§8.2.

Theorem 8.6 extends Theorem 4.2 in Gu, Feng [Gu01d] (it does not require $F_T \neq \emptyset$). The rest of this section is taken from Chidume and Zegeye [ChZ01].

§8.3.

The content of this section is taken from Osilike [Os97d].

For other results on the topic of this Chapter, see the monographs Chang, S.S., Cho, Y.J., Zhou, Y.Y. [CCZ03] and Chidume, C.E. [Chi05]. Theorem 8.9 extends some results from Liang [Lia94] established there in the case of real uniformly convex Banach spaces.

Exercises and Miscellaneous Results

8.1. Let E be a real Banach space and $F : E \to E$ be a strongly accretive (Φ-strongly accretive) operator and let $f \in E$ be fixed. Then $T : E \to E$, defined by

$$Tx = f + (I - F)x, \ x \in E,$$

is strongly pseudocontractive (Φ-strongly pseudocontractive).

8.2. Prove Lemma 8.1 and the assertions in Example 8.1.

8.3. Prove Theorem 8.7 and Theorem 8.10.

8.4. Let $\mathbb{R}$ denote the reals with the usual norm and define $T : \mathbb{R} \to \mathbb{R}$ by

$$Tx = \begin{cases} -1, \ x \in (-\infty, -1) \\ -\sqrt{1 - (x+1)^2}, \ x \in [-1, 0) \\ \sqrt{1 - (x-1)^2}, \ x \in [0, 1] \\ 1, \ x \in (1, \infty) \end{cases}$$

Show that T is m-accretive and has bounded range.

8.5. Let E be a real normed linear space and J be the normalized duality map. A map $A : D(A) \subseteq E \to E$ is called *uniformly accretive* if $\forall x, y \in D(A)$, there exist $j(x - y) \in J(x - y)$ and a strictly increasing function $\Psi : [0, \infty) \to [0, \infty)$ with $\Psi(0) = 0$ such that

$$\langle Ax - Ay, j(x - y) \rangle \geq \Psi(\|x - y\|).$$

The map $T : D(T) \subseteq E \to E$ is called *uniformly pseudocontractive* if $\forall x, y \in D(T)$, there exist $j(x - y) \in J(x - y)$ and a strictly increasing function $\Omega : [0, \infty) \to [0, \infty)$ with $\Omega(0) = 0$ such that

$$\langle Tx - Ty, j(x - y) \rangle \leq \|x - y\|^2 - \Omega(\|x - y\|).$$

(a) Show that the class of uniformly pseudocontractive maps includes the class of strongly pseudocontractive maps and the inclusion is proper;

(b) Show that T is uniformly pseudocontractive if and only if $A = I - T$ is uniformly accretive.

8.6. Show that the sequences $\{\lambda_n\}$ and $\{\theta_n\}$ given by

$$\lambda_n = 1/n, \ \theta_n = (\log \log n)^{-1}, \ n \geq 1, \ n(i) = i^i.$$

are acceptably paired.

8.7. Reich (1980)
Let E be a uniformly smooth Banach space, and let $A \subset E \times E$ be m-accretive. If $0 \in R(A)$, then for each x in E the strong limit $\lim_{t \to \infty} J_t(x)$ exists and belongs to $A^{-1}0$. ($R(A)$ stands for the range of A)

8.8. Chidume and Zegeye (2003)

Let $\{\lambda_n\}$ and $\{b_n\}$ be sequences of nonnegative numbers and $\{\alpha_n\} \subseteq (0,1)$ a sequence satisfying the conditions that $\{\lambda_n\}$ is bounded, $\sum_{n=1}^{\infty} \alpha_n = \infty$ and $b_n \to 0$, as $n \to \infty$. Let the recursive inequality

$$\lambda_{n+1}^2 \le \lambda_n^2 - 2\alpha_n \psi(\lambda_{n+1}) + 2\alpha_n b_n \lambda_{n+1}, \ n = 1, 2, \dots$$

be given, where $\Psi : [0,\infty) \to [0,\infty)$ is a strictly increasing function such that it is positive on $(0,\infty)$ with $\Psi(0) = 0$. Then $\lambda_n \to 0$, as $n \to \infty$.

8.9. Chidume and Zegeye (2003)

Let E be a real normed linear space. Suppose $A : E \to E$ is a uniformly quasi-accretive and uniformly continuous map . For arbitrary $x_1 \in E$ define the sequence $\{x_n\}$ iteratively by

$$x_{n+1} = x_n - \alpha_n A x_n, \ n \ge 1,$$

where $\lim_{n\to\infty} \alpha_n = 0$ and $\sum_{n=0}^{\infty} \alpha_n = \infty$. Then, there exists a constant $d_0 > 0$ such that if $0 < \alpha_n \le d_0$, the sequence $\{x_n\}$ converges strongly to the unique solution of the equation $Ax = 0$.

8.10. Moore and Nnoli (2001)

Let E be a real normed linear space and let $A : E \mapsto 2^E$ be a uniformly continuous and uniformly quasi-accretive multivalued operator with nonempty closed values such that the range of $(I - A)$ is bounded and the inclusion $0 \in Ax$ has a solution $x^* \in E$. Let $\{\alpha_n\}, \{\beta_n\} \subset [0, 1/2)$ be real sequences such that (i) $\lim_{n\to\infty} \alpha_n = \lim_{n\to\infty} \beta_n = 0$, and (ii) $\sum_{n=0}^{\infty} \alpha_n = \infty$. Then the sequence $\{x_n\}$ generated from an arbitrary $x_0 \in E$ by

$$y_n = (1 - \beta_n)x_n + \beta_n \xi_n, \quad \xi_n \in (I - A)x_n, \ n \ge 0,$$
$$x_{n+1} = (1 - \alpha_n)x_n + \alpha_n \eta_n, \quad \eta_n \in (I - A)y_n, \ n \ge 0,$$

converges strongly to x^* as $n \to \infty$.

8.11. Xu, H.K. (1991)

Prove that if E is a smooth Banach space, then the generalized duality mapping J_q is single-valued and

$$J_q(x) = \|x\|^{q-2} J(x), \ x \ne 0.$$

8.12. Xu, H.K. (1991)

Show that in a real q-uniformly smooth Banach space the following geometric inequality holds

$$\|x + y\|^q \le \|x\|^q + q \langle y, j_q(x) \rangle + c_q \|y\|^q,$$

for all $x, y \in E$ and some real constant $c_q > 0$.

9

Error Analysis of Fixed Point Iteration Procedures

Fixed point iteration procedures are mainly designed to be applied in solving concrete nonlinear operator equations, variational equations, variational inequalities etc.

In spite of the great diversity of the theoretical results obtained for the approximation of fixed points, briefly presented in Chapters 1-6 of this book, there is no systematic study of the numerical aspects related to the most recent iteration procedures: Mann, Ishikawa, Mann type and Ishikawa type.

Except for two or three papers by Rhoades [Rho76], [Rh77c] and [Rho91], this study was not systematically approached so far, even if, in some more recent papers, the author tried to draw the attention of researchers on this important numerical topic. This situation is not a natural thing and the incongruous unbalance between theoretical / numerical aspects in the field of approximation of fixed points must be changed at least by empirical studies, in those cases where theoretical results could not be obtained.

Even if Rhoades' opinion [Rho91]: "it is doubtful if any global statement can be made" (with respect to the study of the rate of convergence) should sound discouragingly for researchers, the poor existing results must be theoretically and empirically improved by further studies. The few results presented in Sections 9.2-9.5 could be a possible starting point to such approaches.

The opinion "more numerical work is required to gain additional insight into the [fixed point] iteration schemes", expressed by Rhoades [Rh77c] in an article published thirty years ago, is still valid nowadays.

It is the main aim of this chapter to present both theoretical and empirical results regarding the rate of convergence of the main fixed point iterative methods presented in the book. By comparing some important fixed point iterations, with respect to their rate of convergence, we will also be able to decide about the fastest method for some classes of contractive mappings.

9.1 Rate of Convergence of Iterative Processes

A fixed point theorem is valuable from a numerical point of view if it satisfies several requirements, amongst which we mention (see Rus [Ru79b]):

(a) it is able to provide an error estimate for the iterative process used to approximate the fixed point, and

(b) it can give concrete information on the stability of this procedure or, alternatively, on the data dependence of the fixed point.

As the second requirement was covered satisfactory in Chapter 7, it is the aim of this Chapter to briefly discuss some aspects related to the error estimate or to the rate of convergence of iterative methods.

Only a few fixed point theorems presented in this book do fulfill the two requirements above and, as it can be observed, the error estimate and data dependence of fixed points appear to have been given systematically mainly for *Picard iteration*, in conjunction with various contraction conditions.

Let (X, d) be a certain metric space and let $\{x_n\}_{n=0}^{\infty}$ be a given fixed point iteration that converges to x^*, a fixed point of the operator $T : X \to X$.

Since $x_n \to x^*$ as $n \to \infty$, it results that, for any $\varepsilon > 0$, there exists a positive integer N such that

$$d(x_n, x^*) < \varepsilon \text{ for } n \geq N. \tag{1}$$

If the rank N, depending on ε, on the initial guess x_0 and on the operator T itself, can be practically determined, then (1) serves as a *stopping criterion* for the iterative process.

Example 9.1. As shown by Theorem 2.1, if T is an a-contraction on a complete metric space, then both the *a priori* and the *a posteriori* error estimates

$$d(x_n, x^*) \leq \frac{a^n}{1-a} \cdot d(x_0, x_1), \quad n = 0, 1, 2, \ldots, \tag{2}$$

$$d(x_n, x^*) \leq \frac{a}{1-a} \cdot d(x_{n-1}, x_n), \quad n = 1, 2, \ldots \tag{3}$$

hold, where $\{x_n\}_{n=0}^{\infty}$ is the Picard iteration associated to the operator T, x_0 is the initial guess and x^* is the unique fixed point of T.

Since $0 < a < 1$, from (2), if $d(x_0, Tx_0) \neq 0$, we obtain

$$N = [\log_a(\varepsilon (1 - a) / d(x_0, Tx_0))],$$

where $[x]$ denotes the integer part of x.

This means that, when starting with the initial guess x_0, the N-th Picard iterate x_N approximates x^* with an error less than ε.

So, the *a priori* estimates (2) show how many iterations are needed in order to attain an ε-approximation of the fixed point x^*.

On the other hand, the estimates (3) directly provide a stopping criterion for the iterative process: if we want to obtain x^* with an error less than $\varepsilon > 0$, then we shall stop the iterations at the first step n for which the displacement of two successive iterates verifies

$$d(x_{n-1}, x_n) < \frac{\varepsilon(1-a)}{a}.$$

Together with (3), this guarantees that (1) is satisfied.

Remark. For the contraction mapping theorem (Theorem 2.1), (2) shows that the errors $d(x_n, x^*)$ are decreasing as rapidly as the terms of a geometric progression with ratio a, that is, $\{x_n\}_{n=0}^{\infty}$ converges to x^* at least as rapidly as the geometric series converges to its sum.

Definition 9.1. Let $\{a_n\}_{n=0}^{\infty}$, $\{b_n\}_{n=0}^{\infty}$ be two sequences of positive numbers that converge to a, respectively b. Assume there exists

$$l = \lim_{n \to \infty} \frac{|a_n - a|}{|b_n - b|}. \tag{4}$$

1) If $l = 0$, then it is said that the sequence $\{a_n\}_{n=0}^{\infty}$ converges to a *faster* than the sequence $\{b_n\}_{n=0}^{\infty}$ to b;

2) If $0 < l < \infty$, then we say that the sequences $\{a_n\}_{n=0}^{\infty}$ and $\{b_n\}_{n=0}^{\infty}$ *have the same rate of convergence.*

Remarks.

1) If $l = \infty$, then the sequence $\{b_n\}_{n=0}^{\infty}$ converges *faster* than $\{a_n\}_{n=0}^{\infty}$, that is $b_n - b = o(a_n - a)$.

The concept introduced by Definition 9.1 allows us to compare the rate of convergence of two sequences, and will be useful in the sequel;

2) The concept of rate of convergence given by Definition 9.1 is a relative one, while in literature there exist concepts of absolute rate of convergence, see Ortega and Rheinboldt [ORh70]. However, in the presence of an error estimate of the form (2) or (3), the concept given by Definition 9.1 is much more suitable.

Indeed, the estimate (2) shows that the sequence $\{x_n\}_{n=0}^{\infty}$ converges to x^* faster than any sequence $\{\theta^n\}$ to zero, where $0 < \theta < a$.

Suppose that for two fixed point iterations $\{x_n\}_{n=0}^{\infty}$, and $\{y_n\}_{n=0}^{\infty}$, converging to the same fixed point x^*, the following a priori error estimates

$$d(x_n, x^*) \le a_n, \quad n = 0, 1, 2, \ldots \tag{5}$$

and

$$d(y_n, x^*) \le b_n, \quad n = 0, 1, 2, \ldots \tag{6}$$

are available, where $\{a_n\}_{n=0}^{\infty}$ and $\{b_n\}_{n=0}^{\infty}$ are two sequences of positive real numbers (converging to zero). Then, in view of Definition 9.1, the following concept appears to be very natural.

Definition 9.2. If $\{a_n\}_{n=0}^{\infty}$ converges faster then $\{b_n\}_{n=0}^{\infty}$, then we shall say that the fixed point iteration $\{x_n\}_{n=0}^{\infty}$ *converges faster to* x^* than the fixed point iteration $\{y_n\}_{n=0}^{\infty}$ or, simply, that $\{x_n\}_{n=0}^{\infty}$ *is better* than $\{y_n\}_{n=0}^{\infty}$.

Remarks.

1) Rhoades [Rho76] considered that $\{x_n\}_{n=0}^{\infty}$ is better than $\{y_n\}_{n=0}^{\infty}$ if

$$d(x_n, x^*) \leq d(y_n, x^*), \quad \text{for all } n \in \mathbb{N},$$

see the next section, where some fixed point iteration procedures are compared with respect to the latter concept of rate of convergence.

2) In connection with Q- and/or R-order of convergence, see for example Ortega and Rheinboldt [ORh70], the estimates of the form

$$\| x_{n+1} - x^* \| \leq c \cdot \| x_n - x^* \|^p, \quad c > 0 \tag{7}$$

are precise indicators of the asymptotic rate of convergence of the iteration $\{x_n\}$ at x^*.

Estimates of the form (7) often arise naturally in the study of certain iterative methods, as, for example, the Newton's method, which is in fact a Picard iteration with a particular iteration mapping.

It is also possible to consider estimates of the form (7) in order to define relative concepts of convergence, similar to that in Definition 9.2, but with (5) and (6) derived from an estimation of the form (7).

For example, if T is an a-contraction, then in view of Theorem 2.1, we know that the rate of convergence is expressed by

$$d(x_n, x^*) \leq a \cdot d(x_{n-1}, x^*), \quad n = 1, 2, \ldots,$$

which shows that the convergence rate of the Picard iteration is *linear*.

9.2 Comparison of Some Fixed Point Iteration Procedures for Continuous Functions

It was shown in Section 3.3, Theorem 3.7, that in the class of Lipschitzian and generalized pseudocontractive selfmaps T of a nonempty closed convex subset of a real Hilbert space, we can compare the Picard and Krasnoselskij fixed point iterations with respect to their rate of convergence.

The remarks following Theorem 3.7 express basically (let s be the Lipschitzian constant and r the generalized pseudo-contractiveness constant of T) the fact that, for $s < 1$, the Picard iteration belongs to the family of Krasnoselskij iterations, known to converge to the unique fixed point of T.

Moreover, it is shown by Theorem 3.7 that the fastest Krasnoselskij iteration in that family

1) is faster than the Picard iteration if $r \neq s^2$
and
2) coincides to the Picard iteration, in the case $r = s^2$.

We start this section by presenting some comparison results for the Mann, Ishikawa and Picard iterations in the class of continuous maps.

Theorem 9.1. Let $f : [0,1] \to [0,1]$ be a continuous map, let $\{\alpha_n\}_{n=0}^{\infty}$ and $\{\beta_n\}_{n=0}^{\infty}$ be two sequences satisfying:

$(i) \quad 0 \leq \alpha_n, \beta_n \leq 1; \quad (ii) \quad \lim\limits_{n \to \infty} \alpha_n = 0; \quad (iii) \quad \sum\limits_{n=0}^{\infty} \alpha_n = \infty;$

$(iv) \quad \lim\limits_{n \to \infty} \beta_n = 0.$

Then the Ishikawa sequence given by $x_0 \in J = [0,1]$ and

$$x_{n+1} = (1 - \alpha_n)x_n + \alpha_n f[\beta_n f(x_n) + (1 - \beta_n)x_n], \quad n \geq 0 \qquad (8)$$

converges to a fixed point of f.

Proof. It is well known that f has at least one fixed point. Let's first show that $\{x_n\}_{n=0}^{\infty}$ converges.

The sequence $\{x_n\}$ is contained in $[0,1]$ so it has at least one limit point. For sake of contradiction, assume ξ_1, ξ_2 are two distinct limit points of $\{x_n\}$ and $\xi_1 < \xi_2$. We will show that, as a consequence of the previous assumption, we have $f(x) = x$, for every x in (ξ_1, ξ_2). Let $x^* \in (\xi_1, \xi_2)$.

If $f(x^*) > x^*$, then, by the continuity of the function f, there is a number $\delta \in (0, (x^* - \xi_1)/2)$ such that

$$|x - x^*| < \delta \quad \text{implies} \quad f(x) > x.$$

Since ξ_2 is a limit point of $\{x_n\}$, we can choose an integer N such that $x_N > x^*$ and $\beta_n < \delta/2$, $|x_{n+1} - x_n| < \delta/2$, for all $n \geq N$.

If $x_N \geq x^* + \delta/2$, then $x_{N+1} > x_N - \delta/2 \geq x^*$.

If $x_N < x^* + \delta/2$, then $f(x_N) > x_N$, so that

$$y_N = \beta_N f(x_N) + (1 - \beta_N)x_N > x_N > x^*.$$

Besides $y_N < \delta/2 + (1 - \beta_N)x_N < \delta/2 + x_N$, so that

$$|y_N - x^*| < \delta \quad \text{and} \quad f(y_N) > y_N.$$

Therefore $x_{N+1} - x_N = \alpha_N(f(y_N) - y_N) > 0$, and

$$x_{N+1} > x_N > x^*.$$

We obtain by induction that $x_N > x^*$, for $n \geq N$, contradicting that ξ_1 is a limit point. Similarly, $f(x^*) < x^*$ leads to the contradiction that ξ_2 is a limit point. Therefore every point in the interval (ξ_1, ξ_2) is a fixed point of f.

We will now show that ξ_1 and ξ_2 are not both limit points.

Notice that
$$x_n \notin (\xi_1, \xi_2), \quad \text{for all } n = 1, 2, \ldots$$

since, if $f(x_n) = x_n$ then, by (8), $x_m = x_n$, for all $m > n$ and neither ξ_1 nor ξ_2 could be limit points. Also, by the previous results, it follows that there is a number M such that if $x_M \geq \xi_2$, then $x_n \geq \xi_2 > \xi_1$, for all $n > M$ and ξ_1 is not a limit point. Similarly, if $x_M \leq \xi_1$, then $x_n \leq \xi_1 < \xi_2$, for all $n > M$ and ξ_2 is not a limit point. Either way, $\{x_n\}$ cannot have two distinct limit points. Therefore $\{x_n\}$ converges to its unique limit point, call it ξ.

Suppose $f(\xi) > \xi$. Since $x_n \to \xi$ and f is continuous, with $\varepsilon = (f(\xi) - \varepsilon)/2$ we can find a N such that $n > N$ implies $f(y_N) - x_N > \varepsilon$. Thus

$$\lim_{m \to \infty} (x_{N+m} - x_N) \geq \lim_{m \to \infty} \varepsilon \cdot \sum_{n=N}^{m-1+N} \alpha_n = \infty,$$

a contradiction to the fact that each $x_n \in I$.

The assumption $f(\xi) < \xi$ also leads to a contradiction, so that ξ is a fixed point of f. □

Remark. For nondecreasing functions the hypotheses of Theorem 9.1 can be weakened as bellow.

Theorem 9.2. *Let $f : [0,1] \to [0,1]$ be continuous and nondecreasing, $\{\alpha_n\}_{n=0}^{\infty}$ and $\{\beta_n\}_{n=0}^{\infty}$ satisfying (i) and (iii) in Theorem 9.1.*
Then $\{x_n\}$ given by (8) converges to a fixed point of f.

Proof. Let m, M denote, respectively, the infimum and supremum of the set of fixed points of f in J. For $0 \leq x \leq m$ we get $f(x) > x$, while for $M < x \leq 1$ we get $f(x) < x$.

If p and q are fixed points of f satisfying $m \leq p < q \leq M$ and $f(x) \neq x$ for $x \in (p, q)$, then $f(x) - x$ has constant sign in the interval (p, q). These facts, along with the monotonicity of f, force $\{x_n\}$ to be a monotonic sequence, hence convergent. It remains to show that $\{x_n\}$ tends to a fixed point of f.

Suppose first that $x_0 > M$. Then $\{x_n\}$ is decreasing, $x_n \geq M$ for each n, $\{f(x_n)\}$ is decreasing and $x_n > f(x_n)$ for each n. Thus

$$f(x_n) < y_n = \beta_n f(x_n) + (1 - \beta_n)x_n < x_n.$$

Let $l = \lim_{n \to \infty} x_n$. Then $f(l) = l$. Assume $l > f(l)$. Then $f(l) > f(f(l)) = f^2(l)$, which implies $l > f^2(l)$. Set $\varepsilon = (l - f^2(l))/2$. There exists an integer N such that $x_n - f(y_n) > \varepsilon$ for all $n \geq N$. Hence

$$x_N - x_{N+m} > \varepsilon \sum_{n=N}^{m-1+N} \alpha_n \to \infty,$$

a contradiction. Therefore $l = f(l)$.

For the other choices of x_0, the proof is similar. □

Remark. For any function f, the initial guess x_0 determines which fixed point of f the sequence $\{x_n\}$ will converge. Thus, for some nondecreasing functions f with three distinct fixed points p, q, r satisfying

$$0 \leq p < q < r \leq 1,$$

then $x_0 \in [0, q)$ implies $x_n \to p$, whereas $x_0 \in (q, 1]$ implies $x_n \to r$.

The fixed points p and r are *attractive* fixed points, while q is a *repulsive* fixed point, since the sequence $\{x_n\}$ never converges to q unless $x_0 = q$.

Example 9.1. For $f(x) = 2x^3 - 7x^2 + 8x - 2$ we have $F_T = \{1/2, 1, 2\}$ and only 1 is an attractive fixed point of f.

Definition 9.3. If $\{x_n\}$, $\{z_n\}$ are two iteration schemes which converge to the fixed point p, we shall say that $\{x_n\}$ is *better* than $\{z_n\}$ if

$$|x_n - p| \leq |z_n - p|, \quad \text{for all } n.$$

Theorem 9.3. *Let f, $\{\alpha_n\}$ and $\{\beta_n\}$ satisfying the hypotheses of Theorem 9.2. Then*

(a) $\{x_n\}$ given by (8) is better than $\{z_n\}$ given by $z_0 = x_0$ and

$$z_{n+1} = \alpha_n f(z_n) + (1 - \alpha_n)z_n, \quad n \geq 0. \tag{9}$$

(b) If $w_0 > z_0$, then $w_{n+1} \geq x_{n+1}$, for each n,
where $x_{n+1} = I(x_0, \alpha_n, \beta_n, f)$ and $z_{n+1} = I(z_0, \alpha_n, \beta_n, f)$.
(c) If $\{\gamma_n\}$ satisfies $\beta_n \leq \gamma_n \leq 1$ for each n and $\{t_n\}$ is given by $t_{n+1} = I(x_0, \alpha_n, \gamma_n, f)$, then $\{t_n\}$ is better than $\{x_n\}$.
(d) If $\{\delta_n\}$ satisfies $\alpha_n \leq \delta_n \leq 1$ for each n and $\{z_n\}$ is given by $z_{n+1} = I(x_0, \delta_n, \gamma_n, f)$, then $\{z_n\}$ is better than $\{x_n\}$.

Proof. We shall consider the case $x_0 > M$, where M is defined in the proof of Theorem 9.2 (the other cases are proved similarly).

(a) Let $y_n = (1 - \beta_n)x_n + \beta_n f(x_n)$. As

$$z_1 - x_1 = \alpha_0(f(z_0) - f(y_0)),$$

from $x_0 > M$ we obtain $f(x_0) < x_0$ and hence $y_0 < x_0$. Thus

$$f(y_0) < f(x_0) = f(z_0) \text{ and therefore } z_1 > x_1.$$

Assume now $z_n > x_n$. Then

$$z_{n+1} - x_{n+1} = \alpha_n(f(z_n) - f(y_n)) + (1 - \alpha_n)(z_n - x_n),$$

and so $x_0 > M$ implies $x_n > M$. This means $f(x_n) < x_n$ and hence $y_n < x_n$, which leads to the desired conclusion.

(b) The proof is immediate.

(c) Let $\overline{y}_n = \gamma_n f(t_n) + (1 - \gamma_n)t_n$. Then $x_0 > M$ implies that $\{\gamma_n\}$, $\{t_n\}$ are monotone decreasing in n and x_n, and $t_n \geq M$ for all n. Then

$$x_1 - t_1 = \alpha_0(f(y_0) - f(\overline{y}_0)) \text{ and } y_0 - \overline{y}_0 = (\gamma_0 - \beta_0)(x_0 - f(x_0)) \geq 0,$$

and hence $x_1 \geq t_1$. Assume $x_n \geq t_n$. We have

$$x_{n+1} - t_{n+1} \geq \alpha_n(f(y_n) - f(\overline{y}_n)),$$

and the conclusion follows by

$$\begin{aligned} y_n - \overline{y}_n &= (x_n - t_n) + \beta_n(f(x_n) - x_n) + \gamma_n(t_n - f(t_n)) \geq \\ &\geq (x_n - t_n) + \beta_n(f(x_n) - x_n) + \beta_n(t_n - f(t_n)) = \\ &= (1 - \beta_n)(x_n - t_n) + \beta_n(f(x_n) - f(t_n)) \geq 0. \end{aligned}$$

(d) From $x_0 > M$ we get that $\{x_n\}$ and $\{z_n\}$ are monotone decreasing to M. If we denote

$$\overline{y}_n = \beta_n f(z_n) + (1 - \beta_n)z_n,$$

then

$$x_1 - z_1 = \alpha_0 f(y_0) - \delta_0 f(\overline{y}_0) + (\delta_0 - \alpha_0)x_0 ; \ f(y_0) = f(\overline{y}_0)$$

and $f(x_0) < x_0$, hence $x_1 > z_1$. Assume $x_n > z_n$. Then

$$x_{n+1} - z_{n+1} = (x_n - z_n) + \alpha_n(f(y_n) - y_n) + \delta_n(x_n - f(\overline{y}_n))$$

and $z_n > M$ implies $f(z_n) < z_n$. Therefore $\overline{y}_n < z_n$, which implies $f(\overline{y}_n) < f(z_n)$. Thus $z_n - f(\overline{y}_n) > z_n - f(z_n) > 0$ and

$$x_{n+1} - z_{n+1} \geq x_n - z_n + \alpha_n(f(y_n) - y_n) + \alpha_n(z_n - f(\overline{y}_n)) =$$

$$= (1 - \alpha_n)(x_n - z_n) + \alpha_n(f(y_n) - f(\overline{y}_n)),$$

which shows that $x_{n+1} \geq z_{n+1}$.   □

Remarks.
1) Part (a) in Theorem 9.3 shows that the Ishikawa iteration is better than the Mann iteration;
2) Part (b) shows that the closer the initial guess x_0 is to a fixed point, the better the Ishikawa iteration is;
3) Part (c) and (d) in Theorem 9.3 show that the larger α_n, β_n, the better the iteration scheme is. Since there is an optimum choice, i.e., $\alpha_n = \beta_n = 1$, this shows that the best scheme amongst the Ishikawa iterations (8) for increasing functions is the Picard iteration;
4) For decreasing functions on $[0,1]$ there is no best scheme but, as shown in Section 9.6, some empirical comparisons can however be done.

9.3 Comparing Picard, Krasnoselskij and Mann Iterations in the Class of Lipschitzian Generalized Pseudocontractions

As we proved in Chapter 3, Theorem 3.7, amongst all Krasnoselskij iterations associated to a Lipschitzian generalized pseudocontractive operator T, with $\lambda \in (0, a)$, where a is given by relation (10), there exists one iteration method which is the fastest with respect to the concept of rate of convergence given by Definition 9.2.

Reinterpreting this result, see also the Remarks given after Theorem 3.7, we can say that if $r \neq L^2$, where r and L are the constants of generalized pseudocontractivity, and the Lipschitz constant of T, respectively, then the fastest Krasnoselskij iteration in that family, converges faster than Picard iteration to the unique fixed point of T. The main result of this section compares Krasnoselskij and Mann iterations for the class of mappings mentioned above.

Theorem 9.4. *Let H be a real Hilbert space and K be a nonempty closed convex subset of H. Let $T : K \to K$ be a Lipschitzian and generalized pseudocontractive operator with corresponding constants $L \geq 1$ and $0 < r < 1$.*
Then:
1) T has a unique fixed point p in K;
2) For any $x_0 \in K$ and $\lambda \in (0, a)$, with a given by

$$a = 2(1 - r)/(1 - 2r + L^2),\qquad(10)$$

the Krasnoselskij iteration $\{x_n\}_{n=0}^{\infty} = K(x_0, \lambda, T)$ converges strongly to p;
3) For any $y_0 \in K$ and $\{\alpha_n\}_{n=0}^{\infty}$ in $[0, 1]$ satisfying

$$\sum_{n=1}^{\infty} \alpha_n = \infty,\qquad(11)$$

the Mann iteration $\{y_n\}_{n=0}^{\infty} = M(y_0, \alpha_n, T)$ converges strongly to p;
4) For any Mann iteration converging to p, with $0 \leq \alpha_n \leq b < 1$, there exists a Krasnoselskij iteration that converges faster to p.

Proof. Conclusions 1) and 2) follows by Theorem 3.6 in Section 3.3. Consider now, for all $\lambda \in [0, 1]$, the operator T_λ on K given by

$$T_\lambda x = (1 - \lambda)x + \lambda T x, \quad x \in K.$$

Since $\lambda < a$, it was proved in Section 3.3 that we have

$$\|T_\lambda x - T_\lambda y\| \leq \theta \cdot \|x - y\|, \quad \text{for all } x, y \text{ in } K,\qquad(12)$$

where $0 < \theta = \left[(1 - \lambda)^2 + 2\lambda(1 - \lambda)r + \lambda^2 L^2\right]^{1/2} < 1$.

3) Let $\{y_n\}_{n=0}^{\infty} = M(y_0, \alpha_n, T)$ be the Mann iteration, with the sequence $\{\alpha_n\}_{n=0}^{\infty} \subset [0, 1]$ satisfying (11). Consider t, $0 < t < 1$, and denote

$$a_n = \frac{1}{t}\alpha_n, n = 0, 1, 2, \dots .$$

Then the Mann iteration will be given by

$$y_{n+1} = (1 - ta_n)y_n + ta_n Ty_n , \quad n = 0, 1, 2, \dots .$$

Let p be the unique fixed point of T. We have

$$\|y_{n+1} - p\| = \|(1 - a_n)y_n + a_n[(1 - t)y_n + tTy_n] - p\| \leq$$
$$\leq (1 - a_n)\|y_n - p\| + a_n\|(1 - t)(y_n - p) + t(Ty_n - Tp)\|. \quad (13)$$

Using the properties of T we find that

$$\|t(Ty_n - Tp) + (1 - t)(y_n - p)\|^2 = (1 - t)^2\|y_n - p\|^2 +$$
$$+ 2t(1 - t)\langle Ty_n - Tp, y_n - p\rangle + t^2\|Ty_n - Tp\|^2 \leq$$
$$\leq (1 - t)^2\|y_n - p\|^2 + 2t(1 - t)r\|y_n - p\|^2 + t^2L^2\|y_n - p\|^2 =$$
$$= [(1 - t)^2 + 2t(1 - t)r + t^2L^2]\|y_n - p\|^2. \quad (14)$$

By (13) and (14) we get

$$\|y_{n+1} - p\| \leq \left\{1 - a_n + a_n[(1 - t)^2 + 2t(1 - t)r + t^2L^2]^{1/2}\right\} \cdot \|y_n - p\|$$

$$= \left(1 - (1 - \theta)a_n\right)\|y_n - p\| \leq \prod_{k=1}^{n}\left(1 - (1 - \theta)a_k\right)\|y_1 - p\|, \quad (15)$$

where

$$0 \leq \theta = [(1 - t)^2 + 2t(1 - t)r + t^2L^2]^{1/2} < 1 ,$$

for all t satisfying $0 < t < 2(1 - r)/(1 - 2r + L^2)$.

Since, by (11), $\sum\limits_{n=0}^{\infty} \alpha_n$ diverges, it follows that $\sum\limits_{n=0}^{\infty} a_n$ diverges, too, and in view of the inequality $\theta < 1$ we get

$$\lim_{n \to \infty} \prod_{k=1}^{n} [1 - (1 - \theta)a_k] = 0 ,$$

which by (15) shows that $\{y_n\}$ converges strongly to p.

4) Take $x := x_n$, $y := x_{n-1}$ in (12) to obtain

$$\|x_{n+1} - x_n\| \leq \theta \cdot \|x_n - x_{n-1}\|,$$

which inductively yields $\|x_{n+1} - x_n\| \leq \theta^n\|x_1 - x_0\|$ and then by triangle rule we obtain

$$\|x_{n+k} - x_n\| \leq \theta^n \left(1 + \theta + \cdots + \theta^{k-1}\right) \|x_1 - x_0\|, \tag{16}$$

valid for all $n, k \in \mathbb{N}^*$.

Now letting $k \to \infty$ in (16), we get

$$\|x_n - p\| \leq \frac{\theta^n}{1 - \theta} \|x_1 - x_0\|. \tag{17}$$

Therefore, in view of Definition 9.2, and of previous estimations (16) and (17), in order to compare the Krasnoselskij and Mann iterations, we have to compare

$$\theta^n \quad \text{and} \quad \prod_{k=1}^{n} \left[1 - (1 - \theta)a_k\right].$$

Let $\{y_n\}_{n=0}^{\infty}$ be a certain Mann iteration converging to p, with $\{\alpha_n\}_{n=0}^{\infty}$ satisfying $0 \leq \alpha_n \leq b < 1$. Then $a_k = \alpha_k/t \leq b/t$ (denote b/t by b) and for any $m, 0 < m < 1$, we may find $\theta \in (0, 1)$ such that

$$b(1 - \theta) < 1 - \frac{\theta}{m}.$$

Indeed, to this end it is enough to take $\theta < \dfrac{m(1 - b)}{1 - mb}$. Using the fact that $a_k \leq b$, it results

$$\frac{\theta}{1 - (1 - \theta)a_k} \leq m < 1, \quad \text{for all } k = 1, 2, \ldots,$$

which shows that

$$\lim_{n \to \infty} \frac{\theta}{\displaystyle\prod_{k=1}^{n} \left[1 - (1 - \theta)a_k\right]} \leq \lim_{n \to \infty} m^n = 0,$$

so the Krasnoselskij iteration $\{x_n\}_{n=0}^{\infty} = K(x_0, \theta, T)$ converges faster than the considered Mann iteration, $\{y_n\}_{n=0}^{\infty} = M(y_0, \alpha_n, T)$.

To end the proof we still need to show that the interval $(0, a)$, with a given by (10), and the interval $\left(0, \dfrac{m(1 - b)}{1 - mb}\right)$ have nonempty intersection.

But this is immediate, because, under the hypotheses of the theorem, $0 < \dfrac{m(1 - b)}{1 - mb} < 1$ and $0 < a = \dfrac{2(1 - r)}{1 - 2r + L^2} \leq 1$. $\qquad\square$

Remark.

Part 4) in Theorem 9.4 shows that, in order to approximate the fixed point of a Lipschitzian and generalized pseudo-contractive operator T, it is always more convenient to use a certain Krasnoselskij iteration in the family $\{x_n\}_{n=0}^{\infty}$ given by

$$x_{n+1} = (1 - \lambda)x_n + \lambda T x_n, \quad n = 0, 1, 2, \ldots,$$

with $\lambda \in (0, a)$ and a given by (10).

9.4 Comparing Picard, Mann and Ishikawa Iterations in a Class of Quasi Nonexpansive Maps

We know from the previous chapters that in the class of Zamfirescu operators all important fixed point iterative methods, i.e., Picard iteration (Theorem 2.4), Mann iteration (Theorem 4.10), Ishikawa iteration (Theorem 5.6) and, in particular, Krasnoselskij iteration, are convergent to the unique fixed point of such an operator.

In such situations, it is of theoretical and practical importance to compare these methods in order to establish, if possible, which one converges faster to the unique fixed. The method we shall find, if any, should be preferentially used in applications in order to approximate the fixed points.

The next theorem compares Picard and Mann iterations in the class of Zamfirescu operators.

Theorem 9.5. *Let E be a uniformly convex Banach space, K a closed convex subset of E, and $T : K \to K$ a Zamfirescu operator, i.e., an operator that satisfies (z_1)-(z_3) in Theorem 2.4. Let $\{x_n\}_{n=0}^{\infty}$ be the Picard iteration associated with T and $x_0 \in K$, given by $x_{n+1} = Tx_n$, and $\{y_n\}_{n=0}^{\infty}$ be the Mann iteration given by $y_0 \in K$ and*

$$y_{n+1} = (1 - \alpha_n)y_n + \alpha_n Ty_n , \quad n = 0, 1, 2, \ldots$$

where $\{\alpha_n\}_{n=0}^{\infty}$ is a sequence satisfying

$$(i) \ \alpha_1 = 1; \quad (ii) \ 0 \leq \alpha_n < 1 , \text{ for } n \geq 1; \quad (iii) \ \sum_{n=0}^{\infty} \alpha_n(1 - \alpha_n) = \infty .$$

Then:

1) T has a unique fixed point in E, i.e., $F_T = \{p\}$;

2) The Picard iteration $\{x_n\}$ converges to p for any $x_0 \in K$;

3) The Mann iteration $\{y_n\}$ converges to p for any $y_0 \in K$ and $\{\alpha_n\}$ satisfying (i) - (iii);

4) Picard iteration is faster than any Mann iteration.

Proof. Conclusions 1) - 3) follow by Theorems 2.4 and Theorem 4.10;

4) First of all, we remind, see the proofs of Theorems 2.4 and 4.10, that any Zamfirescu operator satisfies

$$\|Tx - Ty\| \leq \delta \cdot \|x - y\| + 2\delta \cdot \|x - Tx\| , \tag{18}$$

$$\|Tx - Ty\| \leq \delta \cdot \|x - y\| + 2\delta \cdot \|y - Tx\| , \tag{19}$$

for all $x, y \in K$, where δ is given by

$$\delta = \max\left\{\alpha, \frac{\beta}{1 - \beta}, \frac{\gamma}{1 - \gamma}\right\}, \tag{20}$$

and α, β, γ are the contractiveness constants appearing in $(z_1) - (z_3)$. By taking $y := x_n$; $x := p$ in (18) we obtain

$$\|x_{n+1} - p\| \leq \delta \cdot \|x_n - p\|,$$

which inductively yields

$$\|x_{n+1} - p\| \leq \delta^n \cdot \|x_1 - p\|, \quad n \geq 0. \tag{21}$$

Now let $y_0 \in K$ and $\{y_n\}_{n=0}^{\infty}$ be the Mann iteration associated with T, y_0 and the sequence $\{\alpha_n\}$. Then by the definition of Mann iteration we have:

$$\|y_{n+1} - p\| = \|(1 - \alpha_n)y_n + \alpha_n T y_n - [(1 - \alpha_n) + \alpha_n]p\| \leq$$
$$\leq (1 - \alpha_n)\|y_n - p\| + \alpha_n\|T y_n - p\|.$$

Using again (18), this time with $y := y_n$; $x := p$ we get

$$\|T y_n - p\| \leq \delta \cdot \|y_n - p\|$$

and therefore

$$\|y_{n+1} - p\| \leq \left[1 - \alpha_n + \delta\alpha_n\right] \cdot \|y_n - p\|, \quad n = 0, 1, 2, \ldots,$$

which implies that

$$\|y_{n+1} - p\| \leq \prod_{k=1}^{n} \left[1 - \alpha_k + \delta\alpha_k\right] \cdot \|y_1 - p\|, \quad n = 0, 1, 2, \ldots. \tag{22}$$

By (ii), (iii) and the inequality

$$\alpha_n(1 - \alpha_n) < \alpha_n,$$

we obtain that $\sum\limits_{n=0}^{\infty} \alpha_n = \infty$ which implies

$$\prod_{k=1}^{n}(1 - \alpha_k + \delta\alpha_k) \to 0 \text{ as } n \to \infty.$$

Therefore, in view of (21) and (22), in order to compare $\{x_n\}$ and $\{y_n\}$, we must compare the sequences $a_n = \delta^n$ and $b_n = \prod\limits_{k=1}^{n}(1 - \alpha_k + \delta\alpha_k)$. Denote $c_n = a_n/b_n$. Since

$$\frac{c_{n+1}}{c_n} = \frac{\delta}{1 - (1 - \delta)\alpha_{n+1}} < 1,$$

which, by the ratio test implies that $\sum\limits_{n=0}^{\infty} c_n$ converges, we conclude that

$$\lim_{n\to\infty} c_n = \lim_{n\to\infty} \frac{\delta^n}{\prod_{k=1}^{n}(1-\alpha_k+\delta\alpha_k)} = 0.$$

This shows that Picard iteration converges faster than the Mann iteration. $\square$

Remarks.
1) Theorem 9.5 shows that, to efficiently approximate fixed points of Zamfirescu operators, one should always use Picard iteration;
2) The uniform convexity of E is not necessary for the conclusion of Theorem 9.5 to hold, as shown by the next theorem, which also assumes weaker conditions on the sequence $\{\alpha_n\}$.

Theorem 9.6. *Let E be an arbitrary Banach space, K a closed convex subset of E, and $T : K \to K$ an operator satisfying Zamfirescu's conditions. Let $\{y_n\}_{n=0}^{\infty}$ be the Mann iteration associated to T, $y_0 \in K$, and sequence $\{\alpha_n\}$ with $\{\alpha_n\} \subset [0,1]$ satisfying*

$$(iv) \qquad \sum_{n=0}^{\infty} \alpha_n = \infty.$$

Then $\{y_n\}_{n=0}^{\infty}$ converges strongly to the fixed point of T and, moreover, Picard iteration $\{x_n\}_{n=0}^{\infty}$ defined by $x_0 \in K$, converges faster than the Mann iteration.

Proof. We proceed similarly to the proof of Theorem 9.5. $\square$

Remark.
Condition (iv) in Theorem 9.6 is weaker than conditions (i) - (iii) in Theorems 9.5. Indeed, in view of the inequality

$$0 < \alpha_k(1 - \alpha_k) < \alpha_k,$$

valid for all α_k satisfying (i) - (ii), condition (iii) implies (iv).
There also exist values of $\{\alpha_n\}$, e.g., $\alpha_n \equiv 1$, such that (iv) is satisfied but (iii) is not.
Using the same arguments as in proving the previous two theorems, we can compare Mann and Ishikawa iterations in the same class of mappings.

Theorem 9.7. *Let E be an arbitrary Banach space, K be a closed convex subset of E, and $T : K \to K$ be a Zamfirescu operator, that is, an operator that satisfies (z_1)-(z_3) in Theorem 2.4. Let $\{x_n\}$ be the Mann iteration defined by $x_0 \in K$ and $\{\alpha_n\} \subset (0,1)$ satisfying (iv); $\{y_n\}$ be the Ishikawa iteration defined by $y_0 \in K$ and $\{\alpha_n\}, \{\beta_n\}$ satisfying $0 \le \alpha_n, \beta_n < 1$ and (iv).*

Then $\{x_n\}$ and $\{y_n\}$ converges strongly to the unique fixed point of T and, moreover, the Mann iteration converges faster than Ishikawa iteration.

9.5 The Fastest Krasnoselskij Iteration for Approximating Fixed Points of Strictly Pseudo-Contractive Mappings

Let X be a Banach space, K a nonempty closed convex subset of X and $T : K \to K$ a Lipschitzian strictly pseudocontractive mapping. In Chapter 4, Corollaries 4.2 and 4.3, we showed that, in order to approximate the fixed point of T, instead of the Mann iteration, usually considered by many authors, we may use a simpler method, i.e., the Krasnoselskij iterative process.

It is the main aim of this section to show that amongst all Krasnoselskij iterations that converge to the fixed point of such operators, we may select the *fastest* iteration, in some sense. This is indeed a very important achievement in view of concrete applications of fixed point iteration procedures.

The results in this Section open a new important direction of investigation: to analyze all convergence theorems for Mann iteration, Mann-type iteration etc. based on condition (23), in order to decide whether or not this assumption is indeed necessary for the convergence of that iteration and, secondly, to investigate if Krasnoselskij iteration could really replace Mann iteration for those classes of operators.

There are a lot of recent papers in literature devoted to obtaining convergence theorems for the Mann iteration, see Chapter 4 and the list of references in this book, but, as we have seen, the great majority of them are obtained by imposing the following sharp condition on the sequence $\{\alpha_n\}$:

$$\lim_{n \to \infty} \alpha_n = 0. \tag{23}$$

As pointed out in Section 9.7 and also shown by Example 9.2 (or Example 4.3), in most cases condition (23) is not necessary for the convergence of Mann iteration and appears to be an artificial assumption, being tributary to the technique of proof used by the authors.

Example 9.2. Let $X = \mathbb{R}$ with the usual norm, $K = \left[\frac{1}{2}, 2\right]$ and $T : K \to K$ be a function given by $Tx = \frac{1}{x}$, for all x in K. Then:

(a) T is Lipschitzian with constant $L = 4$;

(b) T is strictly pseudocontractive, see Example 4.3 for details;

(c) $Fix\,(T) = \{1\}$, where $Fix\,(T) = \{x \in K|\ Tx = x\}$;

(d) The Picard iteration associated to T does not converge to the fixed point of T, for any $x_0 \in K \setminus \{1\}$;

(e) The Krasnoselskij iteration associated to T converges to the fixed point $p = 1$, for any $x_0 \in K$ and $\lambda \in (0, 1/16)$;

(f) The Mann iteration associated to T with $\alpha_n = \dfrac{n}{2n+1}$, $n \geq 0$ and $x_0 = 2$ converges to 1, the unique fixed point of T (see Example 9.3).

However, $\alpha_n \nearrow \dfrac{1}{2}$ as $n \to \infty$ and so condition (23) is not satisfied.

As we argued in the previous sections of this chapter, when two or more iterative methods are available in order to approximate fixed points of mappings in a certain class, from a computational point of view it is natural to choose a simpler method, when known, in order to avoid complicated computations. On the other hand, it is clear that Krasnoselskij iteration method defined by the initial guess $x_0 \in K$ and

$$x_{n+1} = (1 - \lambda)x_n + \lambda T x_n, \ n \geq 0 \text{ where } \lambda \in [0, 1], \qquad (24)$$

is computationally simpler than the Mann iteration defined by $x_0 \in K$ and

$$x_{n+1} = (1 - \alpha_n)x_n + \alpha_n T x_n, \ n \geq 0,$$

where $\{\alpha_n\}$ is a sequence of real numbers in $[0, 1]$.

Starting from the fact that many papers that were published in the last decade are devoted to the approximation of fixed points of several classes of mappings that include nonexpansive mappings, in Chapter 4, Corollaries 4.2 and 4.3, we showed that, in the case of Lipschitzian strictly pseudo-contractive operators, the Krasnoselskij iteration suffices to approximate fixed points.

By Corollary 4.3, we practically obtain a family $\{x_n^\lambda\}$, $\lambda \in (0, a)$, of Krasnoselskij iterative processes such that each of them could be used to approximate the fixed point p.

A natural question then arises: which Krasnoselskij iteration from the above family, i.e., which λ, would be more suitable to be considered in order to obtain the better method, if any ?

The answer is given by Theorem 9.8. To state it, we use the concept of rate of convergence introduced by Definition 9.2.

Theorem 9.8. *Let X be a Banach space and K a nonempty closed convex subset of X. If $T : K \to K$ is a Lipschitzian (with constant L) and strongly pseudo-contractive operator (with constant k) such that the fixed point set of T, $Fix(T)$, is nonempty, then the Krasnoselskij iteration $\{x_n\} \subset K$ generated by $x_1 \in K$ and (24), with $\lambda \in (0, a)$ and the number a given by*

$$a = \frac{k}{(L + 1)(L + 2 - k)},$$

converges strongly to the (unique) fixed point p of T. Moreover, amongst all Krasnoselskij iterations (24), there exists one which is the fastest one. It is obtained for

$$\lambda_0 = -1 + \sqrt{1 + a}.$$

Proof. We mainly use the arguments presented in the proof of Theorem 4.12.

The proof is now elementary: we have to find λ for which the function

$$q(\lambda) = \frac{1 + (1 - k)\lambda + (L + 1)(L + 2 - k)\lambda^2}{1 + \lambda}$$

attains its minimum value when $\lambda \in (0, a)$, if any. Since $q'(\lambda) = 0$ is equivalent to $\lambda^2 + 2\lambda - a = 0$, we find that $\lambda_0 = -1 + \sqrt{1 + a} \in (0, a)$ is the required value of λ. Then, for any $\lambda \in (0, a)$, $\lambda \neq \lambda_0$, we have $\dfrac{q(\lambda_0)}{q(\lambda)} < 1$ and hence

$$\lim_{n \to \infty} \left(\frac{q(\lambda_0)}{q(\lambda)} \right)^n = 0,$$ which shows that $\{x_n^{\lambda_0}\}$ converges faster than $\{x_n^{\lambda}\}$ to the unique fixed point of T. $\qquad \square$

Remark.

Theorem 9.8 shows that, to efficiently approximate fixed points of Lipschitzian and strictly pseudo-contractive operators, one should always use Krasnoselskij iteration (24) and, more specifically, the one obtained for $\lambda_0 = -1 + \sqrt{1 + a}$.

It is a current tendency in the field of iterative approximation of fixed points to consider more and more complicated fixed point iteration procedures: Ishikawa iteration, Ishikawa iteration with errors, modified Ishikawa iteration etc., see Berinde [Be02c].

Except for some isolated cases, like the case of Lipschitzian pseudo-contractive operators (see Theorem 5.1 in Chapter 5), when it was indeed necessary to consider Ishikawa iteration in order to approximate their fixed points, the use of these complicated iteration procedures is not motivated from a numerical point of view and is not suitable for concrete applications. At most a weak theoretical interest could motivate the numerous papers devoted to this direction of research that appeared in the last decade.

Concluding this Section, at least three problems arise:

1. Give an example, if any, of an operator T for which some Mann iteration converges and no Krasnoselskij iteration converges to the fixed point(s) of T;

2. Try to transpose known convergence results for Mann iteration based on condition (23), to Krasnoselskij iteration, whatever possible;

3. There are recent papers, we quote here Rhoades and Soltuz [RS03a-e], which prove that, for several classes of mappings, Mann iteration is actually equivalent to the more complicated Ishikawa iteration, in the sense that, under certain circumstances, Mann iteration converges (to the fixed point) if and only if Ishikawa converges, too. The challenging problem is then: are Krasnoselskij iteration and Mann iteration equivalent in this sense, for large classes of mappings ?

4. The results regarding the equivalence of fixed point iteration procedures, mentioned before, are actually obtained under a very restrictive assumption (see also Section 5.5): it is always assumed that the initial guesses are identical for all iterations.

A more challenging problem would then be to establish equivalence of various fixed point iteration procedures, without imposing the above restriction.

9.6 Empirical Comparison of Some Fixed Point Iteration Procedures

If, for a given class of mappings, two or more fixed point iteration schemes converge and no analytical information on their rate of convergence is available, then it is of interest, for computational reasons, to know at least empirically, which of these processes appears to be the most efficient.

Let us consider the Mann iteration scheme,

$$x_0 \in [0, 1]$$

$$x_{n+1} = (1 - c_n)x_n + c_n f(x_n), \ n \geq 0,$$

with

$$c_n = [(n + 1)(n + 2)]^{-1/k}, \ k \in \{3, 4, \dots, 8\}$$

for the decreasing functions

$$f(x) = 1 - x^m, \quad g(x) = (1 - x)^m, \quad 1 \leq m \leq 6.$$

The fixed point for each function was first found by the bisection method, accurate to 10 places. Both Mann and Newton-Raphson iteration schemes were used to find each fixed point to within 8 places, using the initial guesses $x_0 = 0.1; \ 0.2; \dots; 0.9$, respectively.

The output of the computations leads to the following observations:

1) Newton converges faster than Mann. This is not surprising, since Mann converges linearly, while Newton is a quadratic method for f smooth enough;

2) However, while Newton converges more rapidly for x_0 near the fixed point, Mann iteration appears to converge somewhat independently of the initial guess. For example, with $m = 4$ or $m = 6$, $k = 4$, Mann scheme converges to the fixed point of f in exactly 8 iterations, for each choice of x_0;

3) The most efficient choice of k is 5, for $m < 3$, and 4, for $m \geq 3$. The number of iterations required increases with the distance from k to 4 or 5.

For f with $m = 2, 3, 4$; $k = 2$ and $x_0 = 0.9$, the Mann scheme needed 400 iterations to find the fixed point accurate to 5 places.

In order to offer a more detailed empirical study of the main fixed point iterative procedures, we designed a program whose input is a certain function, the specific iteration parameters and the initial guess from which to start, and which produces as output a number of iterates, depending on the stopping criterion adopted. The most significant results are given in the following.

Example 9.3. For the decreasing function T in Examples 3.1-3.3 and 9.2, the execution of the program FIXPOINT for some input data leads to the following observations:

1) The Krasnoselskij iteration converges to $p = 1$ for any $\lambda \in (0, 1)$ and any initial guess x_0 (recall that the Picard iteration does not converge for any initial value $x_0 \in [1/2, 2]$ different from the fixed point).

The convergence is slow for λ close enough to 0 (that is, for Krasnoselskij iterations close enough to the Picard iteration) or close enough to 1. The closer to $1/2$, the middle point of the interval $(0,1)$, λ is, the faster it converges.

For $\lambda = 0.5$ the Krasnoselskij iteration converges very fast to $p = 1$, the unique fixed point of T. For example, starting with $x_0 = 1.5$, only 4 iterations are needed in order to obtain p within 6 places: $x_1 = 1.08335$, $x_2 = 1.00325$, $x_3 = 1.000053$, $x_4 = 1$. (Compare these results to that in Example 3.3).

For the same value of λ and $x_0 = 2$, again only 4 iterations are needed to obtain p with the same precision, even though the initial guess is not very close to the fixed point: $x_1 = 1.25$, $x_2 = 1.025$, $x_3 = 1.0003$ and $x_4 = 1$;

2) The speed of Mann and Ishikawa iterations also depends on the position of $\{\alpha_n\}$ and $\{\beta_n\}$ in the interval $(0,1)$.

If we take $\alpha_n = 1/(n+1)$, $\beta_n = 1/(n+2)$ and start with the initial guess $x_0 = 1.5$, then the Mann and Ishikawa iterations converge (slowly) to $p = 1$: after $n = 35$ iterations we get $x_{35} = 1.000155$ for both Mann and Ishikawa iterations.

For $\alpha_n = 1/\sqrt[3]{n+1}$, $\beta_n = 1/\sqrt[4]{n+2}$ we obtain the fixed point within 6 places performing 8 iterations (using the Mann iteration) and, respectively, 9 iterations (using the Ishikawa iteration). Notice that in this case both Mann and Ishikawa iterations converge not monotonically to $p = 1$.

Conditions like $\alpha_n \to 0$ (as $n \to \infty$) or/and $\beta_n \to 0$ (as $n \to \infty$) are usually involved in many convergence theorems presented in this book. The next results show that these conditions are in general not necessary for the convergence of Mann and Ishikawa iterations.

Indeed, taking

$$x_0 = 2, \ \alpha_n = \frac{n}{2n+1} \nearrow \frac{1}{2}, \ \beta_n = \frac{n+1}{2n} \searrow 1/2,$$

we obtain the following results.

For the Mann iteration: $x_1 = 2$, $x_2 = 1.5$, $x_3 = 1.166$, $x_4 = 1.034$, $x_5 = 1.0042$, $x_6 = 1.00397$, $x_7 = 1.000031$, $x_8 = 1.000002$ and $x_9 = 1$.

For the Ishikawa iteration: $x_1 = x_2 = 2$, $x_3 = 1.357$, $x_4 = 1.120$, $x_5 = 1.0289$, $x_6 = 1.0047$, $x_7 = 1.0057$, $x_8 = 1.000054$, $x_9 = 1.00004$ and $x_{10} = 1$.

For all combinations of x_0, λ, α_n and β_n, we notice the following decreasing (with respect to their speed of convergence) chain of iterative methods: Krasnoselskij, Mann, Ishikawa. Consequently, if for a certain operator in the same class, all these methods converge, then we shall use the fastest one (empirically deduced).

Remark. In the case of the function considered in Examples 3.1-3.3, $p = 1$ is a repulsive fixed point of T with respect to the Picard iteration, but, as shown in the preceding example, it is an attractive fixed point with respect to Krasnoselskij, Mann and Ishikawa iterations.

The next example presents a function with two repulsive fixed points with respect to the Picard iteration.

Example 9.4. Let $K = [0, 1]$ and $T : K \to K$ given by $Tx = (1 - x)^6$.

Then T has $p_1 \approx 0.2219$ and $p_1 \approx 2.1347$ as fixed points (obtained with Maple). Both of them are repulsive fixed points with respect to the Picard iteration. However, p_1 is attractive with respect to Krasnoselskij, Mann and Ishikawa iterations, while p_2 stays repulsive.

Here there are some numerical results obtained by running the new version of the program FIXPOINT, to support the previous assertions.

Krasnoselskij iteration: if we start from $x_0 = 2$ and the parameter that defines the iteration is $\lambda = 0.5$, then we obtain $x_1 = 1.5$, $x_2 = 0.757$, $x_3 = 0.379$, $x_4 = 0.2181$, $x_5 = 0.2232$ and $x_6 = 0.2214$;

Mann iteration: if we start from $x_0 = 2$ and the parameter sequence is $\alpha_n = 1/(n + 1)$, then we obtain $x_1 = 1.0$, $x_2 = 0.5$, $x_3 = 0.338$, $x_4 = 0.2748$, $x_5 = 0.2489$ and $x_6 = 0.2378$;

Ishikawa iteration: if we start from $x_0 = 2$ and the parameter sequences are $\alpha_n = 1/(n + 1)$ and $\beta_n = 1/(n + 2)$, then we obtain $x_1 = 0.01$, $x_2 = 0.55$, $x_3 = 0.346$, $x_4 = 0.2851$, $x_5 = 0.2527$ and $x_6 = 0.2392$;

The previous numerical results suggest that Krasnoselskij iteration converges faster than both Mann and Ishikawa iterations. This fact is more clearer illustrated if we choose $x_0 = p_2$, the repulsive fixed point of T: after 20 iterations, Krasnoselskij method gives $x_{20} = 0.2219$, while Mann and Ishikawa iteration procedures give $x_{20} = 0.6346$ and $x_{20} = 0.6347$, respectively. The convergence of Mann and Ishikawa iteration procedures is indeed very slow in this case: after 500 iterations we get $x_{500} = 0.222$ for both methods.

Note that for $x_0 \in \{-2, 3, 4\}$ and the previous values of the parameters λ, α_n and β_n, all three iteration procedures: Krasnoselskij, Mann and Ishikawa, converge to 1, which is not a fixed point of T.

We may infer that, for the function above and, possibly, for all functions possessing similar properties, one can expect that *always* Picard iteration converges faster than Mann or Ishikawa iterations.

The next step would be of course to try to prove (or disprove) this assertion, if possible, but certainly this is not an easy task.

However, sometimes this approach could be successful. It is perhaps important to stress on the fact that the conclusions of Theorems 9.5 and 9.6 were reached in this way: we first observed empirically the behavior of Picard iteration, Mann iteration and Ishikawa iteration for many different sets of initial data and parameters and then tried to prove analytically the observed property.

9.7 Bibliographical Comments

§9.1.

The material included in this section is related to that presented in Berinde [Be02b], [Be02d]. Definitions 9.1 is taken from Berinde [Ber98].

§9.2.

The content of this section, including Theorems 9.1-9.3, is taken from Rhoades [Rho76]. In proving Theorem 9.1 we also used several arguments from the proof in Franks and Mrazec [FrM71].

§9.3.

All results in this section are taken from Berinde [Be04f].

§9.4.

Theorems 9.5 and 9.6 in this section are taken from Berinde [Be04b]. As indicated in the paper Berinde [Be05a], in a similar manner one can prove that in the class of Zamfirescu operators, Mann iteration converges faster than Ishikawa iteration. This was accomplished by Babu and Vara Prasad [BaV06], a result which is stated in Theorem 9.7.

§9.5.

The content of this section is adapted after the paper with the same title Berinde, V. and Berinde, M. [BB05a].

§9.6.

The first empirical study of the fixed point iterative procedures is due to Rhoades [Rh77c], which is the source of the results given at the beginning of this section. The rest of the empirical studies presented were performed by the author and were published for the first time in Berinde [Be02c].

The numerical tests reported in Example 9.4 are given here for the first time. They could suggest new directions of investigation regarding the rate of convergence for those fixed point iterative procedures. We remind that the results demonstrated in the paper [Be04b] and announced in [Be05a] were initially suggested by numerical tests with the program FIXPOINT.

Exercises and Miscellaneous Results

9.1. Let $\{a_n\}$ and $\{b_n\}$ be two sequences of real numbers given by

$$a_n = \frac{1}{n^\alpha}, \, b_n = \frac{1}{2^n}, \, n \geq 1.$$

Find the values of α such that $\{b_n\}$ converges faster than $\{a_n\}$ to zero.

9.2. Let $\{a_n\}$ be a sequence defined by $a_0 \in [-2, +\infty)$ and

$$a_{n+1} = \sqrt{2 + a_n}, \, n \geq 0.$$

Show that $\{a_n\}$ converges to 2 at least as fast as the sequence $\{1/4^n\}$ to zero.

9.3. Let $\{x_n\}$ be given by $x_{n+1} = \dfrac{1}{2}\left(x_n + \dfrac{2}{x_n}\right)$, $n \geq 1$, $x_1 > 0$. Show that $\{x_n\}$ converges to $\sqrt{2}$ faster than any sequence $\left\{1/n^k\right\}_{n \in \mathbb{N}^*}$ to zero, $k \in \mathbb{N}^*$.

9.4. Rhoades (1977)

Let $f : [0,1] \to [0,1]$ be continuous and nondecreasing. Denote $M = \sup\{x | x \in F_f\}$, $m = \inf\{x | x \in F_f\}$ and

$$x_{n+1}^c = c_n f(x_n^c) + (1 - c_n)x_n^c,$$

where $\{c_n\}$ is a sequence in $[0,1]$. Let $\{\alpha_n\}$ be a sequence in $[0,1]$ with $\alpha_0 = 1$ and $\sum\limits_{n=0}^{\infty} \alpha_n = \infty$. For $x_0^\alpha = x_0^\beta$, define the sequences $\{x_n^\alpha\}$, $\{x_n^\beta\}$,$n \geq 0$, where $0 \leq \alpha_n \leq \beta_n \leq 1$. Show that
(a) The sequence $\{x_n^\alpha\}$ converges to a fixed point of f;
(b) If $x_0^\alpha > M$, then $x_n^\alpha \geq x_n^\beta$, for all $n \geq 0$;
(c) If $x_0^\alpha < m$, then $x_n^\alpha \leq x_n^\beta$, for all $n \geq 0$;
(d) If there exists a pair of distinct adjacent fixed points p, q of f satisfying $m \leq p \leq q \leq M$, and $x_0^\alpha \in (p, q)$, then $f(x) > x$ for $x \in (p, q)$ implies $x_n^\alpha \leq x_n^\beta$,$n \geq 0$, and $f(x) < x$ for $x \in (p, q)$ implies $x_n^\alpha \geq x_n^\beta$,$n \geq 0$;
(e) Deduce that for nondecreasing continuous functions, Picard iteration is the best fixed point iteration procedure, in the sense that

$$|f^n(x_0^\alpha) - p| \leq |x_n^\alpha - p|, \quad \text{for all } n \geq 0,$$

where p is the fixed point to which $\{x_n^\alpha\}$ converges.

9.5. Let $T : [0,1] \to [0,1]$ be given by $T(x) = (1 - x)^6$, $x \in [0,1]$.
(a) Show that T has a unique fixed point $p \in [0,1]$;
(b) Prove or disprove the following statements (one can use software packages like Maple, Mathematica etc. if needed):
(b_1) The Picard iteration $\{x_n\}$ converges to p, for any $x_0 \in [0,1]$;
(b_2) The Krasnoselskij iteration $\{y_n\}$ converges to p, for any $y_0 \in [0,1]$ and appropriate parameter λ;
(b_3) The Mann iteration $\{z_n\}$ converges to p, for any $z_0 \in [0,1]$ and an appropriate sequence α_n;
(b_3) The Ishikawa iteration $\{u_n\}$ converges to p, for any $u_0 \in [0,1]$ and appropriate sequences α_n and β_n;
(c) Prove or disprove the following statements (one can use software packages like Maple, Mathematica etc):
(c_1) The Picard iteration $\{x_n\}$, converges to p, for some $x_0 \in [0,1]$;
(c_2) For any Mann iteration $\{z_n\}$ that converges to p, there exists a Krasnoselskij iteration $\{y_n\}$ that converges faster than $\{z_n\}$;
(c_3) For any Ishikawa iteration $\{u_n\}$ that converges to p, there exists a Krasnoselskij iteration $\{y_n\}$ that converges faster than $\{z_n\}$ to p.

References

[AaC02]   Aamri, M., Chaira, K.: Approximation du point fixe et applications faible-
          ment contractantes. Extr. Math., 17, No. 1, 97–110 (2002)
[Ab91a]   Abbaoui, S.: Common fixed points by Ishikawa iterates in metric linear
          spaces. Math. J. Toyama Univ., 14, 147–155 (1991)
[Ab91b]   Abbaoui, S.: Deux theoremes du point fixe. Bull. Soc. Math. Belg. Ser.
          B, 43, No. 2, 117–121 (1991)
[Ab91c]   Abbaoui, S.: Theoremes de point fixe dans un espace uniformement con-
          vexe. Bull. Soc. Math. Belg. Ser. B, 43, No. 1, 75–81 (1991)
[Ac76a]   Achari, J.: On Ciric's non-unique fixed points. Mat. Vesn., N. Ser., 13
          (28), 255–257 (1976)
[Ac76b]   Achari, J.: Some theorems on fixed points in Banach spaces. Math. Semin.
          Notes, Kobe Univ., 4, No. 2, 113–119 (1976)
[Ach78]   Achari, J.: Mean value iteration of a class of mappings in Banach spaces.
          Pure Appl. Math. Sci., 8, No. 1-2, 11–14 (1978)
[Ach87]   Achari, J.: On the existence, uniqueness and approximation of fixed points
          as a generic property. Rev. Anal. Numer. Theor. Approx., 29 (52), No. 2,
          95–98 (1987)
[Aga78]   Agarwal, R.P.: Improved error bounds for the Picard iterates. J. Math.
          Phys. Sci., Madras, 12, 45–48 (1978)
[Aga87]   Agarwal, R.P.: Existence-uniqueness and iterative methods for third-order
          boundary value problems. J. Comput. Appl. Math., 17, 271–289 (1987)
[ACL02]   Agarwal, R.P., Cho, Y.J., Li, J., Huang, N.-J.: Stability of iterative pro-
          cedures with errors approximating common fixed points for a couple
          of quasi-contractive mappings in q-uniformly smooth Banach spaces. J.
          Math. Anal. Appl., 272, No. 2, 435–447 (2002)
[AHC01]   Agarwal, R.P., Huang, N.-J., Cho, Y.J.: Stability of iterative processes
          with errors for nonlinear equations of ϕ-strongly accretive type operators.
          Numer. Funct. Anal. Optimization, 22, No. 5-6, 471–485 (2001)
[AgL84]   Agarwal, R.P., Loi, S.L.: On approximate Picard's iterates for multipoint
          boundary value problems. Nonlinear Anal. TMA, 8, 381–391 (1984)
[AgV85]   Agarwal, Ravi P., Vosmansky, J.: Necessary and sufficient conditions for
          the convergence of approximate Picard's iterates for nonlinear boundary
          value problems. Arch. Math. (Brno), 21, 171–176 (1985)

[AZC00]   Agarwal, R.P., Zhou, H.Y., Cho, Y.J., Kang, S.M.: Ishikawa iterative process with mixed errors for uniformly continuous and strongly pseudo-contractive mappings in Banach spaces. Neural Parallel Sci. Comput., **8**, No. 3-4, 291–298 (2000)

[AR03a]   Agratini, O., Rus, I.A.: Iterates of a class of discrete linear operators via contraction principle. Comment. Math. Univ. Carolin., **44**, No. 3, 555–563 (2003)

[AR03b]   Agratini, O., Rus, I.A.: Iterates of some bivariate approximation process via weakly Picard operators. Nonlinear Anal. Forum, **8**, No. 2, 159–168 (2003)

[AhA00]   Ahmad, Z., Asad, A.J.: Fixed point approximation of weakly commuting mappings in Banach space. Bull. Malays. Math. Sci. Soc. (2), **23**, No. 2, 181–185 (2000)

[AKS98]   Ahmad, Z., Kazmi, K.R., Siddiqui, Z.A.: An Ishikawa iterative algorithm with errors for strongly nonlinear complementarity problem. Far East J. Math. Sci. Special Volume, Part III, 373–382 (1998)

[Ahm05]   Ahmed, M.A.: A characterization of the convergence of Picard iteration to a fixed point for a continuous mapping and an application. Appl. Math. Comput., **169**, No. 2, 1298–1304 (2005)

[AhZ02]   Ahmed, M.A., Zeyada, F.M.: On convergence of a sequence in complete metric spaces and its applications to some iterates of quasi-nonexpansive mappings. J. Math. Anal. Appl., **274**, No. 1, 458–465 (2002)

[Akh90]   Akhiezer, T.A.: Iterative processes that are connected with nonexpansive mappings. (Russian) Teor. Funktsii Funktsional. Anal. i Prilozhen. No. **50**, 17–20 (1988). Translation in J. Soviet Math., **49**, No. 6, 1247–1249 (1990)

[AkK90]   Aksoy, A.G.; Khamsi, M.A.: Nonstandard Methods in Fixed Point Theory. Springer Verlag, Universitext Series, New York (1990)

[Alb96]   Alber, Y.: Metric and generalized projection operators in Banach spaces: properties and applications. In: A. Kartsatos (Ed.), Theory and Applications of Nonlinear Operators of Monotonic and Accretive Type. Marcel Dekker, New York (1996)

[ACZ06]   Alber, Y., Chidume, C.E., Zegeye, H.: Approximating fixed points of total asymptotically nonexpansive mappings. Fixed Point Theory Appl. **2006**, Art. ID 10673, 20 pp. (2006)

[AlG94]   Alber, Y., Guerre-Delabriere, S.: Problems of fixed point theory in Hilbert and Banach spaces. Funct. Differ. Equ., **2**, 5–10 (1994)

[AlG97]   Alber, Y., Guerre-Delabriere, S.: Principle of weakly contractive maps in Hilbert spaces. In: Gohlberg, I., Lyubich, Y. (eds) New Results in Operator Theory. Adv. Appl., 7–22. Birkhauser, Basel (1997)

[AlG01]   Alber, Y., Guerre-Delabriere, S.: On the projection methods for fixed point problems. Analysis, **21**, 17–39 (2001)

[AGZ98]   Alber, Y., Guerre-Delabriere, S., Zelenko, L.: The principle of weakly contractive maps in metric spaces. Commun. Appl. Nonlinear Anal., **5**, No. 1, 45–68 (1998)

[AlN95]   Alber, Y., Notik, A.: On some estimates for projection operator in Banach space. Commun. Appl. Nonlinear Anal., **2**, No. 1, 47–56 (1995)

[AlR94]   Alber, Y., Reich, S.: An iterative method for solving a class of nonlinear operator equations in Banach spaces. Panamer. Math. J., **4**, No. 2, 39–54 (1994)

[ARY03] Alber, Y., Reich, S., Yao, J.-C.: Iterative methods for solving fixed-point problems with nonself-mappings in Banach spaces. Abstr. Appl. Anal. **2003**, No. 4, 193–216 (2003)

[ASe82] Algetmy, A., Sehgal V.M.: A convergence theorem for nonlinear contractions and fixed point. Math. Japon., **27**, 113–116 (1982)

[AGe80] Allgower, E., Georg, K.: Simplicial and continuation methods for approximating fixed points and solutions to systems of equations. SIAM Rev., **22**, 28–85 (1980)

[Als81] Alspach, D.E.: A fixed point free nonexpansive map. Proc. Amer. Math. Soc., **82**, 423–424 (1981)

[AGu96] Altas, I., Gupta, M.M.: Iterative methods for fixed point problems on high performance computing environment. Complex. Int., **3** (electronic) (1996)

[Ame97] Amer, M.A.: Iterative solutions of parameter-dependent nonlinear equations of Hammerstein type. Computers Math. Appl., **34**, No. 9, 123–129 (1997)

[ACD03] de Amo, E., Chitescu, I., Diaz Carillo, M., Secelean, N.A.: A new approximation procedure for fractals. J. Comput. Appl. Math., **151**, 355–370 (2003)

[Ang92] Angelov, V.G.: A continuous dependence of fixed points of ϕ-contractive mappings in uniform spaces. Arch. Math. (Brno), **28**, No. 3-4, 155–162 (1992)

[Ap76a] d'Apuzzo, L.: On the notion of good and special convergence of the method of successive approximations (Italian). Ann. Instit. Univ. Navale Napoli, **45/46**, 123–128 (1976/1977)

[Ap76b] d'Apuzzo, L.: On the convergence of the successive approximation method in metric spaces, Ann. Instit. Univ. Navale Napoli, **45/46**, 99–113 (1976/1977)

[Arg88] Argyros, I.K.: Approximating the fixed points of some nonlinear operator equations. Math. Slovaca, **38**, No. 4, 409–417 (1988)

[Arg95] Argyros, I.K.: Stirling's method and fixed points of nonlinear operator equations in Banach space. Bull. Inst. Math., Acad. Sin., **23**, No. 1, 13–20 (1995)

[Arg99] Argyros, I.K.: A generalization of Ostrowski's theorem on fixed points. Appl. Math. Lett., **12**, No. 6, 77–79 (1999)

[ASz93] Argyros, I., Szidarowsky, F.: The Theory and Applications of Iteration Methods. CRC Press, Boca Raton (1993)

[ASz94] Argyros, I., Szidarowsky, F.: On the convergence of the modified contractions. J. Comput. & Appl. Math., **55**, 183–189 (1994)

[ARH02] Aslam Noor, M., Rasias, T.M., Huang, Z.: Three step iterations for nonlinear accretive operator equations. J. Math. Anal. Appl., **274**, 59–68 (2002)

[Asp67] Asplund, E.: Positivity of duality mappings. Bull. Amer. Math. Soc., **73**, 200–203 (1967)

[Ass88] Assad, N.A.: Approximation for fixed points of multivalued contractive mappings. Math. Nachr., **139**, 207–213 (1988)

[AsK72] Assad, N.A., Kirk, W.A.: Fixed point theorems for set-valued mappings of contractive type, Pacific J. Math., **43**, 553–562 (1972)

[Ath90] Athanasov, Z.S.: Uniqueness and convergence of successive approximations for ordinary differential equations. Math. Japon., **35**, 351–367 (1990)

[At98a]   Atsushiba, S.: Approximating common fixed points of asymptotically non-expansive semigroups by the Mann iterative process. Panamer. Math. J., **8**, No. 4, 45–58 (1998)

[At98b]   Atsushiba, S.: Modulus of convexity and convergence theorems for families of nonexpansive mappings (Japanese). RIMS Kokyuroku 1031, 138–148 (1998)

[Ats99]   Atsushiba, S.: Strong convergence of iterates of nonexpansive mappings and applications. In: Nishizawa, K. (ed.) Convex Analysis and Chaos. The third symposium on nonlinear analysis, NLA '98, Josai University, Saitama, Japan, July 23-25, 1998. Saitama: Josai University, Graduate School of Science, Josai Math. Monogr. 1 (1999)

[Ats03]   Atsushiba, S.: Strong convergence of iterative sequences for asymptotically nonexpansive mappings in Banach spaces. Sc. Math. Japon., **57**, No. 2, 377–388 (2003)

[AST00]   Atsushiba, S., Shioji, N., Takahashi, W.: Approximating common fixed points by the Mann iteration procedure in Banach spaces. J. Nonlinear Convex Anal., **1**, No. 3, 351–361 (2000)

[AT97a]   Atsushiba, S., Takahashi, W.: Approximating common fixed points of nonexpansive semigroups by the Mann iteration process. Ann. Univ. Mariae Curie-Sklodowska, Sect. A, **51**, No. 2, 1–16 (1997)

[AT97b]   Atsushiba, S., Takahashi, W.: Approximating common fixed points of nonexpansive semigroups by the Mann iteration process. Proceedings of Workshop on Fixed Point Theory (Kazimierz Dolny, 1997). Ann. Univ. Mariae Curie-Skodowska Sect. A, **51**, No. 2, 1–16 (1997)

[AtT98]   Atsushiba, S., Takahashi, W.: Approximating common fixed points of two nonexpansive mappings in Banach spaces. Bull. Austral. Math. Soc., **57**, No. 1, 117–127 (1998)

[AT99a]   Atsushiba, S., Takahashi, W.: A weak convergence theorem for nonexpansive semigroups by the Mann iteration process in Banach spaces. In: Takahashi, W. (ed.) et al., Nonlinear Analysis and Convex Analysis. Proceedings of the 1st international conference (NACA98), Niigata, Japan, July 28-31, 1998. World Scientific, Singapore (1999)

[AT99b]   Atsushiba, S., Takahashi, W.: Strong convergence theorems for a finite family of nonexpansive mappings and applications. B. N. Prasad birth centenary commemoration volume Indian J. Math., **41**, No. 3, 435–453 (1999)

[AtT00]   Atsushiba, S., Takahashi, W.: A nonlinear strong ergodic theorem for nonexpansive mappings with compact domains. Math. Japon., **52**, No. 2, 183–195 (2000)

[Bab86]   Babadzhanyan, A.A.: On "constructivity" in a fixed point theorem (Russian). Dokl. Akad. Nauk Arm. SSR, **83**, 147–149 (1986)

[Bab96]   Babadzhanyan, A.A.: "Constructivity" in the fixed-point theorem (Russian). Dopov. Nats. Acad. Nauk. Ukraini, No. **6**, 27–30 (1996)

[BaK02]   Babu, G.V.R., Krishna, M.: Improved convergence rate estimate of Ishikawa iteration process for Lipschitz strongly pseudocontractive maps in a Banach space. Bull. Calcutta Math. Soc., **94**, No. 4, 253–258 (2002)

[BaV06]   Babu, G.V.R., Vara Prasad, K.N.V.V.: Mann iteration converges faster than Ishikawa iteration for the class of Zamfirescu iteration. Fixed Point Theory and Applications, Vol. **2006**, 1-6 (2006); Article ID 49615, DOI 10.1155/FPTA/2006/49615

[BCC98]   Baek, J.H., Cho, Y.J., Chang, S.S.: Iterative process with errors for m-accretive operators. J. Korean Math. Soc., **35**, No. 1, 191–205 (1998)

[BKC99]   Baek, J.H., Kang, S.M., Cho, Y.J., Chang, S.S.: Iterative solutions of non-linear equations for φ-strongly accretive type operators. Panamer. Math. J., **9**, No. 2, 35–50 (1999)

[BaK04]   Bai, C., Kim, J.K.: An implicit iteration process with errors for a finite family of asymptotically quasi-nonexpansive mappings. Nonlinear Funct. Anal. Appl., 9, No. 4, 649–658 (2004)

[Bai01]   Bai, M.: Perturbed iterative process for fixed points of multivalued ϕ-hemicontractive mappings in Banach spaces. Comput. Math. Appl., **41**, No. 1-2, 103–109 (2001)

[Bai75]   Baillon, J.-B.: Un theoreme de type ergodique pour les contraction non-lineaires dans un espace de Hilbert. C.R. Acad. Sci. Paris Ser. I Math., **280**, 1511–1514 (1975)

[Ba76a]   Baillon, J.-B.: Quelques proprietes de convergence asymptotique pour les semigroupes de contractions impaires. C.R. Acad. Sci. Paris Ser. I Math., **283**, 75–78 (1976)

[Ba76b]   Baillon, J.-B.: Quelques proprietes de convergence asymptotique pour les contractions impaires. C.R. Acad. Sci. Paris, **283**, 587–590 (1976)

[Bai78]   Baillon, J.-B.: Comportement asymptotique des iteres de contractions non lineaires dans les espaces L^p. C. R. Acad. Sci. Paris Ser. A, **286**, 157–159 (1978)

[Ba79a]   Baillon, J.-B.: Quelques aspects de la theorie des points fixes dans les espaces de Banach - I. Seminaire d'analyse fonctionnelle 1978-1979, Palaiseau, Expose Numero 7, 13 pages (1979)

[Ba79b]   Baillon, J.-B.: Quelques aspects de la theorie des points fixes dans les espaces de Banach - II. Seminaire d'analyse fonctionnelle 1978-1979, Palaiseau, Expose Numero 8, 32 pages (1979)

[BBr92]   Baillon, J.B., Bruck, R.E.: Optimal rates of asymptotic regularity for av-eraged nonexpansive mappings. In: Fixed Point Theory and Applications, Halifax, NS, 1991, World Sci. Publishing, River Edge, New York (1992)

[BBr96]   Baillon, J.B., Bruck, R.E.: The rate of asymptotic regularity is $O(1/n)$. In: Kartsatos, A.G. (ed.) Theory and Applications of Nonlinear Operators of Accretive and Monotone type. Marcel Dekker, New York (1996)

[BBR78]   Baillon, J.B., Bruck, R.E., Reich, S.: On the asymptotic behavior of non-expansive mappings and semigroups in Banach spaces. Houston J. Math., **4**, 1–9 (1978)

[BaS81]   Baillon, J.B., Schoneberg, R.: Asymptotic normal structure and fixed points of nonexpansive mappings. Proc. Amer. Math. Soc., **81**, 257–264 (1981)

[Ban22]   Banach, S.: Sur les operations dans les ensembles abstraits et leur appli-cations aux equations integrales. Fund. Math., **3**, 133–181 (1922)

[Ban32]   Banach, S.: Theorie des Operations Lineaires. Monografie Matematyczne, Warszawa-Lwow (1932)

[Bao91]   Bao, K.Z.: Iterative approximation of the solution of a locally Lipschitzian equation. Appl. Math. Mech., **12**, No. 4, 409–414 (1991). English trans-lation from Chinese: Appl. Math. Mec, **12**, No.4, 385–389 (1991)

[Bao03]   Bao, Z.Q.: Ishikawa iterative sequences for uniformly quasi-Lipschitzian mappings with errors. (Chinese) Xinan Shifan Daxue Xuebao Ziran Kexue Ban, **28**, No. 3, 367–369 (2003)

[Bar76]   Barbu, V.: Nonlinear Semigroups and Differential Equations in Banach Spaces. Noordhoff, Leyden (1976)

[BaP78]   Barbu, V., Precupanu, Th.: Convexity and Optimization in Banach Spaces. Editura Academiei R.S.R., Bucharest (1978)

[Bar88]   Barnsley, M., Fractals Everywhere. Academic Press, Boston, San Diego etc. (1988)

[Bau96]   Bauschke, H.H.: The approximation of fixed points of compositions of nonexpansive mappings in Hilbert spaces. J. Math. Anal. Appl., **202**, 150–159 (1996)

[Beauz]   Beauzamy, B.: Un cas de convergence des iterees d'une contraction dans une espace uniformement convexe. Unpublished.

[Bea82]   Beauzamy, B.: Introduction to Banach Spaces and their Geometry. Mathematics Studies, vol. **68**, North-Holland, Amsterdam (1982)

[Bea85]   Beauzamy, B., Enflo, P.: Theorème de point fixe et d'approximation. Ark. Mat., **23**, 19–34 (1985)

[Beg01]   Beg, I.: An iteration scheme for asymptotically nonexpansive mappings on uniformly convex metric spaces. Nonlinear analysis and its applications (St. John's, NF, 1999). Nonlinear Anal. Forum, **6**, No. 1, 27–34 (2001)

[Beg03]   Beg, I.: An iteration process for nonlinear mappings in uniformly convex linear metric spaces. Czechoslovak Math. J., **53(128)**, No. 2, 405–412 (2003)

[BAz94]   Beg, I., Azam, A.: On iteration methods for multivalued mappings. Demonstratio Math. **27**, No. 2, 493–499 (1994)

[BAz96]   Beg, I., Azam, A.: Construction of fixed points for generalized nonexpansive mappings. Indian J. Math., **38**, No. 2, 161–170 (1996)

[BeM07]   Berinde, M.: Approximate fixed point theorems. Studia Univ. "Babes-Bolyai" Mathematica, **51**, No. 1, 11–25 (2006)

[BeB07]   Berinde, M., Berinde, V.: On a class of multi-valued weakly Picard mappings. J. Math. Anal. Appl. **326**, 772–782 (2007)

[Ber79]   Berinde, V.: The approximation of fixed points for continuous mappings (Romanian). MSc Thesis, "Babes-Bolyai" Univ. of Cluj-Napoca (1979)

[Be90a]   Berinde, V.: Error estimates in the approximation of fixed points for a class of φ-contractions. Studia Univ. "Babes-Bolyai", **35**, 2, 86–89 (1990)

[Be90b]   Berinde, V.: The stability of fixed points for a class of φ-contractions. Seminar on Fixed Point Theory, 3, 13–20 (1990)

[Be90c]   Berinde, V.: On the solutions of a functional equation using Picard mappings. Studia Univ. Babes-Bolyai Math., **35**, No. 4, 63–69 (1990)

[Ber91]   Berinde, V.: A fixed point theorem of Maia type in K-metric spaces. Seminar on Fixed Point Theory, 3, 7–14 (1991)

[Be92a]   Berinde, V.: Abstract φ-contractions which are Picard mappings. Mathematica, **34(57)**, No. 2, 107–122 (1992)

[Be92b]   Berinde, V.: On the problem of Darboux-Ionescu using a generalized Lipschitz condition. Seminar on Fixed Point Theory, 3, 19–28 (1992)

[Be93a]   Berinde, V.: φ-monotone and φ-contractive operators in Hilbert spaces. Studia Univ. "Babes-Bolyai", **38**, 4, 51–58 (1993)

[Be93b]   Berinde, V.: On an integral equation of Volterra type using a generalized Lipschitz condition. Bul. Stiint. Univ. Baia Mare, **9**, 1–10 (1993)

[Be93c]   Berinde, V.: Generalized contractions in uniform spaces. Bul. Stiint. Univ. Baia Mare, **9**, 45–52 (1993)

[Be93d]   Berinde, V.: Generalized contractions in quasimetric spaces. Seminar on
           Fixed Point Theory, 3, 3–9 (1993)
[Be94a]   Berinde, V.: Generalized contractions in σ-complete lattices. Zb. Rad.
           Prir. Mat. Fac. Univ. Novi Sad, 24, No. 2, 31–38 (1994)
[Be94b]   Berinde, V.: Error estimates for a class of (δ, φ)-contractions. Seminar on
           Fixed Point Theory, 3, 3–9 (1994)
[Be94c]   Berinde, V.: On a homeomorphism theorem. Bul. Stiint. Univ. Baia Mare,
           10, 73–76 (1994)
[Be94d]   Berinde, V.: A fixed point theorem for mappings with contracting orbital
           diameters. Bul. Stiint. Univ. Baia Mare, 10, 29–38 (1994)
[Be95a]   Berinde, V.: A fixed point proof of the convergence of the Newton method.
           In: Proceed. Int. Conf. MicroCAD '94, Univ. of Miskolc, pp. 14–21 (1995)
[Be95b]   Berinde, V.: Generalized contractions and higher order hyperbolic partial
           differential equations. Bul. Stiint. Univ. Baia Mare, 11, 39–54 (1995)
[Be95c]   Berinde, V.: Remarks on the convergence of the Newton-Raphson method.
           Rev. Anal. Numer. Theor. Approx., 24, No. 1-2, 15–21 (1995)
[Be95d]   Berinde, V.: Conditions for the convergence of the Newton method. An.
           Stiint. Univ. Ovidius Constanta Ser. Mat., 3, No. 1, 22–28 (1995)
[Be96a]   Berinde, V.: On some generalized contractive type conditions. Bul. Stiint.
           Univ. Baia Mare, 12, No. 2, 181–190 (1996)
[Be96b]   Berinde, V.: Sequences of operators and fixed points in quasimetric spaces.
           Studia Univ. Babes-Bolyai Math., 41, 23–27 (1996)
[Be97a]   Berinde, V.: Generalized Contractions and Applications (Romanian). Ed-
           itura Cub Press 22, Baia Mare (1997)
[Be97b]   Berinde, V.: On some exit criteria for the Newton method. Novi Sad J.
           Math., 27, No. 1, 19–26 (1997)
[Ber98]   Berinde, V.: On the convergence rate of sequences of real numbers (Ro-
           manian). Gazeta Matematica, 103, No. 4, 146–153 (1998)
[Ber99]   Berinde, V.: A priori and a posteriori error estimates for a class of
           φ-contractions. Bull. for Appl. Math., 90-B, 183–192 (1999)
[Be00a]   Berinde, V.: Error estimates for some Newton-type methods obtained by
           fixed point techniques. In: Proceed. Int. Scientific Conference on Math.
           (Herlany, 1999), Technical Univ. Kosice (2000)
[Be02e]   Berinde, V.: Approximating fixed points of Lipschitzian pseudocontrac-
           tions. In: Mathematics & Mathematics Education (Bethlehem, 2000).
           World Sci. Publishing, River Edge (2002)
[Be02a]   Berinde, V.: Iterative approximations of fixed points for pseudo-
           contractive operators. Seminar on Fixed Point Theory, 3, 209–216 (2002)
[Be02b]   Berinde, V.: On some stability results for fixed point iteration procedures.
           Bul. Stiint. Univ. Baia Mare, 18, No. 1, 7–12 (2002)
[Be02c]   Berinde, V.: Iterative Approximation of Fixed Points. Efemeride, Baia
           Mare (2002)
[Be02d]   Berinde, V.: Weak and almost weak stability of fixed point iteration pro-
           cedures, Preprint, North University of Baia Mare (2002)
[Be03a]   Berinde, V.: On the approximation of fixed points of weak φ-contractive
           operators. Fixed Point Theory, 4, No. 2, 131–142 (2003)
[Be03b]   Berinde, V.: A common fixed point theorem for quasi-contractive type
           mappings. Ann. Univ. Sci. Budap., 46, 81–90 (2003)
[Be03c]   Berinde, V.: On the approximation of fixed points of weak contractive
           mappings. Carpathian J. Math., 19, No. 1, 7–22 (2003)

[Be03d]   Berinde, V.: Summable almost stability of fixed point iteration proce-
           dures. Carpathian J. Math. **19**, No. 2, 81–88 (2003)
[Be03e]   Berinde, V.: On the convergence of Mann iteration for a class of quasi
           contractive operators. Preprint, North University of Baia Mare (2003)
[Be04a]   Berinde, V.: Approximation of fixed points of some nonself generalized
           φ-contractions. Math. Balkanica, **18**, Fasc. 1-2, 85–93 (2004)
[Be04b]   Berinde, V.: Picard iteration converges faster than the Mann iteration in
           the class of quasi-contractive operators. Fixed Point Theory Appl. **2004**,
           No. 2, 97–105 (2004)
[Be04c]   Berinde, V.: On the convergence of Ishikawa iteration for a class of quasi
           contractive operators. Acta Math. Univ. Comen., **73**, No. 1, 119–126
           (2004)
[Be04d]   Berinde, V.: Approximation fixed points of weak contractions using the
           Picard iteration. Nonlinear Analysis Forum, **9**, No. 1, 43–53 (2004)
[Be04e]   Berinde, V.: A common fixed point theorem for nonself mappings. Miskolc
           Math. Notes, **5**, No. 2, 137–147 (2004)
[Be04f]   Berinde, V.: Comparing Krasnoselskij and Mann iterations for Lip-
           schitzian generalized pseudocontractive operators. In: Proceed. of Int.
           Conf. On Fixed Point Theory, Univ. of Valencia, 19-26 July 2003, Yoko-
           hama Publishers (2004)
[Be05a]   Berinde, V.: A convergence theorem for some mean value fixed point
           iterations in the class of quasi contractive operators. Demonstratio Math.,
           38, No. 1, 177–184 (2005)
[Be05b]   Berinde, V.: Error estimates for approximating fixed points of discontin-
           uous quasi-contractions. General Mathematics **13**, No. 2, 23–34 (2005)
[BB05a]   Berinde, V., Berinde, M.: The fastest Krasnoselskij iteration for approxi-
           mating fixed points of strictly pseudo-contractive mappings. Carpathian
           J. Math. **21**, No. 1-2, 13–20 (2005)
[BB05b]   Berinde, V., Berinde, M.: On Zamfirescu's fixed point theorem. Rev.
           Roumaine Math. Pures Appl., **50**, Nos 5-6, 443–453 (2005)
[BeP06]   Berinde, V., Pacurar, M.: A fixed point proof of the convergence o a
           Newton-type method. Fixed Point Theory, **7**, No. 2, 235–244 (2006)
[Bes59]   Bessaga, C.: On the converse of the Banach fixed point principle. Colloq.
           Math., **7**, 41–43 (1959)
[Bet89]   Bethke, M.: Approximation von Fixpunkten streng pseudokontraktiver
           Operatoren. Wiss. Z., Pädagog. Hochsch. "Liselotte Herrmann" Güstrow,
           Math.- Naturwiss. Fak., **27**, No. 2, 263–270 (1989)
[Bet93]   Bethke, M.: Approximation von Fixpunkten kontraktionsartiger Abbil-
           dungen. Rostock: Univ. Rostock, 75 S. (1993)
[BLG97]   Bezanilla Lopez, A., Garcia Juarez, P.: Stability of the iterative process
           $x_{n+1} = Tx_n$ for Kannan maps in metric spaces. Proceedings of the
           3rd International Conference on Approximation and Optimization in the
           Caribbean (Puebla, 1995), 5 pp. (electronic), Benemerita Univ. Auton.
           Puebla, Puebla (1997)
[BLG00]   Bezanilla Lopez, A., Garcia Juarez, P.: Stability of a Sehgal iterative
           process for mappings with Kannan iterates. (Spanish) XXXII National
           Congress of the Mexican Mathematical Society (Spanish) (Guadalajara,
           1999), 109–113. Aportaciones Mat. Comun., **27**, Soc. Mat. Mexicana,
           Mexico (2000)

[Bi56a]   Bielecki, A.: Une remarque sur la methode de Banach-Caccioppoli-Tihonov. Bull. Acad. Sci., **4**, 261–268 (1956)

[Bi56b]   Bielecki, A.: Une remarque sur l'application de la methode de Banach-Caccioppoli-Tihonov dans la theorie de l'equation $s = f(x, y, z, p, q)$. Bull. Acad. Pol. Sci., Cl. III, **4**, 265–268 (1956)

[Bin04]   Binh, T.Q.: Some extensions of contractive mapping theorems. Nonlinear Funct. Anal. & Appl. **9**, No. 4, 659–677 (2004)

[Bog74]   Bogin, J.: On strict pseudo-contractions and a fixed point theorem. Technion Preprint Series No. MT-219, Haifa, Israel (1974)

[Boh71]   Bohl, E.: Zur Iteration bei nichtlinearen Gleichungssystems. Computing, **7**, 53–64 (1971)

[BoW76]   Bolen, J. C., Williams, B.B. On the convergence of successive approximations for quasi-nonexpansive mappings through abstract cones. J. Mathematical and Physical Sci., **10**, No. 3, 271–276 (1976)

[BoB91]   Borwein, D., Borwein, J.M.: Fixed point iterations for real functions. J. Math. Anal. Appl., **157**, 112–126 (1991)

[BRS92]   Borwein J.M., Reich, S., Shafrir, I.: Krasnoselski-Mann iterations in normed spaces. Canad. Math. Bull., **35**, 21–28 (1992)

[Bos78]   Bose, S.C. Weak convergence to the fixed point of an asymptotically non-expansive map. Proc. Amer. Math. Soc., **68**, No. 3, 305–308 (1978)

[BoM78]   Bose, R.K., Mukherjee, R.N.: On fixed points of nonexpansive set-valued mappings. Proc. Amer. Math. Soc., **72**, 97–98 (1978)

[BoM81]   Bose, R.K., Mukherjee, R.N.: Approximating fixed points of some mappings. Proc. Amer. Math. Soc., **82**, 603–606 (1981)

[BoS84]   Bose, R.K., Sahani, D.: Weak convergence and common fixed points of non-expansive mappings of iteration. Indian J. Pure Appl. Math., **15**, 123–126 (1984)

[BoW69]   Boyd, D.W., Wong, J.S.: On nonlinear contractions. Proc. Amer. Math. Soc., **20**, 335–341 (1969)

[Bra00]   Branciari, A.: A fixed point theorem of Banach-Caccioppoli type on a class of generalized metric spaces. Publ. Math. Debrecen, **57**, No. 1-2, 31–37 (2000)

[Bre77]   Brezinski, C.: Acceleration de la Convergence en Analyse Numerique. Lectures Notes in Mathematics, Springer, Berlin, Heidelberg, New York (1977)

[Bre78]   Brezis, H., Browder, F.E.: Nonlinear ergodic theorems. Bull. Amer. Math. Soc., **82**, 959–961 (1978)

[Bri92]   Brimberg, J., Love, R.F.: Local convergence in a generalized Fermat-Weber problem. Ann. Oper. Res., **40**, No. 1-4, 33–66 (1992)

[Brs69]   Brosowski, B.: Fixpunktsatze in der approximation theorie. Mathematica (Cluj), **11**, 195–220 (1969)

[Bro63]   Browder, F.E.: The solvability of nonlinear functional equations. Duke Math. J., **30**, 557–566 (1963)

[Br65a]   Browder, F.E.: Fixed-point theorems for noncompact mappings in Hilbert space. Proc. Nat. Acad. Sci. U.S.A., **53**, 1272–1276 (1965)

[Br65b]   Browder, F.E.: Existence of periodic solutions for nonlinear equations of evolution. Proc. Nat. Acad. Sci. U.S.A., **53**, 1100–1103 (1965)

[Br65c]   Browder, F.E.: Nonexpansive nonlinear operators in Banach spaces. Proc. Nat. Acad. Sci. U.S.A., **54**, 1041–1044 (1965)

[Br67a]   Browder, F.E.: Nonlinear mappings of nonexpansive and accretive type in Banach spaces. Bull. Amer. Math. Soc., **73**, 875–882 (1967)

[Br67b]   Browder, F.E.: Convergence of approximants to fixed points of nonexpansive nonlinear maps in Banach spaces. Arch. Rat. Mech. Anal., **24**, 82–90 (1967)

[Br68a]   Browder, F.E.: On the convergence of successive approximations for nonlinear functional equations. Indagat. Math., **30**, 27–35 (1968)

[Br68b]   Browder, F.E.: Semicontractive and semiaccretive nonlinear mappings in Banach spaces. Bull. Amer. Math. Soc., **74**, 660–665 (1968)

[Br68c]   Browder, F.E.: Nonlinear monotone and accretive operators in Banach spaces. Proc. Nat. Acad. Sci. U.S.A., **61**, 388–393 (1968)

[Bro76]   Browder, F.E.: Nonlinear operators and nonlinear equations of evolution in Banach spaces. Proc. Sympos. Pure Math., 18, Pt. 2, Amer. Math. Soc., Providence, R. I. (1976)

[Bro79]   Browder, F.E.: Remarks on fixed point theorems of contractive type. Nonlinear Anal. TMA, **5**, 657–661 (1979)

[BrP66]   Browder, F.E., Petryshyn, W.V.: The solution by iteration of nonlinear functional equations in Banach spaces. Bull. Amer. Math. Soc., **72**, 571–575 (1966)

[BrP67]   Browder, F.E., Petryshyn, W.V.: Construction of fixed points of nonlinear mappings in Hilbert spaces. J. Math. Anal. Appl., **20**, 197–228 (1967)

[Bru74]   Bruck, R.E.: A strongly convergent iterative method for the solution of $0 \in Ux$ for a maximal monotone operator U in Hilbert space. J. Math. Anal. Appl., **48**, 114–126 (1974)

[Bk78a]   Bruck, R.E.: On the almost-convergence of iterates of a nonexpansive mapping in Hilbert space and the structure of the weak ω-limit set. Israel J. Math., **29**, 1–16 (1978)

[Bk78b]   Bruck, R.E.: On the strong convergence of an averaging iteration for the solution of operator equations involving monotone operators in Hilbert space. J. Math. Anal. Appl., **64**, 319–327 (1978)

[Bk79a]   Bruck, R.E.: A simple proof of the mean ergodic theorem for nonlinear contractions in Banach spaces. Israel J. Math., **32**, 107–116 (1979)

[Bk79b]   Bruck, R.E.: On the convex approximation property and the asymptotic behaviour of nonlinear contractions in Banach spaces. Israel J. Math., **32**, 107–116 (1979)

[BKR82]   Bruck, R.E., Kirk, W.A., Reich, S.: Strong and weak convergence theorems for locally nonexpansive mappings in Banach spaces. Nonlinear Anal. TMA, **6**, 151–155 (1982)

[BKR93]   Bruck, R.E., Kuczumow, T., Reich, S.: Convergence of iterates of asymptotically nonexpansive mappings in Banach spaces with the uniform Opial property. Colloq. Math., **65**, No. 2, 169–179 (1993)

[BrH90]   Bruckner, A.M., Hu, T.: Equicontinuity of iterates of an interval map. Tamkang J. Math., **21**, No. 3, 287–294 (1990)

[Bry68]   Bryant, V.: A remark on a fixed-point theorem for iterated mappings. Amer. Math. Monthly, **75**, No. 4, 399–400 (1968)

[But83]   Butnariu, D.: Computing fixed points for fuzzy mappings. In: Trans. of Prague Conf. on Information Theory, Statistical decision functions, random processes (Prague, 1982), Vol. A, Reidel, Dordrecht (1983)

[But00]   Butnariu, D., Iusen, A.N.: Totally Convex Functions for Fixed Points Computation and Infinite Dimensional Optimization. Kluwer Academic Publishers, Dordrecht (2000)

[Byn01]   Bynum, W.L.: Normal structure coefficients for Banach spaces. Pacific J. Math. **86**, 427–436 (2001)

[Cac30]   Caccioppoli, R.: Un teorema generale sull'esistenza di elementi uniti in una trasformazione funzionale. Rend. Accad. Lincei, **11**, 794–799 (1930)

[Car90]   Carbone, A.: Iterative construction of solutions of functional equations involving multivalued operators in L^p spaces. Jnanabha, vol. 20 (1990)

[CaR77]   Cass, F.P., Rhoades, B.E.: Mercerian theorems via spectral theory. Pacific J. Math., **73**, 63–71 (1977)

[Cat01]   Catinas, E.: On accelerating the convergence of the successive approximations method. Rev. Anal. Numer. Theor. Approx., **30**, 3–8 (2001)

[Cat02]   Catinas, E.: On the superlinear convergence of the successive approximations method. J. Optim. Theory Appl., **113**, No. 3, 473–485 (2002)

[CeR96]   Censor, Y., Reich, S.: Iterations of paracontractions and firmly nonexpansive operators with applications to feasibility and optimization. Optimization, **37**, No. 4, 323–339 (1996)

[Ca97a]   Chang, S.S.: Some problems and results in the study of nonlinear analysis. Nonlinear Anal., **30**, 4197–4208 (1997)

[Ca97b]   Chang, S.S.: On Chidume's open questions and approximate solutions of multivalued strongly accretive mapping equations in Banach spaces. J. Math. Anal. Appl., **216**, No. 1, 94–111 (1997)

[Ca01a]   Chang, S.S.: Some results for asymptotically pseudo-contractive mappings and asymptotically nonexpansive mappings. Proc. Amer. Math. Soc., **129**, No. 3, 845–853 (2001)

[Ca01b]   Chang, S.S.: Iterative approximation problem of fixed points for asymptotically nonexpansive mappings in Banach spaces. Acta Math. Appl., **24**, 236–241 (2001)

[Ca01c]   Chang, S.S.: On the approximating problem of fixed points for asymptotically nonexpansive mappings. Indian J. Pure Appl. Math., **32**, 1–11 (2001)

[Cha06]   Chang, S.S.: Viscosity approximation methods for a finite family of nonexpansive mappings in Banach spaces. J. Math. Anal. Appl., **323**, 1402–1416 (2006)

[CCY02]   Chang, S.S., Cho, Y.J.: On the iterative approximation methods of fixed points for asymptotically contractive type mappings in Banach spaces. In: Fixed Point Theory and Applications. Vol. 3, 33–41. Nova Sci. Publ., Huntington, NY (2002)

[CCK03]   Chang, S.S., Cho, Y.J., Kim, J.K.: The equivalence between the convergence of modified Picard, modified Mann, and modified Ishikawa iterations. Math. Comput. Modelling, **37**, No. 9-10, 985–991 (2003)

[CCK01]   Chang, S.S., Cho, Y.J., Kim, J.K., Kim, K.H.: Iterative approximation of fixed points asymptotically nonexpansive type mappings in Banach spaces. Panamer. Math. J., **11**, 53–63 (2001)

[CCK04]   Chang, S.S., Cho, Y.J., Kim, J.K., Zhou, H.Y.: Random Ishikawa iterative sequence with applications. Stochastic Anal. Appl., **23**, No. 1, 69–77 (2004)

[CCL98]   Chang, S.S., Cho, Y.J., Lee, B.S., Jung, J.S., Kang, S.M.: Iterative approximations of fixed points and solutions for strongly accretive and strongly pseudo-contractive mappings in Banach spaces. J. Math. Anal. Appl., **224**, 149–165 (1998)

[CCZ01]   Chang, S.S., Cho, Y.J., Zhou, H.: Demi-closed principle and weak convergence problems for asymptotically nonexpansive mappings. J. Korean Math. Soc., **38**, No. 6, 1245–1260 (2001)

[CCZ02]   Chang, S.S., Cho, Y.J., Zhou, H.: Iterative Methods for Nonlinear Operator Equations in Banach Spaces. Nova Science Publishers, New York (2002)

[CCZ03]   Chang, S.S., Cho, Y.J., Zhou, Y.Y.: Iterative sequences with mixed errors for asymptotically quasi-nonexpansive type mappings in Banach spaces. Acta Math. Hungar., **100**, No. 1-2, 147–155 (2003)

[CKi03]   Chang, S.S., Kim, J.K.: Convergence theorems of the Ishikawa type iterative sequences with errors for generalized quasi-contractive mappings in convex metric spaces. Appl. Math. Letters, **16**, 535–542 (2003)

[CKJ04]   Chang, S.S., Kim, J.K., Jin, D.S.: Iterative sequences with errors for asymptotically quasi-nonexpansive type mappings in convex metric spaces. Arch. Inequal. Appl., **2**, No. 4, 365–374 (2004)

[CLC04]   Chang, S.S., Lee, H.W.J., Cho, Y.J.: On the convergence of finite steps iterative sequences for asymptotically nonexpansive mappings. Dyn. Contin. Discrete Impuls. Syst. Ser. A Math. Anal., **11**, No. 4, 589–600 (2004)

[CLP02]   Chang, S.S., Lin, L.J., Park, J.Y., Cho, Y.J.: Some convergence theorems of the Ishikawa iterative scheme for accretive and pseudocontractive type mappings in Banach spaces. Commun. Appl. Nonlinear Anal., **9**, No. 2, 69–85 (2002)

[CLC01]   Chang, S.S., Liu, J.A., Cho, Y.J.: On the iterative approximation problems of fixed points for asymptotically nonexpansive type mappings in Banach spaces. Nonlinear Funct. Anal. Appl., **6**, No. 2, 257–270 (2001)

[CPC00]   Chang, S.S., Park, J.Y., Cho, Y.J.: Iterative approximation of fixed points of asymptotically nonexpansive mapping in Banach spaces. Bull. Korean Math. Soc., **37**, 109–119 (2000)

[CTa98]   Chang, S.S., Tan, K.K.: Iteration processes for approximating fixed points of operators of monotone type. Bull. Austral. Math. Soc., **57**, No. 3, 433–445 (1998)

[CZh03]   Chang, S.S., Zhou, Y.Y.: Some convergence theorems for mappings of asymptotically quasi-nonexpansive type in Banach spaces. J. Appl. Math. Comput., **12**, No. 1-2, 119–127 (2003)

[Cha72]   Chatterjea, S.K., Fixed-point theorems. C.R. Acad. Bulgare Sci. **25**, 727–730 (1972)

[CnF97]   Chen, F.Q.: Coupled fixed points for a class of nonlinear operators. Sichuan Shifan Daxue Xuebao Ziran Kexue Ban, **32**, No. 1, 24–30 (1997)

[CYL04]   Chen, J.L., Yang, X., Li, X.Y.: Equivalence between Boyd and Wong's fixed point theorem and Banach's contraction mapping principle. (Chinese). J. Henan Univ. Sci. Technol., Nat. Sci., **25**, No. 1, 90–92 (2004)

[CnS75]   Chen, M.P., Shih, M.-H.: On sequences of quasi-contraction maps and fixed points. Tamkang J. Math., **6**, 291–292 (1975)

[CnS76]   Chen, M.P., Shih, M.-H.: On generalized contractive maps. Math. Japon., **21**, 281–282 (1976)

[CnS79]   Chen, M.P., Shih, M.-H.: Fixed point theorems for point-to-point and point-to-set maps. J. Math. Anal. Appl., **71**, 516–524 (1979)

[CnN03]   Chen, N.: Some fixed point theorems in 2-metric spaces and Mann type iteration. (Chinese) J. Liaoning Univ. Nat. Sci., **30**, No. 4, 311–314 (2003)

[CSZ06]   Chen, R.D., Song, Y., Zhou, H.: Convergence theorems for implicit iteration process for a finite family of continuous pseudocontractive mappings. J. Math. Anal. Appl., **314**, No. 2, 701–709 (2006)

[CnY04]   Chen, R.D., Yao, Y.H.: Approximating fixed points of uniformly ϕ-pseudo contractive mappings by iteration processes. (Chinese) Acta Anal. Funct. Appl., **6**, No. 1, 93–96 (2004)

[CHe04]   Chen, Yan, He, Z.: Three-step iterative approximation method of fixed points for strongly pseudo-contractive mappings in Banach spaces. (Chinese) J. Hebei Univ. Nat. Sci., **24**, No. 1, 15–17 (2004)

[Cen00]   Chen, Y.Z.: Inhomogeneous iterates of contraction mappings and nonlinear ergodic theorems. Nonlinear Anal. **39**, 1–10 (2000)

[Cen02]   Chen, Y.Z.: Stability of positive fixed points of nonlinear operators. Positivity, **6**, 47–57 (2002)

[Cng78]   Cheng, F.-C.: On the Mann iteration process in a uniformly convex Banach space. Kyungpook Math. J., **18**, 189–193 (1978)

[CPM94]   Chiaselotti, G., de Pascale, E., Marino, G.: Nonlinear contraction mappings in $(o) - E$-metric spaces. In: Altomare, F. (ed.) et al. Proceedings of the 2nd Int. Conf. Funct. Anal. Approx. Theory, Acquafredda di Maratea (Potenza), Sept. 14-19, 1992. Palermo: Circolo Matematico di Palermo, Suppl. Rend. Circ. Mat. Palermo, II. Ser. 33 (1994)

[Chi81]   Chidume, C.E.: On the approximation of fixed points of nonexpansive mappings. Houston J. Math., **7**, 345–355 (1981)

[Chi84]   Chidume, C.E.: The solution by iteration of nonlinear equations in certain Banach spaces. J. Nigerian Math. Soc., **3** (1984), 57–62 (1986)

[Chi85]   Chidume, C.E.: On the Ishikawa fixed points iterations for quasi-contractive mappings. J. Nigerian Math. Soc., **4** (1985), 1–11 (1987)

[Ch86a]   Chidume, C.E.: Quasi-nonexpansive mappings and uniform asymptotic regularity. Kobe J. Math., **3**, 29–35 (1986)

[Ch86b]   Chidume, C.E.: An approximation method for monotone Lipschitzian operators in Hilbert spaces. J. Austral. Math. Soc., Ser. A, **41**, 59–63 (1986)

[Ch86c]   Chidume, C.E.: The iterative solution of the equation $f \in x + Tx$ for a monotone operator T in L^p spaces. J. Math. Anal. Appl., **116**, 531–537 (1986)

[Ch86d]   Chidume, C.E.: Iterative construction of fixed points for multivalued operators of the monotone type. Appl. Anal., **23**, 209–218 (1986)

[Chi87]   Chidume, C.E.: Iterative approximation of fixed points of Lipschitzian strictly pseudo-contractive mappings. Proc. Amer. Math. Soc., **99**, 283–288 (1987)

[Chi88]   Chidume, C.E.: Fixed point iterations for certain classes of nonlinear mappings. Appl. Anal., **27**, No. 1-3, 31–45 (1988)

[Ch89a]   Chidume, C.E.: Fixed point iterations for nonlinear Hammerstein equation involving nonexpansive and accretive mappings. Indian J. Pure Appl. Math., **20**, No. 2, 129–135 (1989)

[Ch89b]   Chidume, C.E.: Iterative solution of nonlinear equations of the monotone and dissipative types. Appl. Anal., **33**, No. 1/2, 79–86 (1989)

[Ch90a] Chidume, C.E.: An iterative process for nonlinear Lipschitzian strongly accretive mappings in L^p spaces. J. Math. Anal. Appl., **151**, No. 2, 453–461 (1990)

[Ch90b] Chidume, C.E.: Iterative solution of nonlinear equations of the monotone type in Banach spaces. Bull. Austral. Math. Soc., **42**, No. 1, 21–31 (1990)

[Ch90c] Chidume, C.E.: Iterative methods for nonlinear set-valued operators of the monotone type with applications to operator equations. J. Nigerian Math. Soc., **9**, 7–20 (1990)

[Chi91] Chidume, C.E.: Approximation of fixed points of quasi-contractive mappings in L^p spaces. Indian J. Pure Appl. Math., **22**, No. 4, 273–286 (1991)

[Ch94a] Chidume, C.E.: An iterative method for nonlinear demiclosed monotone-type operators. Dyn. Syst. Appl., **3**, No. 3, 349–355 (1994)

[Ch94b] Chidume, C.E.: Approximation of fixed points of strongly pseudocontractive mappings. Proc. Amer. Math. Soc., **120**, No. 2, 545–551 (1994)

[Chi95] Chidume, C.E.: Iterative solution of nonlinear equations with strongly accretive operators. J. Math. Anal. Appl., **192**, No. 2, 502–518 (1995)

[Ch96a] Chidume, C.E.: Steepest descent approximations for accretive operator equations. Nonlinear Anal. TMA, **26**, No. 2, 299–311 (1996)

[Ch96b] Chidume, C.E.: Iterative solutions of nonlinear equations in smooth Banach spaces. Nonlinear Anal. TMA, **26**, No. 11, 1823–1834 (1996)

[Ch96c] Chidume, C.E.: Steepest solution of nonlinear equations with strongly accretive operators. Nonlinear Anal., **200**, 259–311 (1996)

[Ch98a] Chidume, C.E.: Convergence theorems for strongly pseudo-contractive and strongly accretive maps. J. Math. Anal. Appl., **228**, No. 1, 254–264 (1998)

[Ch98b] Chidume, C.E.: Global iteration schemes for strongly pseudo-contractive maps. Proc. Amer. Math. Soc., **126**, No. 9, 2641–2649 (1998)

[Ch98c] Chidume, C.E.: Iterative solutions of nonlinear equations of the strongly accretive type. Math. Nachr., **189**, 49–60 (1998)

[Chi00] Chidume, C.E.: Iterative methods for nonlinear Lipschitz pseudocontractive operators. J. Math. Anal. Appl., **251**, No. 1, 84–92 (2000)

[Chi01] Chidume, C.E.: Iterative approximation of fixed points of Lipschitz pseudocontractive maps. Proc. Amer. Math. Soc., **129**, No. 8, 2245–2251 (2001)

[Chi02] Chidume, C.E.: Convergence theorems for asymptotically pseudocontractive mappings. Nonlinear Anal. TMA, **49**, 1–11 (2002)

[Chi03] Chidume, C.E.: Iterative algorithms for nonexpansive mappings and some of their generalizations. In: Nonlinear Analysis and Applications: to V. Lakshmikantham on his 80th birthday. vol. 1-2. Kluwer Acad. Publ., Dordrecht (2003)

[Chi04] Chidume, C.E.: Strong convergence theorems for fixed points of asymptotically pseudocontractive semi-groups. J. Math. Anal. Appl., **296**, No. 2, 410–421 (2004)

[Chi05] Chidume, C.E.: Geometric Properties of Banach Spaces and Nonlinear Iterations. International Center for Theoretical Physics, Trieste (in print)

[ChA07] Chidume, C.E., Ali, B.: Approximation of common fixed points for finite families of nonself asymptotically nonexpansive mappings in Banach spaces. J. Math. Anal. Appl., **326**, No. 2, 960–973 (2007)

[ChA93] Chidume, C.E., Aneke, S.J.: Existence, uniqueness and approximation of a solution for a K-positive definite operator equation. Appl. Anal., **50**, No. 3-4, 285–294 (1993)

[ChC05] Chidume, C.E., Chidume, C.O.: Convergence theorems for fixed points of uniformly continuous generalized Φ-hemi-contractive mappings. J. Math. Anal. Appl., **303**, No. 2, 545–554 (2005)

[CC06a] Chidume, C.E., Chidume, C.O.: Iterative approximation of fixed points of nonexpansive mappings. J. Math. Anal. Appl., **318**, No. 1, 288–295 (2006)

[CC06b] Chidume, C.E., Chidume, C.O.: Convergence theorem for zeros of generalized Lipschitz generalized Φ-quasi-accretive operators. Proc. Amer. Math. Soc., **134**, No. 1, 243–251 (2006)

[ChI02] Chidume, C.E., Igbokwe, D.I.: Convergence theorems for asymptotically pseudocontractive maps. Bull. Korean Math. Soc., **39**, No. 3, 389–399 (2002)

[CKZ03] Chidume, C.E., Khumalo, M., Zegeye, H.: Generalized projection and approximation of fixed points of nonself maps. J. Approx. Theory, **120**, 242–252 (2003)

[CLU05] Chidume, C.E., Li, J.L., Udomene, A.: Convergence of paths and approximation of fixed points of asymptotically nonexpansive mappings. Proc. Amer. Math. Soc., **133**, No. 2, 437–480 (2005)

[ChL92] Chidume, C.E., Lubuma, M.-S.: Solution of the Stokes system by boundary integral equations and fixed point iterative schemes. Special issue in honor of Professor James O. C. Ezeilo. J. Nigerian Math. Soc., **11**, No. 3, 1–17 (1992)

[ChM97] Chidume, C.E., Moore, C.: The solution by iteration of nonlinear equations in uniformly smooth Banach spaces. J. Math. Anal. Appl., **215**, No. 1, 132–146 (1997)

[ChM99] Chidume, C.E., Moore, C.: Fixed point iteration for pseudocontractive maps. Proc. Amer. Math. Soc., **127**, No. 4, 1163–1170 (1999)

[ChM00] Chidume, C.E., Moore, C.: Steepest descent method for equilibrium points of nonlinear systems with accretive operators. J. Math. Anal. Appl., **245**, 142–160 (2000)

[ChM01] Chidume, C.E., Moore, C.: Iterative approximation of fixed points of Lipschitz pseudocontractive maps. Proc. Amer. Math. Soc., **129**, No. 8, 2245–2251 (2001)

[CMu01] Chidume, C.E., Mutangadura, S.A.: An example on the Mann iteration method for Lipschitz pseudocontractions. Proc. Amer. Math. Soc., **129**, No. 8, 2359–2363 (2001)

[CNn02] Chidume, C.E., Nnoli, B.V.C.: A necessary and sufficient condition for the convergence of the Mann sequence for a class of nonlinear operators. Bull. Korean Math. Soc., **39**, No. 2, 269–276 (2002)

[COZ03] Chidume, C.E., Ofoedu, E.U., Zegeye, H.: Strong and weak convergence theorems for asymptotically nonexpansive mappings. J. Math. Anal. Appl., **280**, No. 2, 364–374 (2003)

[ChO92] Chidume, C.E., Osilike, M.O.: Iterative solution for nonlinear integral equations of Hammerstein type. Special issue in honor of Professor Chike Obi J. Nigerian Math. Soc., **11**, No. 1, 9–18 (1992)

236   References

[ChO93]   Chidume, C.E., Osilike, M.O.: Fixed point iterations for quasi-contractive maps in uniformly smooth Banach spaces. Bull. Korean Math. Soc., **30**, No. 2, 201–212 (1993)

[ChO94]   Chidume, C.E., Osilike, M.O.: Fixed point iterations for strictly hemi-contractive maps in uniformly smooth Banach spaces. Numer. Funct. Anal. Optimization, **15**, No. 7-8, 779–790 (1994)

[CO95a]   Chidume, C.E., Osilike, M.O.: Approximation methods for nonlinear operator equations of the m-accretive type. J. Math. Anal. Appl., **189**, 225–239 (1995)

[CO95b]   Chidume, C.E., Osilike, M.O.: Ishikawa iteration process for nonlinear Lipschitz strongly accretive mappings. J. Math. Anal. Appl., **192**, No. 3, 727–741 (1995)

[CO95c]   Chidume, C.E., Osilike, M.O.: Approximation methods for nonlinear operator equations of the m-accretive type. J. Math. Anal. Appl., **189**, No. 1, 225–239 (1995)

[ChO98]   Chidume, C.E., Osilike, M.O.: Nonlinear accretive and pseudo-contractive operator equations in Banach spaces. Nonlinear Anal. TMA, **31**, No. 7, 779–789 (1998)

[ChO99]   Chidume, C.E., Osilike, M.O.: Iterative solutions of nonlinear accretive operator equations in arbitrary Banach spaces. Nonlinear Anal. TMA, **36**, No. 7, 863–872 (1999)

[ChO00]   Chidume, C.E., Osilike, M.O.: Equilibrium points for a system involving m-accretive operators. Proc. Edinburgh Math. Soc., **43**, 1–14 (2000)

[CSh05]   Chidume, C.E., Shahzad, N.: Strong convergence of an implicit iteration process for a finite family of nonexpansive mappings. Nonlinear Anal. TMA., **62**, No. 6 (A), 1149–1156 (2005)

[CSZ04]   Chidume, C.E., Shahzad, N., Zegeye, H.: Convergence theorems for mappings which are asymptotically nonexpansive in the intermediate sense. Numer. Funct. Anal. Optim., **25**, No. 3-4, 239–257 (2004)

[ChU06]   Chidume, C.E., Udomene, M.O.: Strong convergence theorems for uniformly continuous pseudocontractive maps. J. Math. Anal. Appl., **323**, No. 1, 88–99 (2006)

[ChZ99]   Chidume, C.E., Zegeye, H.: Approximation of the zeros of m-accretive operators. Nonlinear Anal. TMA, **37B**, No. 1, 81–96 (1999)

[ChZ01]   Chidume, C.E., Zegeye, H.: Global iterative schemes for accretive operators. J. Math. Anal. Appl., **257**, 364–377 (2001)

[ChZ02]   Chidume, C.E., Zegeye, H.: Iterative solution of $0 \in Ax$ for an m-accretive operator A in certain Banach spaces. J. Math. Anal. Appl., **269**, No. 2, 421–430 (2002)

[CZ03a]   Chidume, C.E., Zegeye, H.: Approximate fixed point sequences and convergence theorems for asymptotically pseudocontractive mappings. J. Math. Anal. Appl., **278**, No. 2, 354–366 (2003)

[CZ03b]   Chidume, C.E., Zegeye, H.: On note on two recent papers on approximation of fixed points. Indian J. Pure Appl. Math., **34**, No. 5, 701–703 (2003)

[CZ03c]   Chidume, C.E., Zegeye, H.: Approximation methods for nonlinear operator equations. Proc. Amer. Math. Soc., **131**, No. 8, 2467–2478 (2003)

[CZ03d]   Chidume, C.E., Zegeye, H.: Iterative solution of nonlinear equations of accretive and pseudocontractive types. J. Math. Anal. Appl., **282**, No. 2, 756–765 (2003)

[ChZ04]   Chidume, C.E., Zegeye, H.: Approximate fixed point sequences and convergence theorems for Lipschitz pseudocontractive maps. Proc. Amer. Math. Soc., **132**, No. 3, 831–840 (electronic) (2004)

[CZ05a]   Chidume, C.E., Zegeye, H.: Strong convergence theorems for asymptotically quasi-nonexpansive mappings. Commun. Appl. Nonlinear Anal., **12**, No. 1, 43–50 (2005)

[CZ05b]   Chidume, C.E., Zegeye, H.: Approximation of solutions of nonlinear equations of Hammerstein type in Hilbert space. Proc. Amer. Math. Soc., **133**, No. 3, 851–858 (2005)

[CZA02]   Chidume, C.E., Zegeye, H., Aneke, S.J.: Approximation of fixed points of weakly contractive nonself maps in Banach spaces. J. Math. Anal. Appl., **270**, No. 1, 189–199 (2002)

[CZA03]   Chidume, C.E., Zegeye, H., Aneke, S.J.: Iterative methods for fixed points of asymptotically weakly contractive maps. Appl. Anal., **82**, No. 7, 701–712 (2003)

[CZN99]   Chidume, C.E., Zegeye, H., Ntatin, B.: A generalized steepest descent approximation for the zeros of m-accretive operators. J. Math. Anal. Appl., **236**, No. 1, 48–73 (1999)

[CZP04]   Chidume, C. E., Zegeye, H., Prempeh, E.: Strong convergence theorems for a common fixed point of a finite family of nonexpansive mappings. Commun. Appl. Nonlinear Anal., **11**, No. 2, 25–32 (2004)

[Cho83]   Cho, Y.J.: On successive approximations for nonlinear mappings. Pure Appl. Math. Sci., **17**, No. 1-2, 1–5 (1983)

[Cho00]   Cho, Y.J.: Fixed Point Theory and Applications. In: Proceed. of the Internat. Conf. on Math. Anal. and Applications, Chinju, Korea, August 3-4, 1998. Nova Science, Huntington, New York (2000)

[CFH04]   Cho, Y.J., Fang, Y.P., Huang, N.-J., Hwang, H.J.: Algorithms for systems of nonlinear variational inequalities. J. Korean Math. Soc., **41**, No. 3, 489–499 (2004)

[CFK98]   Cho, Y.J., Fisher, B., Kang, S.M.: Common fixed point theorems for Mann type iterations. Math. Japon., **48**, No. 3, 385–390 (1998)

[CKZ05]   Cho, Y.J., Kang, S.M., Zhou H.: Some control conditions on iterative methods. Commun. Appl. Nonlinear Anal., **12**, No. 2, 27–34 (2005)

[CSJ03]   Cho, Y.J., Sahu, D.R., Jung, J.S.: Approximation of fixed points of asymptotically pseudocontractive mappings in Banach spaces. Southwest J. Pure Appl. Math., No. 2, 49–59 (electronic) (2003)

[CST96]   Cho, Y.J., Sharma, B.K., Thakur, B.S.: Weak convergence theorems for non-Lipschitzian asymptotically nonexpansive mappings in uniformly convex Banach spaces. Commun. Korean Math. Soc., **11**, No. 1, 131–137 (1996)

[CZG04]   Cho, Y.J., Zhou, H., Guo, G.T.: Weak and strong convergence theorems for three-step iterations with errors for asymptotically nonexpansive mappings. Comput. Math. Appl., **47**, No. 4-5, 707–717 (2004)

[CZK01]   Cho, Y.J., Zhou, H., Kang, S.M., Kim, S.S.: Approximations for fixed points of ϕ-hemicontractive mappings by the Ishikawa iterative process with mixed errors. Math. Comput. Modelling, **34**, No. 1-2, 9–18 (2001)

[Chy03]   Choudhury, B.S.: Random Mann iteration scheme. Appl. Math. Lett., **16**, No. 1, 93–96 (2003)

[Chy04]   Choudhury, B.S.: An iteration for finding a common random fixed point. J. Appl. Math. Stochastic Anal. 2004, No. 4, 385–394 (2004)

[CuD65] Chu, S.C., Diaz, J.B.: A fixed point theorem for 'in the large' application of the contraction principle. Atti Acad. Sci. Torino Cl. Sci. Fis. Mat. Natur., **99**, 351–363 (1964/1965)

[Cio90] Cioranescu, I.: Geometry of Banach spaces. Duality Mappings and Nonlinear Problems. Kluwer Academic Publishers (1990)

[Cir71] Ciric, L.B.: Generalized contractions and fixed-point theorems. Publ. l'Inst. Math. (Beograd), **12**, 19–26 (1971)

[Cir74] Ciric, L.B.: A generalization of Banach,s contraction principle. Proc. Amer. Math. Soc., **45**, 267–273 (1974)

[Cir75] Ciric, L.B.: On fixed point theorems in Banach spaces. Publ. Inst. Math., **19(33)**, 43–50 (1975)

[Cir77] Ciric, L.B.: Quasi-contractions in Banach spaces. Publ. Inst. Math., **21(35)**, 41–48 (1977)

[Cir97] Ciric, L.B.: A counterexample to a theorem of Xu. J. Math. Anal. Appl., **213**, 723–725 (1997)

[Cir98] Ciric, L.B.: Common fixed points of nonlinear contractions. Acta Math. Hungar., **80**, No. 1-2, 31–38 (1998)

[Cir99] Ciric, L.B.: Convergence theorems for a sequence of Ishikawa iteration for nonlinear quasi-contractive mappings. Indian J. Pure Appl. Appl. Math., **30**, No. 4, 425–433 (1999)

[Cir03] Ciric, L.B.: Fixed Point Theory. Contraction Mapping Principle. FME Press, Beograd (2003)

[CU03a] Ciric, L.B., Ume, J.S.: On the convergence of the Ishikawa iterates to a common fixed point of multivalued mappings. Demonstratio Math., **36**, No. 4, 951–956 (2003)

[CU03b] Ciric, L.B., Ume, J.S.: Ishikawa iterative process for strongly pseudocontractive operators in arbitrary Banach spaces. Math. Commun., **8**, No. 1, 43–48 (2003)

[CU03c] Ciric, L.B., Ume, J.S.: On the convergence of the Ishikawa iterates associated with a pair of multi-valued mappings. Acta Math. Hungar, **98**, No. 1-2, 1–8 (2003)

[CUm04] Ciric, L.B., Ume, J.S.: Iterative processes with errors for nonlinear equations. Bull. Austral. Math. Soc., 69, No. 2, 177–189 (2004)

[CK03a] Ciric, L.B., Ume, J.S., Khan, M.S.: On the convergence of the Ishikawa iterates to a common fixed point of two mappings. Arch. Math. (Brno), **39**, No. 1, 117–121 (2003)

[CK03b] Ciric, L.B., Ume, J.S., Khan, M.S.: On the convergence of the Ishikawa iterates to a common fixed point of two mappings. Arch. Math. (Brno) **39**, No. 2, 123–127 (2003)

[Col97] Collacao, P., Silva, J.C.E.: A complete comparison of 25 contraction conditions. Nonlinear Anal. TMA, **30**, 471–476 (1997)

[Com95] Combettes, P.L.: Construction d'un point fixe commun une famille de contractions fermes. C. R. Acad. Sci. Paris Sr. I Math., **320**, No. 11, 1385–1390 (1995)

[CoP02] Combettes, P.L., Pennanen, T.: Generalized Mann iterates for constructing fixed points in Hilbert spaces. J. Math. Anal. Appl., **275**, 521–536 (2002)

[Con94] Constantin, A.: On the approximation of fixed points of operators. Bull. Calcutta Math. Soc., **86**, 323–326 (1994)

[Cop55]   Coppel, W.A.: The solution of equations by iteration. Proc. Camb. Philos. Soc., **51**, 41–43 (1955)

[CrP77]   Crandal, M.G., Pazy, A.: On the range of accretive operators. Israel J. Math., **27**, 235–246 (1977)

[Cri76]   Cristescu, R.: Metoda aproximatiilor succesive in spatii liniare ordonate topologice. Stud. Cerc. Mat., **28**, 411–415 (1976)

[Cro02]   Crombez, G.: Finding common fixed points of strict paracontractions by averaging strings of sequential iterations. J. Nonlinear Convex Anal., **3**, No. 3, 345–351 (2002)

[Cro04]   Crombez, G.: Finding common fixed points of a class of paracontractions. Acta Math. Hungar., **103**, No. 3, 233–241 (2004)

[Da93a]   Dai, A.: The convergence of the sequence of Ishikawa iteration for quasi-contractive mappings. J. Nanjing Univ., Math. Biq., **10**, No. 1, 46–53 (1993)

[Da93b]   Dai, A.: On fixed point theorems and the convergence theorem of Ishikawa iteration sequence for some discontinuous operators. J. Nanjing Univ., Math. Biq., **10**, No. 2, 163–171 (1993)

[Dai04]   Dai, H.: Stability of iterative processes with errors for ϕ-pseudocontractive mappings. Sichuan Daxue Xuebao, **41**, No. 3, 462–465 (2004)

[DaD84]   Das, G., Debata, J.P.: On common fixed points of hemicontractive mappings. Indian J. Pure Appl. Math., **15**, No. 7, 713–718 (1984)

[DaD85]   Das, G., Debata, J.P.: Convergence of Ishikawa iteration of quasicontractive mappings. Bull. Inst. Math., Acad. Sin., **13**, 297–302 (1985)

[DaD86]   Das, G., Debata, J.P.: Fixed points of quasi-nonexpansive mappings. Indian J. Pure Appl. Math., **17**, 1263–1269 (1986)

[DaD88]   Das, G., Debata, J.P.: Fixed points on unit interval using infinite matrices. J. Indian Math. Soc., New Ser., **53**, No. 1-4, 167–176 (1988)

[DMS95]   Das, G., Manjari Swain, M.: A convergence theorem of nonexpansive mappings in Hilbert spaces. Proc. Nat. Acad. Sci. India Sect. A, **65**, No. 4, 445–453 (1995)

[DMS00]   Das, G., Manjari Swain, M.: Approximating fixed points of nonexpansive mappings. Proc. Natl. Acad. Sci. India Sect. A Phys. Sci., **70**, No. 3, 295–308 (2000)

[DSW81]   Das, K.M., Singh, S.P., Watson, B.: A note on Mann iteration for quasi-nonexpansive mappings. Nonlinear Anal. TMA, **5**, No. 6, 675–676 (1981)

[DBM76]   De Blasi, F.S., Myjak, J.: Sur la convergence des approximations successives pour les contractions non lineaires dans un espace de Banach. C.R. Acad. Sci. Paris, **283**, 185–187 (1976)

[Dei74]   Deimling, K.: Zeros of accretive operators. Manuscripta Math., **13**, 365–374 (1974)

[Dei85]   Deimling, K.: Nonlinear Functional Analysis. Springer Verlag, Berlin (1985)

[Dg93a]   Deng, L.: On Chidume's open questions. J. Math. Anal. Appl., **174**, No. 2, 441–449 (1993)

[Dg93b]   Deng, L.: An iterative process for nonlinear Lipschitzian and strongly accretive mappings in uniformly convex and uniformly smooth Banach spaces. Acta Appl. Math., **32**, No. 2, 183–196 (1993)

[Dng94]   Deng, L.: Iteration processes for nonlinear Lipschitzian strongly accretive mappings in L_p spaces. J. Math. Anal. Appl., **188**, No. 1, 128–140 (1994)

[Dng95]   Deng, L.: Iterative approximation of Lipschitz strictly pseudo-contractive mappings in uniformly smooth Banach spaces. Nonlinear Anal. TMA, **24**, No. 7, 981–987 (1995)

[Dng96]   Deng, L.: Convergence of the Ishikawa iteration process for nonexpansive mappings. J. Math. Anal. Appl., **199**, No. 3, 769–775 (1996)

[Dng99]   Deng, L.: The Ishikawa iteration process for nonexpansive mappings in uniformly convex Banach spaces (Chinese). Xinan Shifan Daxue Xuebao Ziran Kexue Ban, **24**, No. 2, 142–144 (1999)

[DDi92]   Deng, L., Ding, X.P.: Ishikawa's iterations of real Lipschitz functions. Bull. Austral. Math. Soc., **46**, No. 1, 107–113 (1992)

[DD94a]   Deng, L., Ding, X.P.: Iterative process for Lipschitz local strictly pseudo-contractive mappings. Appl. Math. Mech., Engl. Ed., **15**, No. 2, 119–123 (1994)

[DD94b]   Deng, L., Ding, X.P.: Iterative construction of fixed points for multivalued operators of the monotone type in uniformly smooth Banach spaces. Appl. Math. Mech. (English Ed.), **15**, No. 10, 897–902 (1994)

[DDi95]   Deng, L., Ding, X.P.: Iterative approximation of Lipschitz strictly pseudo-contractive mappings in uniformly smooth Banach spaces. Nonlinear Anal. TMA, **24**, No. 7, 981–987 (1995)

[DJi00]   Deng, L., Jiang, X.Y.: Fixed points in the Ishikawa iteration process for nonexpansive mappings (Chinese). Xinan Shifan Daxue Xuebao Ziran Kexue Ban, **25**, No. 1, 1–3 (2000)

[DL00a]   Deng, L., Li, S.H.: Ishikawa iteration process with errors for nonexpansive mappings in uniformly convex Banach spaces. Int. J. Math. Math. Sci., **24**, No. 1, 49–53 (2000)

[DL00b]   Deng, L., Li, S.H.: The Ishikawa iteration process for nonexpansive mappings in uniformly convex Banach spaces (Chinese). Chinese Ann. Math. Ser. A, **21**, No. 2, 159–164 (2000). Translation in Chinese J. Contemp. Math., **21**, No. 2, 127–132 (2000)

[DXi04]   Deng, L., Xia, X.: Modified Ishikawa iterative sequences with errors for uniformly quasi-Lipschitzian mappings. (Chinese) Acta Anal. Funct. Appl., **6**, No. 2, 146–149 (2004)

[dPM93]   de Pascale, E., Marino, G., Pietramala, P.: The use of the E-metric spaces in the search for fixed points. Matematiche, **48**, No. 2, 367–376 (1993)

[DeY98]   Deutsch, F., Yamada, I.: Minimizing certain convex functions over the intersection of the fixed point sets of nonexpansive mappings. Numer. Funct. Anal. Optim., **19**, No. 1, 2, 33–56 (1998)

[Dha88]   Dhage, B.C.: On approximating common fixed points of some mappings. J. Math. Phys. Sci., **22**, No. 6, 775–788 (1988)

[DhS92]   Dhage, B.C., Sharma, S.: Approximate common fixed points of some quasi-contraction mappings. Indian J. Pure Appl. Math., **23**, No. 11, 763–771 (1992)

[DiM67]   Diaz, J.B., Metcalf, F.T.: On the structure of the set of subsequential limit points of successive approximations. Bull. Amer. Math. Soc., **73**, 516–519 (1967)

[DiM69]   Diaz, J.B., Metcalf, F.T.: On the set of subsequential limit points of successive approximations. Trans. Amer. Math. Soc., **135**, 459–485 (1969)

[Die75]   Diestel, J.: Geometry of Banach Spaces. Selected Topics. Lectures Notes in Mathematics, Vol. 485, Springer Verlag (1975)

[DiL81]   Di Lena, G.: Convergenza globale del metodo delle approssimazioni successive in $\mathbb{R}^m$ per una classe di funzioni. Boll. Unione Mat. Ital., V. Ser. A, **18**, 235–241 (1981)

[DiL85]   Di Lena, G.: Global convergence of the method of successive approximations on S^1. J. Math. Anal. Appl., **106**, 196–201 (1985)

[DLM86]   Di Lena, G., Messano, B.: Global convergence and global plus-convergence of the method of successive approximations in closed subsets of $\mathbb{R}^h$ (Italian). Rend. Mat. Appl., VII. Ser., **6**, No. 1/2, 199–214 (1986)

[DLM94]   Di Lena, G., Messano, B., Roux, D.: On the successive approximations method for isotone functions. Boll. Unione Mat. Ital., VII. Ser., A, **8**, No. 2, 169–180 (1994)

[DLM97]   Di Lena, G., Messano, B., Roux, D.: Rigid sets and nonexpansive mappings. Proc. Amer. Math. Soc., **125**, No. 12, 3575–3580 (1997)

[DMZ87]   Di Lena, G., Messano, B., Zitarosa, A.: On the global convergence of the generalized successive approximation method (Italian). Ric. Mat., **36**, No. 2, 278–288 (1987)

[DMZ88]   Di Lena, G., Messano, B., Zitarosa, A.: On the generalized successive approximation method. Calcolo, **25**, No. 3, 249–267 (1988)

[DMZ95]   Di Lena, G., Messano, B., Zitarosa, A.: Results related to the interactive process $x_{n+1} = f(x_n, x_{n-1})$ (Italian). Ric. Mat., **44**, No. 1, 109–130 (1995)

[DLP78]   Di Lena, G., Peluso, R.I.: A characterization of global convergence for fixed point iteration in $\mathbb{R}^1$. Pubbl., Ser. III, Ist. Appl. Calcolo 133, 11 p. (1978)

[DLP81]   Di Lena, G., Peluso, R.I.: Sulla convergenza del metodo delle approssimazioni successive in $\mathbb{R}^1$. Calcolo, **17**, 313–319 (1981)

[Din81]   Ding, X.P.: Iteration method to construct fixed points of nonlinear mappings (Chinese). Math. Numer. Sin., **3**, 285–295 (1981)

[Din88]   Ding, X.P.: Iteration processes for nonlinear mappings in convex metric spaces. J. Math. Anal. Appl., **132**, No. 1, 114–122 (1988)

[Di93a]   Ding, X.P.: Weak contractor directions and weak directional contractions for a set valued operator with a closed range. Zb. Rad. Prirod.-Mat. Fak. Ser. Mat., **23**, No. 1, 39–50 (1993)

[Di93b]   Ding, X.P.: Approximating fixed points of asymptotically quasinonexpansive mappings by Ishikawa iteration. Sichuan Shifan Daxue Xuebao Ziran Kexue Ban, **16**, No. 4, 43–49 (1993)

[Di96a]   Ding, X.P.: Approximation of fixed points for monotone multivalued operators in uniformly smooth Banach spaces. J. Sichuan Normal Univ. (Sichuan Shifan Daxue Xuebao Ziran Kexue Ban), **19**, No. 3, 1–9 (1996)

[Di96b]   Ding, X.P.: Iterative process with errors to locally strictly pseudocontractive maps in Banach spaces. Comput. Math. Appl., **32**, No. 10, 91–97 (1996)

[Di96c]   Ding, X.P.: Iterative solutions of nonlinear equations involving accretive and dissipative operators. Sichuan Shifan Daxue Xuebao Ziran Kexue Ban, **19**, No. 6, 1–12 (1996)

[Di97a]   Ding, X.P.: Iterative solution of equation $f \in x + Tx$ for an accretive operator T in uniformly smooth Banach spaces. Indian J. Pure Appl. Math., **28**, No. 1, 13–21 (1997)

[Di97b]   Ding, X.P.: Iterative process with errors of nonlinear equations involving m-accretive operators. J. Math. Anal. Appl., **209**, No. 1, 191–201 (1997)

[Di97c]   Ding, X.P.: Iterative processes with errors for finding fixed points of multi-valued monotone operators (Chinese). J. Sichuan Normal Univ. (Sichuan Shifan Daxue Xuebao Ziran Kexue Ban), **20**, No. 2, 54–59 (1997)

[Di97d]   Ding, X.P.: Iterative process with errors to nonlinear ϕ-strongly accretive operator equations in arbitrary Banach spaces. Comput. Math. Appl., **33**, No. 8, 75–82 (1997)

[Din98]   Ding, X.P.: Iteration process with errors to nonlinear equations in arbitrary Banach spaces. Acta Math. Sinica (N.S.), **14**, suppl., 577–584 (1998)

[DiD94]   Ding, X.P., Deng, L.: Iterative solution of nonlinear equations of the monotone and dissipative types in uniformly smooth Banach spaces. J. Sichuan Normal Univ. (Sichuan Shifan Daxue Xuebao Ziran Kexue Ban), **17**, No. 1, 43–48 (1994)

[DiD96]   Ding, X.P., Deng, L.: The iterative solution of the equation $f \in x + Tx$ for a monotone operator T in uniformly smooth Banach spaces. Chinese J. Math., **24**, No. 4, 307–314 (1996)

[DiZ00]   Ding, X.P., Zhang, H.L.: Iterative process to ϕ-hemicontractive operator and ϕ-strongly accretive operator equations. Appl. Math. Mech. (English Ed.), **21**, No. 11, 1256–1263 (2000)

[Dja90]   Djafari Rouhani, B.: Asymptotic behaviour of almost nonexpansive sequences in a Hilbert space. J. Math. Anal. Appl., **151**, No. 1, 226–235 (1990)

[Dja95]   Djafari Rouhani, B.: Asymptotic behaviour of firmly nonexpansive sequences. Proc. Amer. Math. Soc., **123**, No. 3, 771–777 (1995)

[DTW01]   Djafari Rouhani, B., Tarafdar, E., Watson, P.J.: Fixed point theorems, coincidence theorems and variational inequalities. In: Hadjisavvas, N. (ed.) et al. Generalized Convexity and Generalized Monotonicity. Proceedings of the 6th International Symposium, Samos, Greece, September 1999. Springer, Berlin (2001)

[DoB96]   Dominguez Benavides, T.(ed.): Recent Advances on Metric Fixed Point Theory. Universidad de Sevilla, Ciencias, 48 (1996)

[DFR03]   Donchev, T., Farkhi, E., Reich, S.: Fixed set iterations for relaxed Lipschitz multimaps. Nonlinear Anal., **53**, No. 7-8, 997–1015 (2003)

[Dot70]   Dotson, W.G.: On the Mann iterative process. Trans. Amer. Math. Soc., **149**, 65–73 (1970)

[Dot71]   Dotson, W.G.: Mean ergodic theorems and iterative solution of linear functional equations. J. Math. Anal. Appl., **34**, 141–150 (1971)

[Dot72]   Dotson, W.G.: Fixed points of quasi-nonexpansive mappings. J. Austral. Math. Soc., **13**, 167–170 (1972)

[Dot78]   Dotson, W.G.: An iterative process for nonlinear monotonic nonexpansive operators in Hilbert space. Math. Comp., **32**, No. 151, 223–225 (1978)

[DoM68]   Dotson, W.G., Mann, W.R.: A generalized corollary of the Browder-Kirk fixed point theorem. Pacific J. Math., **26**, 455–459 (1968)

[Dow77]   Downing, D.J.: Fixed-point theorems and surjectivity results for nonlinear mappings in Banach spaces. Ph.D. Thesis, Iowa State University, Ames (1977)

[DuG82]   Dugundji, J., Granas, A.: Fixed Point Theory. Monografie Matematycne, Warsazawa (1982)

[Dun73]   Dunn, J.C.: On recursive averaging processes and Hilbert space extensions of the contraction mapping principle. J. Franklin Inst., **295**, 117–133 (1973)

[Dun78]   Dunn, J.C.: Iterative construction of fixed points for multivalued opera-
          tors of the monotone type. J. Funct. Anal., **27**, 38–50 (1978)
[Dun79]   Dunn, J.C.: A relaxed Picard iteration process for set-value operators of
          the monotone type. Proc. Amer. Math. Soc., **73**, 319–327 (1979)
[Dzi93]   Dzitac, I.: Solving on multiprocessors of nonlinear fixed point systems via
          the asynchronous iteration method (in Romanian). Anal. Univ. Oradea,
          Seria Matematica, **3**, 97–102 (1993)
[Eav71]   Eaves, B.C.: Computing Kakutani fixed points. SIAM J. Appl. Math., **21**,
          236–244 (1971)
[Eav72]   Eaves, B.C.: Homotopies for computation of fixed points. Math. Program-
          ming, **3**, 1–22 (1972)
[Eav76]   Eaves, B.C.: A short course in solving equations with PL-homotopies.
          SIAM-AMS Proceedings, **9**, 73–143 (1976)
[EaS72]   Eaves, B.C., Saigal, R.: Homotopies for computation of fixed points on
          unbounded regions. Math. Programming, **3**, 225–237 (1972)
[Ede66]   Edelstein, M.: A remark on a theorem of M. A. Krasnoselski. Amer. Math.
          Monthly, **73**, 509–510 (1966)
[Ede72]   Edelstein, M.: The construction of an asymptotic center with a fixed-point
          property. Bull. Amer. Math. Soc., **78**, 206–208 (1972)
[Ede82]   Edelstein, M.: On fixed and periodic point under contractive mappings.
          J. London Math. Soc., **25**, 139–144 (1982)
[EdO78]   Edelstein, M., O'Brian, R.C.: Nonexpansive mappings, asymptotic regu-
          larity and successive approximations. J. London Math. Soc., **17**, No. 3,
          547–554 (1978)
[Em82a]   Emmanuele, G.: Convergence of the Mann-Ishikawa iterative process for
          nonexpansive mappings. Nonlinear Anal. TMA, **6**, No. 10, 1135–1141
          (1982)
[Em82b]   Emmanuele, G.: A remark on my paper "Convergence of the Mann-
          Ishikawa iterative process for nonexpansive mappings". Nonlinear Anal.
          TMA, **7**, No. 5, 473–474 (1982)
[Emm84]   Emmanuele, G.: A remark on a paper: "Common fixed points of nonex-
          pansive mappings by iteration" [Pacific J. Math. 97 (1981), No. 1, 137–
          139; MR0638181 (82k:47076)] by P. K. F. Kuhfittig. Pacific J. Math., **110**,
          No. 2, 283–285 (1984)
[Emm85]   Emmanuele, G.: Asymptotic behavior of iterates of nonexpansive map-
          pings in Banach spaces with Opial's condition. Proc. Amer. Math. Soc.,
          94, 103–109 (1985)
[En77a]   Engl, H.W.: Weak convergence of Mann iteration for nonexpansive map-
          pings without convexity assumptions. Boll. Unione Mat. Ital., V. Ser., A
          14, No. 3, 471–475 (1977)
[En77b]   Engl, H.W.: Schwache Konvergenz asymptotisch regulärer Iterationsver-
          fahren bei Fixpunktgleichungen nichtexpansiver Funktionen. Z. angew.
          Math. Mech., **57**, 272–273 (1977)
[EnL01]   Engl, H.W., Leitao, A.: A Mann iterative regularization method for el-
          liptic Cauchy problems. Numer. Funct. Anal. Optimization, **22**, No. 7-8,
          861–884 (2001)
[Eva76]   Evans, C.: Homotopies for computation of fixed points. Math. Program-
          ming, SIAM, **3**, 1–22, (1976)

[EvZ85]   Evhuta, N.A., Zabrejko, P.P.: On the convergence of the successive approximations in Samojlenko's method for finding periodic solutions (Russian). Dokl. Akad. Nauk. BSSR, **29**, No. 1, 15–18 (1985)

[Fal96]   Falkowski, B.J.: On the convergence of Hillam's iteration scheme. Math. Mag. **69**, 299–303 (1996)

[FKH02]   Fang, Y.-P., Kim, J.K., Huang, N.-J.: Stable iterative procedures with errors for strong pseudocontractions and nonlinear equations of accretive operators without Lipschitz assumption. Nonlinear Funct. Anal. Appl., **7**, No. 4, 497–507 (2002)

[FeD79]   Feathers, G., Dotson, W.G.: A nonlinear theorem of ergodic type. II. Proc. Amer. Math. Soc., **73**, 37–39 (1979)

[FWD79]   Feathers, G., Wayne Pace, J., Dotson, W.G.: A nonlinear theorem of ergodic type. Proc. Amer. Math. Soc., **73**, 35–36 (1979)

[FeN05]   Feng, X.Z., Ni, R.X.: A convergence theorem for modified Reich-Takahashi iterative sequences with random errors for asymptotically non-expansive mappings. (Chinese). Qufu Shifan Daxue Xuebao Ziran Kexue Ban, **31**, No. 1, 19–24 (2005)

[For80]   Forster, W. (ed.): Numerical solution of highly nonlinear problems. Fixed point algorithms and complementary problems. North-Holland, Amsterdam (1980)

[Frn02]   Franklin, J.N., Methods of Mathematical Economics: Linear and Nonlinear Programing, Fixed Point Theorems. (Rev & corr. ed.). SIAM, Philadelphia (2002)

[FrM71]   Franks, R.L., Mrazec, R.P.: A theorem on mean-value iterations. Proc. Amer. Math. Soc., **30**, 324–326 (1971)

[FrK67]   Frum-Ketkov, R.L.: Mappings into a Banach space sphere. Dokl. Akad. Nauk SSSR, **175**, 1229–1231 (1967)

[Fuc77]   Fuchssteiner, B.: Iterations and fixpoints. Pacific J. Math., **68**, No. 1, 73–80 (1977)

[Gal89]   Gal, Sorin G.: A construction of monotonically convergent sequences from successive approximations in certain Banach spaces. Numer. Math., **56**, 67–71 (1989)

[Gan80]   Ganguly, A.: On common fixed point of two mappings. Math. Semin. Notes, Kobe Univ., **8**, 343–345 (1980)

[Gan91]   Ganguly, D.K., Bandyopadhyay, D.: Some results on fixed point theorem using infinite matrix of regular type. Soochow J. Math., **17**, 269–285 (1991)

[GB96a]   Ganguly, D.K., Bandyopadhyay, D.: Fixed point theorems for multifunctions. J. Nat. Phys. Sci., **9-10**, 77–86 (1996)

[GB96b]   Ganguly, D.K., Bandyopadhyay, D.: Approximation of fixed points in Banach space by iteration processes using infinite matrices. Soochow J. Math., **22**, No. 3, 395–403 (1996)

[GaB98]   Ganguly, D.K., Bandyopadhyay, D.: Fixed point and summability method on iteration in Banach spaces. Kyungpook Math. J., **38**, No. 2, 235–243 (1998)

[GCG04]   Gao, G.L., Chen, D.Q., Guo, J.T., Wu, C.Y.: Modified Mann iterative schemes with errors. (Chinese). J. Hebei Norm. Univ., Nat. Sci. Ed., **28**, No. 2, 113–115,119 (2004)

[GZC03] Gao, G.L., Zhou, H., Chen, D.Q.: Multi-step iterations for asymptotically nonexpansive mappings. Acta Anal. Funct. Appl., **5**, No. 2, 119–124 (2003)

[GKK01] Garcia Falset, J., Kaczor, W., Kuczumow, T., Reich, S.: Weak convergence theorems for asymptotically nonexpansive mappings and semigroups. Nonlinear Anal., **43**, 377–401 (2001)

[GaK72] Gatica, J.A., Kirk, W.A.: Fixed point theorems for Lipschitzian pseudocontractive mappings. Proc. Amer. Math. Soc., **36**, 111–115 (1972)

[GeC03] Ge, C.S.: Iterative approximation of fixed points of Φ-hemicontractive operators in Banach spaces (Chinese). Pure Appl. Math. (Xi'an), **19**, No. 2, 173–178 (2003)

[GeL75] Genel, A., Lindenstrauss, J.: An example concerning fixed points. Israel J. Math., **22**, 81–86 (1975)

[Gho80] Ghosh, M.K.: A note on a theorem of Rhoades. Math. Sem. Notes, Kobe Univ., **8**, 505–507 (1980)

[Gh95a] Ghosh, M.K.: Approximating common fixed points of families of quasi-nonexpansive mappings. Ganita, **46**, No. 1-2, 53–58 (1995)

[Gh95b] Ghosh, M.K.: Approximating common fixed points of families of quasi-nonexpansive mappings. Ganita, **46**, No. 1-2, 53–58 (1995)

[GhD92] Ghosh, M.K., Debnath, L.: Approximation of the fixed points of quasi-nonexpansive mappings in a uniformly convex Banach space. Appl. Math. Lett., **5**, No. 3, 47–50 (1992)

[GhD95] Ghosh, M.K., Debnath, L.: Approximating common fixed points of families of quasi-nonexpansive mappings. Internat. J. Math. Math. Sci., **18**, No. 2, 287–292 (1995)

[GD97a] Ghosh, M.K., Debnath, L.: Convergence of Ishikawa iterates of generalized nonexpansive mappings. Internat. J. Math. Math. Sci., **20**, No. 3, 517–520 (1997)

[GD97b] Ghosh, M.K., Debnath, L.: Convergence of Ishikawa iterates of quasi-nonexpansive mappings. J. Math. Anal. Appl., **207**, No. 1, 96–103 (1997)

[GiK96] Gillespie, A., Kannan, R.: Iterative process for finding common fixed points of nonlinear mappings. In: Kartsatos, A.G. (ed.) Theory and Applications of Nonlinear Operators of Accretive and Monotone Type. Lect. Notes Pure Appl. Math. 178. Marcel Dekker, New York (1996)

[Gin82] Gindac, F.: Successive approximations in uniform spaces. (Romanian) Stud. Cerc. Mat., **34**, No. 5, 416–424 (1982)

[Gin89] Gindac, F.: On the method of successive approximations. (Romanian) Bul. Inst. Politehn. Bucureşti Ser. Transport. Aeronave, **51**, 3–8 (1989)

[GbK83] Goebel, K., Kirk, W.A.: Iteration processes for nonexpansive mappings. Contemp. Math., **21**, 115–123 (1983)

[GbK90] Goebel, K., Kirk, W.A.: Topics in Metric Fixed Point Theory. Cambridge University Press, Cambridge (1990)

[GKS78] Goebel, K., Kirk, W.A., Shimi, T.S.: A fixed point theorem in uniformly convex spaces. Boll. Un. Mat. Ital. A, **15**, 67–75 (1978)

[GbR84] Goebel, K., Reich, S.: Uniform Convexity, Hyperbolic Geometry and Nonexpansive Mappings. Marcel Dekker, New York (1984)

[Goh65] Gohde, D.: Zum Prinzip der kontraktiven Abbildung. Math. Nachr., **30**, 251–258 (1965)

[GoL82] Goncharov, G.M., Lubashevskii, V.K.: Calculation of general fixed points for generalized-nonexpanding operators by the iteration method. (Russian) Functional analysis, No. 19, 58–68, Ulyanovsk. Gos. Ped. Inst., Ulyanovsk (1982)

[GoL84] Goncharov, G.M., Lubashevskii, V.K.: Convergence of iterations to the common fixed point of aggregates of operators. (Russian). In: Functional analysis, No. 23, 62–68, Ulyanovsk. Gos. Ped. Inst., Ulyanovsk, (1984)

[Gor89] Gornicki, J.: Weak convergence theorems for asymptotically nonexpansive mappings in uniformly convex Banach spaces. Comment. Math. Univ. Carolin., **30**, 249–252 (1989)

[Gor91] Gornicki, J.: Nonlinear ergodic theorems for asymptotically nonexpansive mappings in Banach spaces satisfying Opial's condition. J. Math. Anal. Appl., **161**, 440–446 (1991)

[Gra02] Graca, M.M.: Acceleration of nonhyperbolic sequences of Mann. Int. J. Math. Math. Sci. **32**, No. 9, 565–572 (2002)

[GrD03] Granas, A., Dugundji, J.: Fixed Point Theory. Springer, New York (2003)

[Gro72] Groetsch, C.W.: A note on segmenting Mann iterates. J. Math. Anal. Appl., **40**, 369–372 (1972)

[Gr74a] Groetsch, C.W.: A nonstationary iterative process for nonexpansive mappings. Proc. Amer. Math. Soc., **43**, 155–158 (1974)

[Gr74b] Groetsch, C.W.: Summation methods associated with an iteration. Nanta Math., **7**, No. 2, 13–16 (1974)

[Gro77] Groetsch, C.W.: Some aspects of Mann's iterative method for approximating fixed points. In: Fixed points: Algorithms and Applications (Proc. First Internat. Conf., Clemson Univ., Clemson, S.C., 1974). Academic Press, New York (1977)

[Gu99a] Gu, F.: An iterative process for a class of nonlinear mappings satisfying a generalized Lipschitz condition (Chinese). Math. Appl., **12**, No. 3, 44–48 (1999)

[Gu99b] Gu, F.: Iterative approximation of solutions to a class of nonlinear operator equations (Chinese). Pure Appl. Math., **15**, No. 2, 93–98 (1999)

[Gu01a] Gu, F.: Some strong convergence theorems for Ishikawa iterative sequence of certain nonlinear operators (Chinese). Acta Math. Sci. Ser. A Chin. Ed., **21**, No. 1, 102–109 (2001)

[Gu01b] Gu, F.: Iteration processes for approximating fixed points of operators of monotone type. Proc. Amer. Math. Soc., **129**, No. 8, 2293–2300 (2001)

[Gu01c] Gu, F.: Ishikawa iterative approximation of fixed points for ϕ-strongly pseudo-contractive mappings (Chinese). J. Eng. Math., Xi'an, **18**, No. 1, 63–67 (2001)

[Gu01d] Gu, F.: Iterative process for certain nonlinear mappings with Lipschitz condition (Chinese). Appl. Math. Mech. (English Ed.), **22**, No. 12, 1458–1467 (2001). Translated from Appl. Math. Mech., **22**, No. 12, 1309–1316 (2001)

[Gu01e] Gu, F.: Convergence theorems of Φ-pseudo contractive type mappings in normed linear spaces. Northeast. Math. J., **17**, No. 3, 340–346 (2001)

[Gu02a] Gu, F.: On the convergence problems of Ishikawa and Mann iterative processes with error for Φ-pseudo contractive type mappings. Chin. Ann. Math., Ser. A, **23**, No. 1, 49–54 (2002)

[Gu02b]   Gu, F.: Convergence problems of the Ishikawa and Mann iterative processes with error for Φ-pseudo-contractive type mappings. (Chinese) Chinese Ann. Math. Ser. A, **23**, No. 1, 49–54 (2002). Translation in Chinese J. Contemp. Math., **23**, No. 1, 39–44 (2002)

[Gu03a]   Gu, F.: Convergence and stability of the Ishikawa iteration procedures with mixed errors for nonlinear equations of the accretive type. (Chinese) Sichuan Shifan Daxue Xuebao Ziran Kexue Ban, **26**, No. 3, 257–260 (2003)

[GuD03]   Gu, F., Du, X.F.: Iterative approximations of fixed points for asymptotically pseudo-contractive mappings in normed linear spaces. Acta Analysis Functionalis Applicata, **5**, No. 2, 125–131 (2003)

[GG00a]   Gu, F., Gao, W.: Approximating fixed points of Φ-hemicontractive mappings by the Ishi-kawa iteration process in normed linear spaces (Chinese). Pure Appl. Math., **16**, No. 2, 50–54 (2000)

[GG00b]   Gu, F., Gao, W.: Ishikawa iterative approximation with errors for fixed points of multivalued operators of monotone type (Chinese). Natur. Sci. J. Harbin Normal Univ., **16**, No. 3, 13–17 (2000)

[GHL00]   Gu, F., Han, Y., Liu, C.P.: Iterative approximation of fixed points for multi-valued Φ-pseudo contractive type mappings in uniformly smooth Banach spaces (Chinese). Heilongjiang Daxue Ziran Kexue Xuebao, **17**, No. 4, 11–13 (2000)

[GHL01]   Gu, F., Han, Y., Liu, C.P.: Some strong convergence theorems for Ishikawa iterative sequence of certain nonlinear operators (Chinese). Acta Math. Sci., Ser. A, Chin. Ed., **21**, No. 1, 102–109 (2001)

[GLu04]   Gu, F., Lu, J.: Stability of Mann and Ishikawa iterative processes with errors for a class of nonlinear variational inclusions problem. Math. Commun., **9**, No. 2, 149–159 (2004)

[GuQ00]   Gu, F., Qin, Y.X.: Iterative approximation of solutions for m-accretive operator equations in Banach spaces (Chinese). Natur. Sci. J. Harbin Normal Univ., **16**, No. 5, 8–11 (2000)

[GSi83]   Guay, M.D., Singh, K.L.: Fixed points of asymptotically regular mappings. Mat. Vesnik, **35**, 101–106 (1983)

[GSi84]   Guay, M.D., Singh, K.L.: Convergence of sequences of iterates for a pair of mappings. J. Math. Phy. Sci., **18**, 461–472 (1984)

[GSi86]   Guay, M.D., Singh, K.L.: A nonstationary iteration process for nonexpansive type mappings. Bull. Math. Soc. Sci. Math. Repub. Soc. Roum., Nouv. Ser., **30**(78), 117–121 (1986)

[GSW84]   Guzzardi, R., Singh, S.P., Watson, B.: Convergence of the sequence of iterates of nonexpansive mappings (a survey). Numerical methods of approximation theory, Vol. 7 (Oberwolfach, 1983), 99–104. Internat. Schriftenreihe Numer. Math., 67, Birkhuser, Basel (1984)

[Gwi78]   Gwinner, J.: On the convergence of some iteration processes in uniformly convex Banach spaces. Proc. Amer. Math. Soc., **71**, 29–35 (1978)

[HJ90a]   Ha, K.S., Jung, J.S.: Convergence of approximants in Banach spaces. In: Differential Equations: Stability and Control, Proc. Int. Conf., Colorado Springs/CO (USA) 1989, Lect. Notes Pure Appl. Math. 127 (1990)

[HJ90b]   Ha, K.S., Jung, J.S.: Strong convergence theorems for accretive operators in Banach spaces. J. Math. Anal. Appl., **147**, 330–339 (1990)

[Had77]   Hadzic, O.: Osnovi teorije nepokretne tacke (Fundamental Elements of Fixed Point Theory). Institut za Matematiku, Novi Sad (1977)

[Hal67]   Halpern, B.: Fixed points of nonexpanding maps. Bull. Amer. Math. Soc., **73**, 957–961 (1967)

[HZK02]   Hao, J., Zhang, L., Kang, S.M.: Fixed point iteration for quasi-contractive mappings. In: Fixed Point Theory and Applications. Vol. 3, 71–81, Nova Sci. Publ., Huntington, NY (2002)

[HAK05]   Hao, J., An, Z., Kang, S.M., Kim, H.K.: Iterative approximation of fixed points for a class of generalized nonexpansive mappings. Commun. Appl. Nonlinear Anal., **12**, No. 1, 69–75 (2005)

[Har87]   Harder, A.M.: Fixed point theory and stability results for fixed points iteration procedures. PhD Thesis, University of Missouri-Rolla (1987)

[HH88a]   Harder, A.M., Hicks, T.L.: A stable iteration procedure for nonexpansive mappings. Math. Japon., **33**, No. 5, 687–692 (1988)

[HH88b]   Harder, A.M., Hicks, T.L.: Stability results for fixed point iteration procedures. Math. Japon., **33**, No. 5, 693–706 (1988)

[HH88c]   Harder, A.M., Hicks, T.L.: Fixed point theory and iteration procedures. Indian J. Pure Appl. Math., **19**, 17–26 (1988)

[HRo73]   Hardy, G.E., Rogers, T.D.: A generalization of a fixed point theorem of Reich. Canad. Math. Bull., **16**, No. 2, 201–206 (1973)

[HeC03]   He, C.: Convergence problem of iterative asymptotically nonexpansive type mappings in Banach spaces. (Chinese) Gongcheng Shuxue Xuebao, **20**, No. 3, 75–81 (2003)

[HeS03]   He, C., Sun, Z.H.: The convergence problem of iteration with errors for asymptotically nonexpansive type mappings. (Chinese) Sichuan Daxue Xuebao, **40**, No. 2, 199–203 (2003)

[HeC07]   He, H., Chen, R.D.: Viscosity approximation to common fixed points of nonexpansive semigroups in Hilbert spaces. Int. J. Math. Analysis, **1**, No. 2, 73–78 (2007)

[HKT92]   Heinkenschloss, M., Kelley, C.T., Tran, H.T.: Fast algorithms for nonsmooth compact fixed-point problems. SIAM J. Numer. Anal., **29**, No. 6, 1769–1792 (1992)

[HKr96]   Herceg, D., Krejic, N.: Convergence results for fixed point iterations in ℝ. Comput. Math. Appl., **31**, No. 2, 7–10 (1996)

[Hks78]   Hicks, T.L.: On locating fixed points of a function. Math. Japon., **22**, 557–564 (1978)

[HK77a]   Hicks, T.L., Kubicek, J.R.: On the Mann iteration process in Hilbert space. J. Math. Anal. Appl., **59**, 498–504 (1977)

[HK77b]   Hicks, T.L., Kubicek, J.R.: Nonexpansive mappings in locally convex spaces. Canad. Math. Bull., **20**, No. 4, 455–461 (1977)

[Hig96]   Higham, N.J.: Accuracy and stability of numerical algorithms. SIAM, Philadelphia (1996)

[HiR79]   Hicks, T.L., Rhoades, B.E. , A Banach type fixed point theorem. Math. Japon., **24**, No. 3, 327–330 (1979)

[Hil73]   Hillam, B.P.: Fixed point iterations and infinite matrices, and subsequential limit points of fixed point sets. PhD Thesis, University of California, Riverside (1973)

[Hil75]   Hillam, B.P.: A generalization of Krasnoselski's theorem on the real line. Math. Mag., **48**, 167–168 (1975)

[Hil76]   Hillam, B.P.: A characterization of the convergence of successive approximations. Amer. Math. Monthly, **83**, No. 4, 273 (1976)

[HiH03]   Hirano, N., Huang, Z.: Convergence theorems for multivalued Φ-hemicontractive operators and Φ-strongly accretive operators. Comput. Math. Appl., **46**, 1461–1471 (2003)

[Hoa91]   Hoang, T.: Computing fixed points by global optimization methods. In: Thera, M.A.; Baillon, J.B.(eds.) Fixed Point Theory and Applications, Proceed. of Int. Conf. Univ. d'Aix-Marseille I, 5-8 June 1989. Pitman Research Notes in Mathematics Series. Longman Sc. (1991)

[HNL78]   Hoang, T., Nguyen, V.T., Le Dung, M.: Un nouvel algorithme de point fixe. C. R. Acad. Sci. Paris, Ser. A, **286**, 783–785 (1978)

[HuC98]   Hu, C.S.: Convergence of Mann's iteration processes for a class of nonlinear operators in L_p (Chinese). Math. Appl., **11**, No. 1, 101–105 (1998)

[HuC99]   Hu, C.S.: Convergence of iterative processes for asymptotically hemicontractive mappings in P-uniformly convex Banach spaces (Chinese). Math. Appl., **12**, No. 3, 72–76 (1999)

[HuC00]   Hu, C.S.: Convergence theorems for AP-iteration processes for asymptotically hemicontractive mappings in L^p (Chinese). J. Systems Sci. Math. Sci., **20**, No. 1, 40–46 (2000)

[HC04a]   Hu, C.S.: Strong convergence of approximated sequences for asymptotically nonexpansive mappings in Banach spaces. (Chinese) Acta Math. Sci. Ser. A Chin. Ed., **24**, No. 2, 216–222 (2004)

[HC04b]   Hu, C.S.: Iterative approximation problems of fixed points for asymptotically nonexpansive type mappings in Banach spaces (Chinese). J. Math., Wuhan Univ. **24**, No. 6, 675–679 (2004)

[HHe01]   Hu, T., Heng, W.-S.: Iterative procedures to approximate fixed points. Indian J. Pure Appl. Math., **32**, No. 2, 267–270 (2001)

[HHu97]   Hu, T., Huang, J.C.: Iteration of fixed points on hypersurfaces. Chin. Ann. Math., Ser. B, **18**, No. 4, 423–428 (1997)

[HHR97]   Hu, T., Huang, J.C., Rhoades, B.E.: A general principle for Ishikawa iterations for multi-valued mappings. Indian J. Pure Appl. Math., **28**, No. 8, 1091–1098 (1997)

[HuY80]   Hu, T., Yang, G.: Generalized iteration process. Tamkang J. Math., **11**, 135–140 (1980)

[HJC02]   Huang, J.C.: Approximating common fixed points of infinite asymptotically nonexpansive mappings. Far East J. Math. Sci., **6**, No. 2, 113–123 (2002)

[Hu98a]   Huang, J.C.: On the convergence of the iteration methods to a common fixed point for a pair of mappings. Publ. Math. Debrecen, **53**, No. 1-2, 59–67 (1998)

[Hu98b]   Huang, J.C.: Iteration processes for nonlinear multi-valued mappings in convex metric spaces. Tamsui Oxf. J. Math. Sci., **14**, 19–24 (1998)

[Hua01]   Huang, J.C.: Convergence theorems of the sequence of iterates for a finite family asymptotically nonexpansive mappings. Int. J. Math. Math. Sci., **27**, No. 11, 653–662 (2001)

[Hu02a]   Huang, J.C.: On common fixed points of asymptotically hemicontractive mappings. Indian J. Pure Appl. Math., **33**, No. 7, 1121–1135 (2002)

[Hu02b]   Huang, J.C.: Common fixed points of asymptotically hemicontractive mappings. Indian J. Pure Appl. Math., **33**, No. 12, 1811–1825 (2002)

[Hua03]   Huang, J.C.: Convergence and stability of iterative procedures with errors for a couple of quasi-contractive mappings in q-uniformly smooth Banach spaces. Far East J. Math. Sci., **10**, No. 2, 121–146 (2003)

[Hu04a]   Huang, J.C.: Common fixed points iteration processes for a finite family of asymptotically nonexpansive mappings. Georgian Math. J., **11**, No. 1, 83–92 (2004)

[Hu04b]   Huang, J.C.: On common fixed points iteration processes for generalized asymptotically contractive and generalized hemi-contractive mappings. Panamer. Math. J., **14**, No. 1, 55–67 (2004)

[HJ01a]   Huang, J.L.: Ishikawa iteration process with errors for the sequence of the nonexpansive mappings (Chinese). J. Sichuan Norm. Univ., Nat. Sci., **24**, No. 1, 42–44 (2001)

[HJ01b]   Huang, J.L.: Ishikawa iteration process with errors for nonexpansive mappings. Int. J. Math. Math. Sci., **27**, No. 7, 413–417 (2001)

[HJL03]   Huang, J.L.: Ishikawa iterative sequences for asymptotically quasi-nonexpansive mappings. Sichuan Shifan Daxue Xuebao Ziran Kexue Ban, **26**, No. 1, 10–12 (2003)

[HBa99]   Huang, N.-J., Bai, B.R.: A perturbed iterative procedure for multivalued pseudo-contractive mapping and multivalued accretive mapping in Banach spaces. Comput. Math. Appl., **37**, 7–15 (1999)

[HCK00]   Huang, N.-J., Cho, Y.J., Kang, S.M., Hwang, H.J.: Ishikawa and Mann iterative processes with errors for set-valued strongly accretive and ϕ-hemicontractive mappings. Math. Comput. Modelling, **32**, No. 7-8, 791–801 (2000)

[HCL00]   Huang, N.-J., Cho, Y.J., Lee, B.S., Jung, J.S.: Convergence of iterative processes with errors for set-valued pseudocontractive and accretive type mappings in Banach spaces. Comput. Math. Appl., **40**, No. 10-11, 1127–1139 (2000)

[HGH02]   Huang, N.-J., Gao, C.J., Huang, X.P.: New iteration procedures with errors for multivalued Φ-strongly pseudocontractive and Φ-strongly accretive mappings. Comput. Math. Appl., **43**, 1381–1390 (2002)

[HLa04]   Huang, N.-J., Lan, H.Y.: A new iterative approximation of fixed points for asymptotically contractive type mappings in Banach spaces. Indian J. Pure Appl. Math., **35**, No. 4, 441–453 (2004)

[HLi03]   Huang, N.-J., Li, J.: Iterative approximating with errors of common fixed points for a couple of asymptotically nonexpansive type mappings in Banach spaces. In: Fixed Point Theory and Applications (Chinju/Masan, 2001), 121–136. Nova Sci. Publ., Hauppauge, NY (2003)

[HKG01]   Huang, X.P., Kim, J.K., Gao, C.J., Huang, N.-J.: Ishikawa iterative procedures with errors for multi-valued mappings in Banach spaces. Nonlinear Funct. Anal. Apl., **6**, 57–68 (2001)

[HuY95]   Huang, Y.: A remark on an iteration theorem of B. E. Rhoades. Soochow J. Math., **21**, No. 1, 121–123 (1995)

[HJe97]   Huang, Y., Jeng, J.-C.: Approximating fixed points by iteration processes. Indian J. Pure Appl. Math., **28**, No. 2, 129–138 (1997)

[HZ97a]   Huang, Z.: A remark on Ishikawa iteration theorem of M. O. Osilike. Soochow J. Math., **23**, No. 1, 113–114 (1997)

[HZ97b]   Huang, Z.: A new convergence result for fixed-point iteration in bounded intervals of $\mathbb{R}^n$. Comput. Math. Appl., **34**, No. 12, 33–36 (1997)

[HZ98a]   Huang, Z.: Approximating fixed points of Φ-hemicontractive mappings by the Ishikawa iteration process with errors in uniformly smooth Banach spaces. Comput. Math. Appl., **36**, No. 2, 13–21 (1998)

[HZ98b]   Huang, Z.: Fast algorithms for fixed points and nonlinear equations. MS Thesis, University of Utah (1998)

[HZ99a]   Huang, Z.: Mann and Ishikawa iterations with errors for asymptotically nonexpansive mappings. Comput. Math. Appl., **37**, No. 3, 1–7 (1999)

[HZ99b]   Huang, Z.: A generalization of Ishikawa fixed point iteration theorems. J. Nanjing Univ., Math. Biq., **16**, No. 2, 179–183 (1999)

[HZ00a]   Huang, Z.: Iterative process with errors for fixed points of multivalued Φ-hemicontractive operators in uniformly smooth Banach spaces. Comput. Math. Appl., **39**, No. 3-4, 137–145 (2000)

[HZ00b]   Huang, Z.: Iterative approximation of Φ-hemicontractive mappings without Lipschitz assumption. Numer. Math. J. Chinese Univ. (English Ser.), **9**, No. 2, 193–203 (2000)

[HZ00c]   Huang, Z.: Stability results for generalized contractive mappings. Numer. Math. J. Chinese Univ. (English Ser.), **9**, No. 1, 83–90 (2000)

[HZ00d]   Huang, Z.: Almost T-stability of the iteration procedures with errors for strongly pseudocontractions in q-uniformly smooth Banach spaces without continuity assumption. Yokohama Math. J., **48**, No. 1, 71–82 (2000)

[HZ00e]   Huang, Z.: Iterative solution of nonlinear equations of m-accretive type in arbitrary Banach spaces. J. Nanjing Univ., Math. Biq., **17**, No. 1, 27–34 (2000)

[HgZ01]   Huang Z.: Ishikawa iterative process in uniformly smooth Banach spaces. Appl. Math. Mech. (English Ed.), **22**, No. 11, 1306–1310 (2001). Translated from Appl. Math. Mech. (Chinese), **22**, No. 11, 1177–1180 (2001)

[HKS99]   Huang, Z., Khachiyan, L., Sikorski, K.: Approximating fixed points of weakly contracting mappings. J. Complexity, **15**, No. 2, 200–213 (1999)

[HSi98]   Huang, Z., Sikorski, K.: Interior ellipsoid algorithm for fixed points. Technical Report UUCS-98-006, Department of Computer Science, University of Utah (1998)

[HZZ03]   Hui, S.R., Zhang, G.W., Zhou, F.C.: The range of perturbed m-accretive operators and maximal monotone operators. (Chinese). Dongbei Shida Xuebao, **35**, No. 2, 7–10 (2003)

[Hum80]   Humphreys, M.: Algorithms for fixed points of nonexpansive operators. PhD Thesis, University of Missouri-Rolla (1980)

[HwC98]   Hwang, H.J., Cho, Y.J.: Iterative process with errors for m-accretive operators in Banach spaces. Nonlinear Anal. Forum, **3**, 75–88 (1998)

[Igb02]   Igbokwe, D.I.: Approximation of fixed points of asymptotically demicontractive mappings in arbitrary Banach spaces. J. Inequal. Pure Appl. Math., **3**, No. 1, Paper No. 3 (Electronic) (2002)

[Igb03]   Igbokwe, D.I.: Iterative construction of fixed points of asymptotically pseudocontractive maps. Panamer. Math. J., **13**, No. 4, 83–97 (2003)

[IKS88]   Imdad, M., Khan, M.S., Sessa, S.: On sequences of contractive mappings and their fixed points. Int. J. Math. Math. Sci., **11**, No. 3, 527–533 (1988)

[ImO03]   Imoru, C.O., Olatinwo, M.O.: On the stability of Picard and Mann iteration processes. Carpathian J. Math., **19**, No. 2, 155–160 (2003)

[Isa92]   Isac, G.: Complementarity Problems. Lecture Notes in Mathematics 1528, Springer-Verlag, Berlin (1992)

[IsL04]   Isac, G., Li, J. : The convergence property of Ishikawa iteration schemes in noncompact subsets of Hilbert spaces and its applications to complementarity theory. Comput. Math. Appl., **47**, No. 10-11, 1745–1751 (2004)

252   References

[Ish74]   Ishikawa, S.: Fixed points by a new iteration method. Proc. Amer. Math. Soc., **44**, No. 1, 147–150 (1974)

[Ish76]   Ishikawa, S.: Fixed points and iteration of a nonexpansive mapping in a Banach space. Proc. Amer. Math. Soc., **59**, No. 1, 65–71 (1976)

[Ish77]   Ishikawa, S.: Fixed points and iteration of a Kannan's mapping in a Banach space. Keio Eng. Rep., **30**, 29–34 (1977)

[Ish79]   Ishikawa, S.: Common fixed points and iteration of commuting nonexpansive mappings. Pacific J. Math., **80**, 493–501 (1979)

[Ist73]   Istratescu, V.I.: Introducere in teoria punctelor fixe (Introduction to the Fixed Points Theory). Editura Academiei R.S.R, Bucuresti (1973)

[Ist81]   Istratescu, V.I.: Fixed Point Theory. An introduction. D. Reidel Publishing Company, Dordrecht (1981)

[Iva76]   Ivanov, A.A., Fixed points of metric space mappings (in Russian). Isledovaniia po topologii. II., pp. 5-102, Akademia Nauk, Moskva (1976)

[Jac97]   Jachymski, J.R.: An extension of A. Ostrowski's theorem on the round-off stability of iterations. Aequationes Math., **53**, No. 3, 242–253 (1997)

[Jac99]   Jachymski, J.R.: On iterative equivalence of some classes of mappings. Ann. Math. Sil., **13**, 149–165 (1999)

[Jac04]   Jachymski, J.R., Jozwik, I.: On Kirk's asymptotic contractions. J. Math. Anal. Appl., **300**, 147–159 (2004)

[Ja77a]   Jaggi, D.S.: Fixed point theorems for orbitally continuous functions. Mat. Vesnik, **1(14)(29)**, No. 2, 129–135 (1977)

[Ja77b]   Jaggi, D.S.: Fixed point theorems for orbitally continuous functions. II. Indian J. Math., **19**, 113–118 (1977)

[Jan67]   Janos, L.: A converse of Banach's contraction principle. Proc. Amer. Math. Soc., **18**, No. 2, 287–289 (1967)

[Jbi91]   Jbilou, K., Sadok, H.: Some results about vector extrapolation methods and related fixed-point iterations. J. Comput. Appl. Math., **36**, No. 3, 385–398 (1991)

[JeW97]   Jensen, S., Wieczorek, A.: Convergence of Mann iteration processes for nonexpansive operators in equi-connected metric spaces. Numer. Funct. Anal. Optim., **18**, No. 5-6, 609–621 (1997)

[Jeo97]   Jeong, J.U.: Ishikawa and Mann iteration methods for strongly accretive operators. Korean J. Comput. Appl. Math., **4**, No. 2, 417–425 (1997)

[Jeo02]   Jeong, J.U.: Stability of the Mann and Ishikawa type iteration procedures for nonlinear equations involving m-accretive operators. Int. Math. J., **2**, No. 7, 641–649 (2002)

[Jer91]   Jerome, J.W.: Numerical approximation of PDE system fixed-point maps via Newton's method. J. Comput. Appl. Math., **38**, No. 1-3, 211–230 (1991)

[JCK00]   Jiang, G.-J., Chun, S., Kim, Ki Hong: Iterative approximation of fixed points for asymptotically demicontractive mappings. Nonlinear Funct. Anal. Appl., **5**, No. 2, 15–21 (2000)

[JLi94]   Jiang, Y.-L., Liu, W.S.: An approximation method for equations involving Lipschitz strongly accretive mappings. (Chinese). Xi'an Jiaotong Daxue Xuebao, **28**, No. 9, 69–73, 82 (1994)

[JXR96]   Jiang, Y.-L., Xu, Z.B., Roach, G. F.: On conditions of weak convergence of nonlinear contraction semigroups and of iterative methods for accretive operators in Banach spaces. Nonlinear Anal., **27**, No. 4, 387–396 (1996)

[JLK04]   Jin, L., Liu, Z., Kang, S.M.: Ishikawa iterative processes with errors for nonlinear ϕ-strongly accretive operator equations. In: Fixed Point Theory and Applications. Vol. 5, 33–39. Nova Sci. Publ., Hauppauge, NY (2004)

[Ji00a]   Jin, M.: Ishikawa iteration process with errors for nonexpansive mappings in a uniformly convex Banach spaces (Chinese). J. Sichuan Norm. Univ., Nat. Sci., **23**, No. 3, 250–252 (2000)

[Ji00b]   Jin, M.: The Ishikawa iteration process with errors for fixed points of nonexpansive mappings. (Chinese). Xinan Shifan Daxue Xuebao Ziran Kexue Ban, **25**, No. 1, 4–6 (2000)

[Ji00c]   Jin, M.: The construction and convergence of Mann iterative sequences for nonexpansive mappings with boundary conditions. (Chinese). Xinan Shifan Daxue Xuebao Ziran Kexue Ban, **25**, No. 3, 239–241 (2000)

[Ji02a]   Jin, M.: A stability problem for the Ishikawa iteration procedure with errors for strongly pseudocontractive mappings. (Chinese). Sichuan Daxue Xuebao, **39**, No. 5, 800–804 (2002)

[Ji02b]   Jin, M.: Ishikawa iterative approximation with errors for solutions of accretive operator equations. (Chinese). Sichuan Shifan Daxue Xuebao Ziran Kexue Ban, **25**, No. 4, 373–375 (2002)

[Ji03a]   Jin, M.: Iterative approximation for strictly pseudocontractive mappings. (Chinese). Sichuan Daxue Xuebao, **40**, No. 6, 1019–1021 (2003)

[Ji03b]   Jin, M.: Ishikawa iterative processes with error for Φ-pseudocontractive mappings. (Chinese). Sichuan Daxue Xuebao, **40**, No. 2, 208–211 (2003)

[Ji03c]   Jin, M.: Ishikawa iteration process with errors in Banach spaces. (Chinese). Xinan Shifan Daxue Xuebao Ziran Kexue Ban, **28**, No. 3, 358–361 (2003)

[JiD02]   Jin, M., Deng, L.: Strong stability of the Ishikawa iteration procedure with errors for strongly pseudocontractive operators. (Chinese). Acta Anal. Funct. Appl., **4**, No. 2, 164–168 (2002)

[JHu00]   Jin, M., Huang, J.L.: Mann iteration process with errors for fixed points of nonexpansive mappings. (Chinese). J. Henan Norm. Univ. Nat. Sci., **28**, No. 3, 124–125 (2000)

[JLu04]   Jin, M., Liu, Q.K.: Nonlinear quasi-variational inclusions involving generalized m-accretive mappings. Nonlinear Funct. Anal. Appl., **9**, No. 3, 485–494 (2004)

[Jin00]   Jin, W.X.: Iterative approximation of fixed points of strongly pseudocontractive mappings. (Chinese). J. Nanjing Norm. Univ. Nat. Sci. Ed., **23**, No. 1, 15–18 (2000)

[Joh72]   Johnson, G.G.: Fixed points by mean value iteration. Proc. Amer. Math. Soc., **34**, 193–194 (1972)

[Jor94]   Jorgensen, N.: Finding fixpoints in finite function spaces using neededness analysis and chaotic iteration. Static analysis. (Namur, 1994), 329-345, Lecture Notes in Comput. Sci., 864. Springer, Berlin (1994)

[JuJ01]   Jung, J.S.: Iterative approximation for perturbed m-accretive operator equations in arbitrary Banach spaces. Commun. Appl. Nonlinear Anal., **8**, No. 1, 51–62 (2001)

[JuJ02]   Jung, J.S.: Convergence of nonexpansive iteration processes in Banach spaces. J. Math. Anal. Appl., **273**, 153–159 (2002)

[JuJ05]   Jung, J.S.: Iterative approaches to common fixed points of nonexpansive mappings in Banach spaces. J. Math. Anal. Appl., **302**, No. 2, 509–520 (2005)

[JCA05] Jung, J.S., Cho, Y.J., Agarwal, R.P.: Iterative schemes with some control conditions for a family of finite nonexpansive mappings in Banach spaces. Fixed Point Theory Appl., **2005**, No. 2, 125–135 (2005).

[JCL00] Jung, J.S., Cho, Y.J., Lee, B.S.: Asymptotic behavior of nonexpansive iterations in Banach spaces. Commun. Appl. Nonlinear Anal., 7, No. 1, 63–76 (2000)

[JCS00] Jung, J.S., Cho, Y.J.: Sahu, D.R.: Existence and convergence for fixed points of non-Lipschitzian mappings in Banach spaces without uniform convexity. Commun. Korean Math. Soc., 15, No. 2, 275–284 (2000)

[JC02a] Jung, J.S., Cho, Y.J., Zhou, H.: Iterative methods with mixed errors for perturbed m-accretive operator equations in arbitrary Banach spaces. Math. Comput. Modelling, 35, No. 1-2, 55–62 (2002)

[JC02b] Jung, J.S., Cho, Y.J., Zhou, H.: Iterative processes with mixed errors for nonlinear equations with perturbed m-accretive operators in Banach spaces. Appl. Math. Comput., **133**, 389–406 (2002)

[JKi96] Jung, J.S., Kim, T.H.: Strong convergence theorems for multivalued nonexpansive mappings in Banach spaces. Nonlinear Anal. Forum, **2**, 49–55 (1996)

[JKi97] Jung, J.S., Kim, T.H.: Convergence of approximate sequences for compositions of nonexpansive mappings in Banach spaces. Bull. Korean Math. Soc., **34**, No. 1, 93–102 (1997)

[JKi98] Jung, J.S., Kim, T.H.: Strong convergence of approximating fixed points for nonexpansive nonself-mappings in Banach spaces. Kodai Math. J., **21**, No. 3, 259–272 (1998)

[JKS95] Jung, J.S., Kim, S.S.: Strong convergence theorems for nonexpansive nonself-mappings in Banach spaces. Nonlinear Anal. Forum, **1**, 31–42 (1995)

[JK98a] Jung, J.S., Kim, S.S.: Strong convergence theorems for nonexpansive nonself-mappings in Banach spaces. Nonlinear Anal. TMA, **33**, No. 3, 321–329 (1998)

[JK98b] Jung, J.S., Kim, S.S.: Strong convergence theorems for nonexpansive nonself-mappings in Banach spaces. In: Tangmanee, E. (ed.) et al. Proceedings of the Second Asian Mathematical Conference 1995, Nakhon Ratchasima, Thailand, October 17-20, 1995. World Scientific, Singapore (1998)

[JMo01] Jung, J.S., Morales, C.H.: The Mann process for perturbed m-accretive operators in Banach spaces. Nonlinear Anal. TMA, **46**, No. 2, 231–243 (2001)

[JPP97] Jung, J.S., Park, J.S., Park, E.H.: Convergence of approximating fixed points for nonexpansive nonself-mappings in Banach spaces. Commun. Korean Math. Soc., **12**, No. 2, 275–285 (1997)

[JSa98] Jung, J.S., Sahu, D.R.: Approximating fixed points of asymptotically nonexpansive mappings. Nonlinear Anal. Forum, **3**, 41–52 (1998)

[JSa03] Jung, J.S., Sahu, D.R.: Dual convergences of iteration processes for nonexpansive mappings in Banach spaces. Czechoslovak Math. J., **53(128)**, No. 2, 397–404 (2003)

[JST98] Jung, J.S., Sahu, D.R., Thakur, B.S.: Strong convergence theorems for asymptotically nonexpansive mappings in Banach spaces. Commun. Appl. Nonlinear Anal., **5**, No. 3, 53–69 (1998)

[KaG00] Kalantari, B., Gerlach, J.: Newton's method and generation of a determinantal family of iterations. J. Comput. Appl. Math., **116**, 195–200 (2000)

[Kal03] Kalinde, A.K.: Corrigendum: "Iterative solutions of generalized strongly pseudo-accretive type nonlinear equations in Banach spaces" [Far East J. Math. Sci. (FJMS) 4 (2002), No. 2, 221–233; MR1902959]. Far East J. Math. Sci. (FJMS), **10**, No. 3, 367–369 (2003)

[KaR92] Kalinde, A.K., Rhoades, B.E.: Fixed point Ishikawa iterations J. Math. Anal. Appl., **170**, No. 2, 600–606 (1992)

[KCZ04] Kang, J.I., Cho, Y.J., Zhou, H.: Convergence theorems of the iterative sequences for nonexpansive mappings. Commun. Korean Math. Soc., **19**, No. 2, 321–328 (2004)

[KLW03] Kang, S.M., Liu, Z., Wang, L.: Convergence and stability of Ishikawa iterative processes with errors for a pair of quasi-contractive mappings in uniformly convex Banach spaces. Far East J. Math. Sci. (FJMS), **9**, No. 1, 105–119 (2003)

[KZC02] Kang, S.M., Zhou, H., Cho, Y.J.: Ishikawa iterative process with mixed errors for Lipschitzian and strongly pseudo-contractive mappings in Banach spaces. In: Fixed point theory and applications. Vol. 3, 99–111. Nova Sci. Publ., Huntington, NY (2002)

[Kng91] Kang, Z.B.: Iterative approximation of the solution of locally Lipschitzian equation. Appl. Math. Mech. (English Ed.), **12**, No. 4, 409–414 (1991). Translated from (Chinese) Appl. Math. Mech., **12**, No. 4, 385-389 (1991)

[Kan71] Kaniel, S.: Construction of a fixed point for contractions in Banach space. Israel J. Math., **9**, 535–540 (1971)

[Knn68] Kannan, R.: Some results on fixed points. Bull. Calcutta Math. Soc., **10**, 71–76 (1968)

[Knn71] Kannan, R.: Some results on fixed points. III. Fund. Math., **70**, 169–177 (1971)

[Knn73] Kannan, R.: Construction of fixed points of a class of nonlinear mappings. J. Math. Anal. Appl., **41**, 430–438 (1973)

[Knt39] Kantorovich, L.: The method of successive approximations for functional equations. Acta Math., **71**, 63–67 (1939)

[Kar79] Karlovitz, L.A.: Geometric methods in the existence and construction of fixed points of nonexpansive mappings. In: Constructive Approaches to Mathematical Models (Proc. Conf. in honor of R. J. Duffin, Pittsburgh, Pa., 1978), pp. 413–420. Academic Press, New York London Toronto, Ont. (1979)

[KaG77] Karamardian, S., Garcia, C.B. (eds.): Fixed points. Algorithms and Applications. Proceed. 1st Intern. Conf. on Computing Fixed Points with Applications, Clemson University, June 26-28, 1976. Academic Press, New York etc. (1977)

[Kas78] Kasahara, S.: Fixed point iterations using linear mappings. Math. Sem. Notes Kobe Univ., **6**, No. 1, 87–90 (1978)

[Kas79] Kasahara, S.: Fixed point iterations via linear mappings. Math. Sem. Notes Kobe Univ., **7**, No. 2, 401–408 (1979)

[KKo88] Kassay, G., Kolumban, I.: Remarks on local stability of fixed points. Itinerant Seminar on Functional Equations, Approximation and Convexity (Cluj-Napoca, 1988), 191–196. Preprint, 88-6, Univ. "Babes-Bolyai", Cluj-Napoca (1988)

256   References

[Kat67]   Kato, T.: Nonlinear semigroups and evolution equations. J. Math. Soc.
          Japan, **19**, 508–520 (1967)
[KaR80]   Kaucher, E., Rump, S.M.: Generalized iteration methods for bounds of
          the solution of fixed point operator-equations. Computing, **24**, No. 2-3,
          131–137 (1980)
[KLY76]   Kellog, R.B., Li, T.Y, Yorke, J.: A constructive proof of the Brouwer
          fixed point theorem and computational results. SIAM J. Numer. Anal.,
          13, 473–483 (1976)
[KhH01]   Khan, A.R., Hussain, N.: Iterative approximation of fixed points of non-
          expansive maps. Sci. Math. Japon., **54**, No. 3, 503–511 (2001)
[Kha86]   Khan, L.A.: On the convergence of Mann iterates to a common fixed point
          of two mappings. J. Pure Appl. Sc., **5**, No. 1, 57–58 (1986)
[Kha87]   Khan, L.A.: On a fixed point theorem for iterates in locally convex spaces.
          J. Nat. Sci. Math., **27**, No. 1, 1–5 (1987)
[Kha88]   Khan, L.A.: Fixed point theorems for Mann iterates in metrisable linear
          topological spaces. Math. Japon., **33**, No. 2, 247–251 (1988)
[Kha89]   Khan, L.A.: Fixed points by Ishikawa iterates in metric linear spaces.
          Math. Rep., Toyama Univ., **12**, 57–63 (1989)
[Kh90a]   Khan, L.A.: Common fixed point results by iterations using linear map-
          pings. J. Pure Appl. Sci., Bahawalpur, **9**, No. 2, 43–45 (1990)
[Kh90b]   Khan, L.A.: Extensions of some fixed point theorems of Kannan and Wong
          to paranormed spaces. J. Math., Punjab Univ., **23**, 77–82 (1990)
[Kha92]   Khan, L.A.: Common fixed point results for iterations in metric linear
          spaces. Stud. Sci. Math. Hung., **27**, No. 1-2, 143–146 (1992)
[KHA96]   Khan, M.S., Hussain, N., Aslam Noor, M.: Mann iterative construction
          of fixed points in locally convex spaces. J. Nat. Sci. Math., **36**, No. 2,
          155–159 (1996)
[KIS86]   Khan, M.S., Imdad, M., Sessa, S.: A coincidence theorem in linear normed
          spaces. Libertas Math., **6**, 83–94 (1986)
[Kha00]   Khan, S.H.: Estimating common fixed points of two nonexpansive map-
          pings by strong convergence. Nihonkai Math. J., **11**, No. 2, 159–165 (2000)
[Kh04a]   Khan, S.H.: On iterative convergence of resolvents of accretive operators.
          Demonstratio Math., **37**, No. 2, 407–417 (2004)
[Kh04b]   Khan, S.H.: Iterative convergence of resolvents of maximal monotone op-
          erators perturbed by the duality map in Banach space., Acta Math. Acad.
          Paedagog. Nyhzi. (N.S.), **20**, 45–51 (2004)
[KT01a]   Khan, S.H., Takahashi, W.: Approximating common fixed points of two
          asymptotically nonexpansive mappings. Sci. Math. Japon., **53**, No. 1,
          143–148 (2001)
[KT01b]   Khan, S.H., Takahashi, W.: Iterative approximation of fixed points of
          asymptotically nonexpansive mappings with compact domains. Panamer.
          Math. J., **11**, No. 1, 19–24 (2001)
[KhP86]   Khanh, P.Q.: Remarks on fixed points theorems based on iterative ap-
          proximations. Polish Acad. Sc., Preprint 361 (1986)
[KhP89]   Khanh, P.Q.: Fixed points by Ishikawa iterates in metric linear spaces.
          Math. Rep. Toyama Univ., **12**, 57–63 (1989)
[KiT03]   Kikkawa, M., Takahashi, W.: Approximating fixed points of infinite non-
          expansive mappings by the hybrid method. J. Optim. Theory Appl., **117**,
          No. 1, 93–101 (2003)

[KiT04]   Kikkawa, M., Takahashi, W.: Approximating fixed points of nonexpansive mappings by the block iterative method in Banach spaces. Int. J. Comput. Numer. Anal. Appl., **5**, No. 1, 59–66 (2004)

[KmK01] Kim, E.S., Kirk, W.A.: A note on Picard iterates of nonexpansive mappings. Ann. Pol. Math., **76**, No. 3, 189–196 (2001)

[KGE98] Kim, G.E.: Strong convergence to fixed points of non-Lipschitzian mappings in Banach spaces. RIMS Kokyuroku, **1071**, 99–104 (1998)

[KGE00] Kim, G.E.: Approximating fixed points of λ-firmly nonexpansive mappings in Banach spaces. Int. J. Math. Math. Sci., **24**, No. 7, 441–448 (2000)

[KKi98]   Kim, G.E., Kim, T.H.: Strong convergence to fixed points of non-Lipschitzian mappings in Banach spaces. Kodai Math. J., **21**, No. 3, 259–272 (1998)

[KKi01]   Kim, G.E., Kim, T.H.: Mann and Ishikawa iterations with errors for non-Lipschitzian mappings in Banach spaces. Comput. Math. Appl., **42**, No. 12, 1565–1570 (2001)

[KKT02] Kim, G.E., Kiuchi, H., Takahashi, W.: Weak and strong convergence theorems for nonexpansive mappings. Sci. Math. Japon., **56**, No. 1, 133–141 (2002)

[KKT04] Kim, G.E., Kiuchi, H., Takahashi, W.: Weak and strong convergences of Ishikawa iterations for asymptotically nonexpansive mappings in the intermediate sense. Sci. Math. Japon., **60**, No. 1, 95–106 (2004)

[KT96a] Kim, G.E., Takahashi, W.: Strong convergence theorems for nonexpansive nonself-mappings in Banach spaces. Nihonkai Math. J., **7**, No. 1, 63–72 (1996)

[KT96b] Kim, G.E., Takahashi, W.: Approximating common fixed points of nonexpansive semigroups in Banach spaces. Sci. Math. Jpn., **63**, No. 1, 31–36 (2006)

[KKi88]   Kim, H.S., Kim, T.H.: Weak convergence theorems for mappings of asymptotically nonexpansive type in Banach space. Math. Japon., **33**, No. 3, 431–438 (1988)

[KJH02] Kim, J.H.: Iterative processes with mixed errors for nonlinear equations involving m-accretive mappings. In: Fixed Point Theory and Applications. Vol. 3, 113–124. Nova Sci. Publ., Huntington, NY (2002)

[KJL03] Kim, J.K., Jang, S.M., Liu, Z.: Convergence theorems and stability problems of Ishikawa iterative sequences for nonlinear operator equations of the accretive and strongly accretive operators. Commun. Appl. Nonlinear Anal., **10**, No. 3, 85–98 (2003)

[KKK04] Kim, J.K., Kim, Ki Hong, Kim, K.S.: Convergence theorems of modified three-step iterative sequences with mixed errors for asymptotically quasi-nonexpansive mappings in Banach spaces. Panamer. Math. J. **14**, No. 1, 45–54 (2004)

[KLK04] Kim, J.K., Liu, Z., Kang, S.M.: Almost stability of Ishikawa iterative schemes with errors for φ-strongly quasi-accretive and φ-hemicontractive operators. Commun. Korean Math. Soc., **19**, No. 2, 267–281 (2004)

[KLN04] Kim, J.K., Liu, Z., Nam, Y.M., Chun, S.A.: Strong convergence theorems and stability problems of Mann and Ishikawa iterative sequences for strictly hemi-contractive mappings. J. Nonlinear Convex Anal., **5**, No. 2, 285–294 (2004)

[KSH00] Kim, K.W., Shin, S.S., Hwang, S.-Y.: An Ishikawa type iteration for generalized contraction mappings on metric spaces., Panamer. Math. J., **10**, No. 1, 17–23 (2000)

[KTH01] Kim, T.H.: Approximation of common fixed points for a family of non-Lipschitzian self-mappings. In: Nonlinear analysis and convex analysis (Japanese) (Kyoto, 2000). Sūrika isekikenkyūsho Kōkyūroku No. 1187 (2001)

[KHu99] Kim, T.H., Hur, M.-D.: Approximation of common fixed points of a family of self-mappings. Nonlinear Anal. Forum, **4**, 1–13 (1999)

[KJ97a] Kim, T.H., Jung, J.S.: Approximating fixed points of nonlinear mappings in Banach spaces. Ann. Univ. Mariae Curie-Skodowska Sect. A, **51**, No. 2, 149–165 (1997)

[KJ97b] Kim, T.H., Jung, J.S.: Fixed point theorems for non-Lipschitzian mappings in Banach spaces. Math. Japon., **45**, No. 1, 61–67 (1997)

[KJu98] Kim, T.H., Jung, J.S.: An Ishikawa type iteration scheme in complete metric spaces. Numer. Funct. Anal. Optimization, **19**, No. 5-6, 557–563 (1998)

[KJu00] Kim, T.H., Jung, J.S.: Remarks on approximation of fixed points of strictly pseudocontractive mappings. Bull. Korean Math. Soc., **37**, No. 3, 461–475 (2000)

[KKE00] Kim, T.H., Kim, E.S.: Remarks on approximation of fixed points of strictly pseudocontractive mappings. Bull. Korean Math. Soc., **37**, No. 3, 461–475 (2000)

[KKG98] Kim, T.H., Kim, G.E.: Iterative algorithms for approximating common fixed points of hemi-relaxed Lipschitz semigroups. Panamer. Math. J., **8**, No. 4, 81–88 (1998)

[KXu98] Kim, T.H., Xu, H.K.: Some Hilbert space characterizations and Banach space inequalities. Math. Inequal. Appl., **1**, No. 1, 113–121 (1998)

[KXu07] Kim, T.H., Xu, H.K.: Robustness of Mann's algorithm for nonexpansive mappings. J. Math. Anal. Appl., **327**, No. 2, 1105–1115 (2007)

[KTW01] Kimura, Y., Takahashi, W.: Weak convergence to common fixed points of countable nonexpansive mappings and its applications. J. Korean Math. Soc., **38**, No. 6, 1275–1284 (2001)

[KTT05] Kimura, Y., Takahashi, W., Toyoda, M.: Convergence to common fixed points of a finite family of nonexpansive mappings. Arch. Math. 84, No. 4, 350–363 (2005)

[Kir65] Kirk, W.A.: A fixed point theorem for mappings which do not increase distances. Amer. Math. Monthly, **72**, 1004–1006 (1965)

[Kir70] Kirk, W.A.: Remarks on pseudo-contractive mappings., Proc. Amer. Math. Soc., **25**, 820–823 (1970)

[Kir71] Kirk, W.A.: On successive approximations for nonexpansive mappings in Banach spaces. Glasgow Math. J., **12**, 6–9 (1971)

[Kir79] Kirk, W.A.: A fixed point theorem for local pseudo-contraction in uniformly convex spaces. Manuscripta Math., **30**, 89–102 (1979)

[Kir81] Kirk, W.A.: Fixed point theory for nonexpansive mappings. In: Fixed Point Theory (Sherbrooke, Que., 1980), 484–505, Lecture Notes in Math. 886. Springer, Berlin New York (1981)

[Kir82] Kirk, W.A.: Krasnoselskij 's iteration process in hyperbolic space. Numer. Funct. Anal. Optim., **4**, No. 4, 371–381 (1981/82)

[Kir83]   Kirk, W.A.: Fixed point theory for nonexpansive mappings. II. Contemp. Math., **18**, 121–140 (1983)

[Kir89]   Kirk, W.A.: An iteration process for nonexpansive mappings with applications to fixed point theory in product spaces. Proc. Amer. Math. Soc., **107**, No. 2, 411–415 (1989)

[Kir91]   Kirk, W.A.: An application of a generalized Krasnoselski-Ishikawa iteration process. In: Thera, M.A., Baillon, J.B.(eds.) Fixed Point Theory and Applications, Proceed. of Int. Conf. Univ. d'Aix-Marseille I, 5-8 June 1989. Pitman Research Notes in Mathematics Series, Longman Sc. (1991)

[Kir97]   Kirk, W.A.: Remarks on approximation and approximate fixed points in metric fixed point theory. Ann. Univ. Mariae Curie-Skodowska Sect. A, **51**, 167-178 (1997)

[Kk00a]   Kirk, W.A.: Nonexpansive mappings and asymptotic regularity. Nonlinear Anal. **40**, 323–332 (2000)

[Kk00b]   Kirk, W.A.: Nonexpansive mappings and asymptotic regularity. Lakshmikantham's legacy: a tribute on his 75th birthday. Nonlinear Anal. TMA, **40**, No. 1-8, 323–332 (2000)

[Kir03]   Kirk, W.A.: Fixed points of asymptotic contractions. J. Math. Anal. Appl., **277**, 645–650 (2003)

[KMS98]   Kirk, W.A., Martinez Yanez, C., Shin, S.S.: Asymptotically nonexpansive mappings. Nonlinear Anal., **33**, 1–12 (1998)

[KMo80]   Kirk, W.A., Morales, C.: Fixed point theorems for local strong pseudo-contractions. Nonlinear Anal. TMA, **4**, 363–368 (1980)

[KMo81]   Kirk, W.A., Morales, C.: On the approximation of fixed points of locally nonexpansive mappings. Canad. Math. Bull., **24**, No. 4, 441–445 (1981)

[KSa00]   Kirk, W.A., Saliga, L.: Some results on existence and approximation in metric fixed point theory. J. Comput. Appl. Math., **113**, 141–152 (2000)

[KSc77]   Kirk, W.A., Schoneberg, R.: Some results on pseudocontractive mappings. Pacific J. Math., **71**, No. 1, 89–100 (1977)

[KSi99]   Kirk, W.A., Sims, B.: Convergence of Picard iterates of nonexpansive mappings. Bull. Polish Acad. Sci., **47**, 147–155 (1999)

[KSi01]   Kirk, W.A., Sims, B.: Handbook of Metric Fixed Point Theory. Kluwer Academic Publishers (2001)

[KiL97]   Kiwiel, K.C., Lopuch, B.: Surrogate projection methods for finding fixed points of firmly nonexpansive mappings. SIAM J. Optim., **7**, 1084–1102 (1997)

[Kob81]   Kobayashi, K.: On the strong convergence of the Cesàro means of contractions in Banach spaces. II. In: Nonlinear Functional Analysis. Proceedings of a Symposium held at the Research Institute for Mathematical Sciences, Kyoto University, Kyoto, October 13-15, 1980. Kyoto University, Research Institute for Mathematical Sciences. II (1981)

[KbN81]   Kohlberg, E., Neyman, A.: Asymptotic behavior of nonexpansive mappings in normed linear spaces. Israel J. Math., **38**, No. 4, 269–275 (1981)

[Kbh01]   Kohlenbach, U.: On the computational content of the Krasnoselski and Ishikawa fixed point theorems. In: Blanck, J. et all. (eds) Proceedings of the Fourth Workshop on Computability and Complexity in Analysis. Springer LNCS vol. 2064, 119-145. Springer, Berlin (2001)

[Kbh03]   Kohlenbach, U.: Uniform asymptotic regularity for Mann iterates. J. Math. Anal. Appl., **279**, 531–544 (2003)

260   References

[Kbh05]   Kohlenbach, U.: Some computational aspects of metric fixed-point theory. Nonlinear Anal. TMA, **61A**, No. 5, 823–837 (2005)

[KbL03]   Kohlenbach, U., Leustean, L.: Mann iterates of directionally nonexpansive mappings in hyperbolic spaces. Abstr. Appl. Anal., No. 8, 449–477 (2003)

[KoW92]   Koparde, P.V., Waghmode, B.B.: Fixed point theorem for a strictly pseudocontractive mapping in Hilbert space. Math. Stud., **61**, No. 1-4, 13–17 (1992)

[Kos62]   Koshelev, A.I.: On the convergence of the method of successive approximations for quasilinear elliptic equations. Dokl. Akad. Nauk SSSR, **142**, 1007–1010 (1962)

[Ko81a]   Koshelev, A.I.: Quasilinear elliptic degenerate equations and convergence of an iteration process. I (Russian). Izv. Vyssh. Uchebn. Zaved. Mat., No. **10**, 31–40 (1981)

[Ko81b]   Koshelev, A.I.: Quasilinear elliptic degenerate equations and convergence of an iteration process. II (Russian). Izv. Vyssh. Uchebn. Zaved. Mat. No. **11**, 21–28 (1981)

[Ko82a]   Koshelev, A.I.: Krasnoselskii's iteration process in hyperbolic space. Numer. Funct. Anal. Optim., **4**, 371–381 (1982)

[Ko82b]   Koshelev, A.I.: Quasilinear elliptic degenerate equations and convergence of an iteration process. III (Russian). Izv. Vyssh. Uchebn. Zaved. Mat., No. 7, 30–39 (1982)

[KoV81]   Koshelev, V.N.: Estimation of mean error for a discrete successive-approximation scheme. (Russian). Problems Inform. Transmission, **17**, No. 3, 161–171 (1981). Translated from Problemy Peredachi Informatsii, **17**, No. 3, 20–33 (1981)

[Kra55]   Krasnoselskij, M.A.: Two remarks on the method of successive approximations (Russian). Uspehi Mat. Nauk., **10**, No. 1 (63), 123–127 (1955)

[KB92a]   Kruppel, M., Bethke, M.: An iteration method for mappings of uniformly monotone type. Zesz. Nauk. Politech. Rzesz. 103, Mat. Fiz. 16, Mat., **12**, 67–72 (1992)

[KB92b]   Kruppel, M., Bethke, M.: An iteration method for mappings of uniformly accretive type in L_p-spaces. Zesz. Nauk. Politech. Rzesz. 103, Mat. Fiz. 16, Mat., **12**, 73–78 (1992)

[KGo94]   Kruppel, M., Gornicki, J.: An ergodic theorem for asymptotically nonexpansive mappings. Proc. R. Soc. Edinb., Sect. A, **124**, No. 1, 23–31 (1994)

[KMA97]   Kubiaczyk, I., Mostafa Ali, N.: On the convergence of the Ishikawa iterates to a common fixed point for a pair of multi-valued mappings. Acta Math. Hungar., **75**, No. 3, 253–257 (1997)

[Kuh80]   Kuhfitting, P.K.F.: The mean-value iteration for set-valued mappings. Proc. Amer. Math. Soc., **80**, No. 3, 401–405 (1980)

[Kuh81]   Kuhfitting, P.K.F.: Common fixed points of nonexpansive mappings by iteration. Pacific J. Math., **97**, No. 1, 137–139 (1981)

[Khn80]   Kuhn, G.: Generalized non-expansive mappings: approximation of fixed points. Atti Accad. Sci. Torino Cl. Sci. Fis. Mat. Natur., **114**, 93–101 (1980)

[Khn68]   Kuhn, H.W.: Simplicial approximation of fixed points. Proc. Nat. Acad. Sci., **61**, 1238–1242 (1968)

[Lak79]   Lakshmikantham, V. (Ed.): Applied Nonlinear Analysis. Academic Press, New York (1979)

[LaH02]   Lan, H.Y., Huang, N.-J.: Approximation of fixed points for asymptotically nonexpansive mappings in Banach spaces. (Chinese). Sichuan Daxue Xuebao, **39**, No. 5, 785–788 (2002)

[LHC04]   Lan, H.Y., Huang, N.-J., Cho, Y.J.: A new method for nonlinear variational inequalities with multi-valued mappings. Arch. Inequal. Appl., **2**, No. 1, 73-84 (2004)

[LLH02]   Lan, H.Y., Lee, B.S., Huang, N.-J.: Iterative approximation with mixed errors of fixed points for a class of asymptotically nonexpansive type mappings. Nonlinear Anal. Forum, **7**, No. 1, 55–65 (2002)

[LLL04]   Lan, H.Y., Liu, Q.K., Li, J.: Iterative approximation for a system of nonlinear variational inclusions involving generalized m-accretive mappings. Nonlinear Anal. Forum, **9**, No. 1, 33–42 (2004)

[LaD73]   Lami Dozo, E.: Multivalued nonexpansive mappings and Opial's condition. Proc. Amer. Math. Soc., **38**, 286–292 (1973)

[Lea83]   Leader, S.: Equivalent Cauchy sequences and contractive fixed points in metric spaces. Studia Math., **76**, No. 1, 63–67 (1983)

[Lee95]   Lee, Y.S.: Convergence of nonexpansive mappings. Commun. Korean Math. Soc., **10**, No. 2, 301–304 (1995)

[LMi86]   Lefebvre, O., Michelot, C.: Calcul d'un point fixe d'une application prox par la methode des approximations successives; conditions de convergence finie. C. R. Acad. Sci., Paris, Ser. I, **303**, 905–908 (1986)

[Lem96]   Lemaire, B.: Stability of the iteration for nonexpansive mappings. Serdica Math. J., **22**, No. 3, 229–238 (1996)

[Lem97]   Lemaire, B.: Which fixed point does the iteration method select ?. In: Gritzmann, P. (ed.) et al. Recent Advances in Optimization, Proceed. 8th French-German Conf. on Optimization. Trier, Germany, July 21-26, 1996, Lect. Notes Econ. Math. Syst. 452. Springer, Berlin (1997)

[Lev85]   Levi, L.: Fixed points of generalized nonexpansive multivalued mappings. Instit. Lombardo Accad. Sci. Lett. Rend. A, **116** (1982), 343–349 (1985)

[LiK01]   Li, G., Kim, J.K.: Demiclosedness principle and asymptotic behavior for nonexpansive mappings in metric spaces. Appl. Math. Lett., **14**, 645–649 (2001)

[LiH03]   Li, H.M.: The convergence and stability of Ishikawa iteration with errors for generalized Lipschitz accretive operator equations. (Chinese). Sichuan Shifan Daxue Xuebao Ziran Kexue Ban, **26**, No. 2, 116–119 (2003)

[LiH04]   Li, H.M.: Convergence and stability of Ishikawa iterative sequences for locally strongly pseudo contractive mappings. (Chinese). Sichuan Shifan Daxue Xuebao Ziran Kexue Ban, **27**, No. 3, 238–241 (2004)

[LiH05]   Li, J., Huang, N.-J.: Approximating random common fixed point of random set-valued strongly pseudo-contractive mappings. J. Appl. Math. Comput., **17**, No. 1-2, 329–341 (2005)

[LHH05]   Li, J., Huang, N.-J., Hwang, H.J., Cho, Y.J.: Stability of iterative procedures with errors for approximating common fixed points of quasi-contractive mappings. Appl. Anal., **84**, No. 3, 253–267 (2005)

[LLH02]   Li, J., Lan, H.Y., Huang, N.-J.: A new iterative approximation of fixed points for asymptotically demi-contractive mappings. (Chinese). J. Liaoning Norm. Univ. Nat. Sci., **25**, No. 1, 7–11 (2002)

[Li04a]   Li, Y.J.: Convergence of the Ishikawa iterative sequence with errors for Lipschitz accretive operators in Banach spaces. (Chinese). Acta Sci. Natur. Univ. Sunyatseni, **43**, No. 4, 10–13 (2004)

[Li04b]   Li, Y.J.: Equivalence of Mann and Ishikawa iteration methods in Banach spaces. (Chinese). Acta Sci. Natur. Univ. Sunyatseni, **43**, No. 1, 5–7 (2004)

[Li04c]   Li, Y.J.: Ishikawa iterative process with errors for strongly pseudocontractive mappings in Banach spaces. Commun. Korean Math. Soc., **19**, No. 3, 461–467 (2004)

[Li04d]   Li, Y.J.: Ishikawa iterative sequence with errors for Lipschitzian ϕ-strongly accretive operators in arbitrary Banach spaces. Far East J. Appl. Math., **16**, No. 2, 161–170 (2004)

[LiS03]   Li, Y.J., Shu, X.B.: The convergence of Ishikawa iterative sequence with errors for k-subaccretive operators in Banach spaces. Far East J. Math. Sci., **10**, No. 1, 75–86 (2003)

[LYL98]   Li, Y.Q., Liu, L.W.: Iterative processes for Lipschitz strongly accretive operators. Acta Math. Sinica, **41**, 845–850 (1998)

[Lia94]   Liang, Z.: Iterative solution of nonlinear equations involving m-accretive operators in Banach spaces. J. Math. Anal. Appl., **188**, 410–416 (1994)

[Lie81]   Liepinsh, A.K.: On the convergence of iteration processes of quasinonexpanding mappings in metric spaces. (Russian) Topological spaces and their mappings, pp. 75–87, 177, 183, Latv. Gos. Univ., Riga (1981)

[Lim77]   Lim, T.-C.: Fixed point theorems for mappings of nonexpansive type. Proc. Amer. Math. Soc., **66**, No. 1, 69–74 (1977)

[Lim85]   Lim, T.-C.: On fixed point stability for set-valued contractive mappings with applications to generalized differential equations. J. Math. Anal. Appl., **110**, No. 2, 436–441 (1985)

[Lim94]   Lim, T.-C., Xu, H.K.: Fixed point theorems for asymptotically nonexpansive mappings. Nonlinear Anal. TMA, **22**, No. 11, 1345–1355 (1994)

[LTX95]   Lin, P.-K., Tan, K.K., Xu, H.K.: Demiclosedness principle and asymptotic behavior for asymptotically nonexpansive mappings. Nonlinear Anal. TMA, **24**, No. 6, 929–946 (1995)

[LTs79]   Lindenstrauss, J., Tsafiri, J.: Classical Banach Spaces, vol. 2. Springer Verlag, Berlin (1979)

[Lns77]   Lions, P.-L.: Approximation de points fixes de contractions. C.R. Acad. Sci. Ser. A-B Paris, **284**, 1357–1359 (1977)

[Li837]   Liouville, J.: Sur le developpment des fonction ou parties de fonctions de fonctions en series..., Second Memoire. Journ. de Math., **2**, 16–35 (1837)

[LDL00]   Liu, G., Deng, L., Li, S.H.: Approximating fixed points of nonexpansive mappings. Int. J. Math. Math. Sci., **24**, No. 3, 173–177 (2000)

[LiJ02]   Liu, J.A.: Some convergence theorems of implicit iterative process for nonexpansive mappings in Banach spaces. Math. Commun., **7**, No. 2, 113–118 (2002)

[LiJ04]   Liu, J.R., Cui, Y.L.: Three-step iterative algorithms for a new class of generalized mixed nonlinear implicit quasi-variational inclusions in Banach spaces. (Chinese). Math. Appl. (Wuhan), **17**, No. 2, 197–202 (2004)

[LiK84]   Liu, K.: Iterative method for constructing zeros of accretive sets. MSc. Thesis, Xi'an Jaiotong University (1984)

[LLS93]   Liu, L.S.: Mann iteration processes for constructing solutions of strongly monotone operator equations in Banach spaces (Chinese). J. Eng. Math., Xi'an, **10**, No. 4, 117–121 (1993)

[LLS94]   Liu, L.S.: On approximation theorems and fixed point theorems for nonself mappings in infinite dimensional Banach spaces. J. Math. Anal. Appl., **188**, No. 2, 541–551 (1994)

[LL95a]   Liu, L.S.: Fixed points of local strictly pseudo-contractive mappings using Mann and Ishikawa iteration with errors. Indian J. Pure Appl. Math., **26**, No. 7, 649–659 (1995)

[LL95b]   Liu, L.S.: Ishikawa and Mann iteration process with errors for nonlinear strongly accretive mappings in Banach spaces. J. Math. Anal. Appl., **194**, 114–125 (1995)

[LL95c]   Liu, L.S.: Ishikawa iteration methods for a solution of nonlinear Lipschitzian strongly accretive operator equations in uniformly smooth Banach spaces (Chinese). J. Qufu Norm. Univ., Nat. Sci., **21**, No. 1, 1–5 (1995)

[LLS98]   Liu, L.S.: Ishikawa-type and Mann-type iterative processes with errors for constructing solutions of nonlinear equations involving m-accretive operators in Banach spaces. Nonlinear Anal., **34**, 307–317 (1998)

[LL00a]   Liu, L.S.: Ishikawa and Mann iterative processes with errors for nonlinear operator equations of strongly accretive and strongly pseudo-contractive type in Banach spaces. In: Fixed Point Theory and Applications (Chinju, 1998). Nova Sci. Publ., Huntington (2000)

[LL00b]   Liu, L.S.: Iterative method for solutions and coupled quasi-solutions of nonlinear integro-differential equations of mixed type in Banach spaces. Nonlinear Anal. TMA, **42**, No. 4, 583–598 (2000)

[LLS01]   Liu, L.S.: Approximation theorems and fixed point theorems for various classes of 1-set-contractive mappings in Banach spaces. Acta Math. Sin. (Engl. Ser.), **17**, No. 1, 103–112 (2001)

[LZh98]   Liu, L.S, Zhang, H.Q.: Solutions of mixed monotone operator equations and their application to nonlinear integral equations (Chinese). J. Eng. Math., Xi'an, **15**, No. 3, 17–24 (1998)

[LZK01]   Liu, L.S., Zhang, X.Y., Kim, J.K.: Existence of global solutions of initial value problem for nonlinear impulsive integro-differential equations of mixed type in Banach spaces. Nonlinear Funct. Anal. Appl., **6**, No. 3, 313–327 (2001)

[LiW94]   Liu, L.W.: An iterative process for Lipschitzian strongly accretive mappings in L_p 1. Numer. Math., Nanjing, **16**, No. 3, 264–270 (1994)

[LiW97]   Liu, L.W.: Approximation of fixed points of a strictly pseudocontractive mapping. Proc. Amer. Math. Soc., **125**, No. 5, 1363–1366 (1997)

[LiW98]   Liu, L.W.: On Mann and Ishikawa iteration processes of strongly pseudo-contractive mappings. Sci. Math., **1**, No. 2, 189–193 (1998)

[LiW00]   Liu, L.W.: Strong convergence of iteration methods for equations involving accretive operators in Banach spaces. Nonlinear Anal. TMA, **42**, No. 2, 271–276 (2000)

[LXi02]   Liu, L.W., Xiao, H.: On the stability of iteration procedures for strongly pseudocontractive mappings. (Chinese) Acta Anal. Funct. Appl., **4**, No. 2, 158–163 (2002)

[LiQ87]   Liu, Q.: On Naimpally and Singh's open questions. J. Math. Anal. Appl., **124**, 157–164 (1987)

[LQ90a]   Liu, Q.: A convergence theorem of the sequence of Ishikawa iterates for quasi-contractive mappings. J. Math. Anal. Appl., **146**, 301–305 (1990)

[LQ90b]   Liu, Q.: The convergence theorems of the sequence of Ishikawa iterates for hemicontractive mappings. J. Math. Anal. Appl., **148**, 55–62 (1990)

[LiQ92]   Liu, Q.: A convergence theorem for Ishikawa iterates of continuous generalized nonexpansive maps. J. Math. Anal. Appl., **165**, 305–309 (1992)

[LiQ96]   Liu, Q.: Convergence theorems of the sequence of iterates for asymptotically demicontractive and hemicontractive mappings. Nonlinear Anal. TMA, **26**, No. 11, 1835–1842 (1996)

[LQ01a]   Liu, Q.: Iterative sequence for asymptotically nonexpansive type mappings in Banach spaces. J. Math. Anal. Appl., **256**, 1–7 (2001)

[LQ01b]   Liu, Q.: Iterative sequences for asymptotically quasi-nonexpansive mappings with error member. J. Math. Anal. Appl., **259**, No. 1, 18–24 (2001)

[LiQ02]   Liu, Q.: Iteration sequences for asymptotically quasi-nonexpansive mappings with an error member in a uniform convex Banach space. J. Math. Anal. Appl., **266**, 468–471 (2002)

[LXe00]   Liu, Q., Xue, L.: Convergence theorems of iterative sequences for asymptotically non-expansive mapping in a uniformly convex Banach space. J. Math. Res. Exposition, **20**, No. 3, 331–336 (2000)

[LiZ98]   Liu, Z.: On the structure of the set of subsequential limit points of a sequence of iterates. J. Math. Anal. Appl., **222**, No. 1, 297–304 (1998)

[LiZ99]   Liu, Z.: On Park's open questions and some fixed-point theorems for general contractive type mappings. J. Math. Anal. Appl., **234**, No. 1, 165–182 (1999)

[LAK04]   Liu, Z., An, Z., Kang, S.M., Ume, J.S.: Convergence and stability of the three-step iterative schemes for a class of general quasivariational-like inequalities. Int. J. Math. Math. Sci. 2004, No. 69-72, 3849–3857 (2004)

[LAL04]   Liu, Z., An, Z., Li, Y.J., Kang, S.M.: Iterative approximation of fixed points for ϕ-hemicontractive operators in Banach spaces. Commun. Korean Math. Soc., **19**, No. 1, 63–74 (2004)

[LBK02]   Liu, Z., Bounias, M., Kang, S.M.: Iterative approximation of solutions to nonlinear equations of Φ-strongly accretive operators in Banach spaces. Rocky Mountain J. Math., **32**, No. 3, 981–997 (2002)

[LFK03]   Liu, Z., Feng, C., Kang, S.M., Kim, Kun Ho: Convergence and stability of modified Ishikawa iterative procedures with errors for some nonlinear mappings. Panamer. Math. J., **13**, No. 4, 19–33 (2003)

[LFK04]   Liu, Z., Feng, C., Kang, S.M., Ume, J.S.: Approximating fixed points and common fixed points of quasi-nonexpansive mappings. Commun. Appl. Nonlinear Anal., **12**, No. 1, 59–68 (2005)

[LK01a]   Liu, Z., Kang, S.M.: Convergence and stability of the Ishikawa iteration procedures with errors for nonlinear equations of the φ-strongly accretive type. Neural Parallel Sci. Comput., **9**, No. 1, 103–117 (2001)

[LK01b]   Liu, Z., Kang, S.M.: Stability of Ishikawa iteration methods with errors for strong pseudocontractions and nonlinear equations involving accretive operators... Math. Comput. Modelling, **34**, No. 3-4, 319–330 (2001)

[LK01c]   Liu, Z., Kang, S.M.: Convergence theorems for Φ-strongly accretive and Φ-hemicontractive operators. J. Math. Anal. Appl., **253**, 35–49 (2001)

[LK01d]   Liu, Z., Kang, S.M.: Iterative approximation of fixed points for Φ-hemicontractive operators in arbitrary Banach spaces. Acta Sci. Mat. (Szeged), **67**, 821–831 (2001)

[LK03a]   Liu, Z., Kang, S.M.: Stable and almost stable iteration schemes for non-linear accretive operator equations in arbitrary Banach spaces. Panamer. Math. J., **13**, No. 1, 91–102 (2003)

[LK03b]   Liu, Z., Kang, S.M.: Iterative solutions of nonlinear equations with Φ-strongly accretive operators in uniformly smooth Banach spaces. Computers Math. Appl., **45**, 623–634 (2003)

[LK04a]   Liu, Z., Kang, S.M.: Iterative process with errors for nonlinear equations of local ϕ-strongly accretive operators in arbitrary Banach spaces. Int. J. Pure Appl. Math., **12**, No. 2, 229–246 (2004)

[LK04b]   Liu, Z., Kang, S.M.: On general principles of Ishikawa iterative scheme with errors of multi-valued mappings in normed linear spaces. Int. J. Pure Appl. Math., **17**, No. 2, 189–199 (2004)

[LK04c]   Liu, Z., Kang, S.M.: Convergence and stability of perturbed three-step iterative algorithm for completely generalized nonlinear quasivariational inequalities. Appl. Math. Comput., **149**, No. 1, 245–258 (2004)

[LKC04]   Liu, Z., Kang, S.M., Cho, Y.J.: Convergence and almost stability of Ishikawa iterative scheme with errors for m-accretive operators. Comput. Math. Appl., **47**, No. 4-5, 767–778 (2004)

[LKU01]   Liu, Z., Kang, S.M., Ume, J.S.: Iterative solutions of K-positive definite operator equations in real uniformly smooth Banach spaces. Int. J. Math. Math. Sci., **27**, No. 3, 155–160 (2001)

[LKU02]   Liu, Z., Kang, S.M., Ume, J.S.: Error bounds of the iterative approxima-tions of Ishikawa iterative schemes with errors for strictly hemicontractive and strongly quasiaccretive operators. Commun. Appl. Nonlinear Anal., **9**, No. 4, 33–46 (2002)

[LKU03]   Liu, Z., Kang, S.M., Ume, J.S.: General principles for Ishikawa iterative process for multi-valued mappings. Indian J. Pure Appl. Math., **34**, No. 1, 157–162 (2003)

[LKS03]   Liu, Z., Kang, S.M., Shim, S.H.: Almost stability of the Mann itera-tion with errors for strictly hemi-contractive operators in smooth Banach spaces. Int. J. Math. Math. Sci., **40**, No. 1, 29–40 (2003)

[LKC01]   Liu, Z., Kim, J.K., Chun, S.-A.: Convergence theorems and stability prob-lems of Ishikawa iterative processes with errors for quasi-contractive map-pings. Commun. Appl. Nonlinear Anal., **8**, No. 3, 69–79 (2001)

[LKC02]   Liu, Z., Kim, J.K., Chun, S.-A.: Iterative approximation of fixed points for generalized asymptotically contractive and generalized hemicontractive mappings. Panamer. Math. J., **12**, No. 4, 67–74 (2002)

[LKH04]   Liu, Z., Kim, J.K., Hyun, H.: Convergence theorems and stability of the Ishikawa iteration procedures with errors for strong pseudocontractions and nonlinear equations involving accretive operators. In: Fixed Point Theory and Appl. Vol. 5, 79–95. Nova Sci. Publ., Hauppauge, (2004)

[LKK03]   Liu, Z., Kim, J.K., Kang, S.M.: Necessary and sufficient condi-tions for convergence of Ishikawa iterative schemes with errors to ϕ-hemicontractive mappings. Commun. Korean Math. Soc., **18**, No. 2, 251–261 (2003)

[LKK02]   Liu, Z., Kim, J.K., Kim, Ki Hong: Convergence theorems and stability problems of the modified Ishikawa iterative sequences for strictly succes-sively hemicontractive mappings. Bull. Korean Math. Soc., **39**, No. 3, 455–469 (2002)

266   References

[LKU02] Liu, Z., Kim, J.K., Ume, J.S.: Characterizations for the convergence of
         Ishikawa iterative processes with errors in normed linear spaces. J. Non-
         linear Convex Anal., **3**, No. 1, 59–66 (2002)
[LLK00] Liu, Z., Lee, J., Kim, J.K.: On Meir-Keeler type contractive mappings
         with diminishing orbital diameters. Nonlinear Funct. Anal. Appl., **5**, No.
         1, 73–83 (2000)
[LNK02] Liu, Z., Nam, Y.M., Kim, J.K., Ume, J.S.: Stability of Ishikawa iterative
         schemes with errors for nonlinear accretive operators in arbitrary Banach
         spaces. Nonlinear Funct. Anal. Appl., **7**, No. 1, 55–67 (2002)
[LUm02] Liu, Z., Ume, J.S.: Stable and almost stable iteration schemes for strictly
         hemi-contractive operators in arbitrary Banach spaces. Numer. Funct.
         Anal. Optim., **23**, No. 7-8, 833–848 (2002)
[LUK03] Liu, Z., Ume, J.S., Kang, S.M.: Convergence and stability and almost sta-
         bility of the Ishikawa iteration procedures with errors for quasi-contractive
         mappings in q-uniformly smooth Banach spaces. In: Fixed Point Theory
         and Applications (Chinju/Masan, 2001), 171–188. Nova Sci. Publ., Haup-
         pauge, NY (2003)
[LU04a] Liu, Z., Ume, J.S., Kang, S.M.: Stability of Noor iterations with errors
         for generalized nonlinear complementarity problems. Acta Math. Acad.
         Paedagog. Nyhzi. (N.S.) **20**, No. 1, 53–61 (electronic) (2004)
[LU04b] Liu, Z., Ume, J.S., Kang, S.M.: Convergence and almost stability of
         Ishikawa iteration method with errors for strictly hemi-contractive op-
         erators in Banach spaces. J. Korea Soc. Math. Educ. Ser. B Pure Appl.
         Math., **11**, No. 4, 293–308 (2004)
[LWS05] Liu, Z., Wang, L., Shim, S.H., Kang, S.M.: The equivalence of Mann
         and Ishikawa iteration methods with errors for Lipschitzian φ-strongly
         accretive operators. Int. J. Pure Appl. Math., **18**, No. 1, 61-72 (2005)
[LWK03] Liu, Z., Wang, L., Kang, S.M., Kim, Kang Hak: Convergence of three-
         step iteration methods for quasi-contractive mappings. Commun. Appl.
         Nonlinear Anal., **10**, No. 4, 41–47 (2003)
[LXC01] Liu, Z., Xu, Y.G., Cho, Y. J.: Iterative solution of nonlinear equations
         with ϕ-strongly accretive operators. Arch. Math. (Basel), **77**, No. 6, 508–
         516 (2001)
[LZC01] Liu, Z., Zhang, L., Cho, Y.J.: Convergence and stability of Ishikawa iter-
         ative methods with errors for quasi-contractive mappings in q-uniformly
         smooth Banach spaces. Appl. Anal., **79**, No. 1-2, 277–292 (2001)
[LZK01] Liu, Z., Zhang, L., Kang, S.M.: Iterative solutions to nonlinear equations
         of the accretive type in Banach spaces. East Asian Math. J., **17**, No. 2,
         265–273 (2001)
[LZK02] Liu, Z., Zhang, L., Kang, S.M.: Convergence theorems and stability results
         for Lipschitz strongly pseudocontractive operators. Int. J. Math. Math.
         Sci., **31**, No. 10, 611–617 (2002)
[LZL02] Liu, Z., Zhao, Y.L., Lee, B.S.: Convergence and stability of the three-step
         iterative process with errors for strongly pseudocontractive operators in
         L_p spaces. Nonlinear Anal. Forum, **7**, No. 1, 15–22 (2002)
[LZK05] Liu, Z., Zhu, B.B., Kang, S.M., Lee, S.K.: The equivalence among mod-
         ified Mann iteration, modified Ishikawa iteration and modified multistep
         iteration. Int. J. Pure Appl. Math., **18**, No. 1, 73-87 (2005)
[Maa68] Maia, M.G.: Un'osservazione sulle contrazioni metriche. Rend. Sem. Mat.
         Univ. Padova, **40**, 139–143 (1968)

[Mai07]   Mainge, P.-E.: Approximation methods for common fixed points of non-expansive mappings in Hilbert spaces. J. Math. Anal. Appl., **325**, 469-479 (2007)

[Mai76]   Maiti, M., Achari, J., Pal, T.K.: Mappings having common fixed points. Pure Appl. Math. Sci., **3**, 101–104 (1976)

[Mai81]   Maiti, M., Babu, A.C.: On subsequential limit points of a sequence of iterates. Proc. Amer. Math. Soc., **82**, 377–381 (1981)

[Mai85]   Maiti, M., Babu, A.C.: On subsequential limit points of a sequence of iterates. II. J. Austral. Math. Soc., Ser. A, **38**, 118–129 (1985)

[MaG89]   Maiti, M., Ghosh, M.K.: Approximating fixed points by Ishikawa iterates. Bull. Austral. Math. Soc., **40**, No. 1, 113–117 (1989)

[MaS93]   Maiti, M., Saha, B.: Approximating fixed points of nonexpansive and generalized nonexpansive mappings. Int. J. Math. Math. Sci., **16**, No. 1, 81–86 (1993)

[Man53]   Mann, W.R.: Mean value methods in iteration. Proc. Amer. Math. Soc., **44**, 506–510 (1953)

[Man79]   Mann, W.R.: Averaging to improve convergence of iterative processes. Functional analysis methods in numerical analysis, Spec. Sess., AMS, St. Louis 1977, Lect. Notes Math. 701 (1979)

[Mar92]   Marino, G.: Approximating fixed points for nonexpansive maps in Hilbert spaces. Approximation theory, spline functions and applications (Maratea, 1991), 405–409, NATO Adv. Sci. Inst. Ser. C, Math. Phys. Sci., 356. Kluwer Acad. Publ., Dordrecht (1992)

[MPS94]   Marino, G., Pietramala, P., Singh, S.P.: Convergence of approximating fixed point sets for multivalued mappings. J. Math. Sci., Delhi, **28**, 117–130 (1994)

[MPT90]   Marino, G., Pietramala, P., Trombetta, G.: Convergence of approximating fixed point sets. Ann. Univ. Ferrara Sez. VII (N.S.), **36**, 195–206 (1990)

[MaT92]   Marino, G., Trombetta, G.: On approximating fixed points of nonexpansive maps. Indian J. Math., **34**, 91–98 (1992)

[MaX06]   Marino, G., Xu, H.K.: A general iterative method for nonexpansive mappings in Hilbert spaces. J. Math. Anal. Appl., **318**, No. 1, 43–52 (2006)

[Mrk73]   Markin, J.T.: Continuous dependence of fixed point sets. Proc. Amer. Math. Soc., **38**, 545–547 (1973)

[Mrk76]   Markin, J.T.: A fixed point stability theorem for nonexpansive set valued mappings. J. Math. Anal. Appl., **54**, No. 2, 441–443 (1976)

[Mrs76]   Maruster, St.: Quasi-nonexpansivity and two classical methods for solving nonlinear equations. Proc. Amer. Math. Soc., **62**, No. 1, 119–123 (1976)

[Mrs77]   Maruster, St.: The solution by iteration of nonlinear equations in Hilbert spaces. Proc. Amer. Math. Soc., **63**, No. 1, 69–73 (1977)

[Mrs81]   Maruster, St.: Metode numerice in rezolvarea ecuatiilor neliniare. Editura Tehnica, Bucuresti (1981)

[Ms77a]   Massa, S.: Approximation of fixed points by Cesaro's means of iterates. Rend. Ist. Mat. Univ. Trieste, **9**, 127–133 (1977)

[Ms77b]   Massa, S.: On a method of successive approximations. Atti Accad. Naz. Lincei, VIII. Ser., Rend., Cl. Sci. Fis. Mat. Nat., **62**, 584–587 (1977)

[Ms77c]   Massa, S.: On the approximation of fixed points for quasi nonexpansive mappings. Ist. Lombardo Accad. Sci. Lett. Rend. A, **111**, 188–193 (1977)

[Mas78]   Massa, S.: Convergence of an iterative process for a class of quasi-nonexpansive mappings. Boll. U.M.I., **51-A**, 154–158 (1978)

[Mas82]   Massa, S.: Fixed point approximation for quasi-nonexpansive mappings. Matematiche, **37**, No. 1, 3–7 (1982)

[Mas83]   Massa, S.: Opial spaces, asymptotic centers and fixed points (Italian). Rend. Semin. Mat. Fis. Milano, **53**, 35–47 (1983)

[MKu03]   Matsushita, S., Kuroiwa, D.: Approximation of fixed points of nonexpansive nonself mappings. Sc. Math. Japon., **57**, No. 1, 171–176 (2003)

[MKu04]   Matsushita, S., Kuroiwa, D.: Strong convergence of averaging iterations of nonexpansive nonself-mappings. J. Math. Anal. Appl., **294**, 206–214 (2004)

[MTa04]   Matsushita, S., Takahashi, W.: Weak and strong convergence theorems for relatively nonexpansive mappings in Banach spaces. Fixed Point Theory and Applications 2004, **1**, 37–47 (2004)

[MeK69]   Meir, A., Keeler, E.: A theorem on contraction mappings. J. Math. Anal. Appl., **28**, 326–329 (1969)

[Mel96]   Melentsov, A. A.: Iterative nets of weakly contractive operators. (Russian) Acta Comment. Univ. Tartu. Math., No. **1**, 9–12 (1996)

[Mes92]   Meszaros, J.: A comparison of various definitions of contractive type mappings. Bull. Calcutta Math. Soc., **84**, No. 2, 167–194 (1992)

[Mey80]   Meyer, P.W.: Die Anwendung von verallgemeinerten Normen zur Fehlerabschaetzung bei Iterationsverfahren. Dissertation, Mathematisch-Naturwissenschaft-liche Fakultät der Universität Düsseldorf. (1980)

[MiP83]   Miczko, A., Palczewski, B.: On convergence of successive approximations of some generalized contraction mappings. Ann. Pol. Math., **40**, No. 3, 213–232 (1983)

[Mik91]   Mikhlin, S.G.: Error analysis in numerical processes. Wiley-Interscience Series in Pure and Applied Mathematics. John Wiley & Sons, Chichester (1991)

[Min62]   Minty, G.J.: Monotone (nonlinear) operators in Hilbert space. Duke Math. J., **29**, 341–346 (1962)

[Mis95]   Mishra, S.N.: Fixed point Ishikawa iteration in a convex metric space. C. R. Math. Acad. Sci., Soc. R. Can., **17**, No. 4, 153–158 (1995)

[MKa96]   Mishra, S.N., Kalinde, A.K.: Iterative construction of fixed points. Numer. Funct. Anal. Optimization, **17**, No. 5-6, 639–647 (1996)

[MKa98]   Mishra, S.N., Kalinde, A.K.: A note on an Ishikawa type iteration scheme. Demonstratio Math., **31**, No. 3, 587–594 (1998)

[Mis85]   Misiurewicz, M.: Chaos almost everywhere. In: Iteration theory and its functional equations (Lochau, 1984), 125-130, Lecture Notes in Math., 1163. Springer, Berlin (1985)

[Miy83]   Miyazaki, K.-I.: Iteration methods for common fixed points of nonexpansive mappings. Proc. Japan Acad. Ser. A Math. Sci., **59**, No. 3, 75–78(1983)

[MKM83]   Miyazaki, K.-I., Kawatani, T., Miyaura, S.: Some ergodic theorems for a finite family of nonexpansive mappings. Bull. Kyushu Inst. Tech. Math. Natur. Sci. No., **30**, 31–35 (1983)

[Mol89]   Moloney, J.: Some fixed point theorems. Glas. Mat., **24**, 59–76 (1989)

[Mol94]   Moloney, J.: Construction of a sequence strongly converging to a fixed point of an asymptotically non-expansive mapping. J. Math. Anal. Appl., **182**, No. 3, 589–593 (1994)

[MoW95]   Moloney, J., Weng, X. L.: A fixed point theorem for demicontinuous pseudocontractions in Hilbert space. Stud. Math., **116**, No. 3, 217–223 (1995)

[Moo90]   Moore, C.: Iterative approximation of the solution to a K-accretive operator equation in certain Banach spaces. Indian J. Pure Appl. Math., **21**,
No. 12, 1087–1093 (1990)

[Moo91]   Moore, C.: A fixed point iteration process for Hammerstein equations
involving angle-bounded operators. Bol. Soc. Mat. Mex., II. Ser., **36** , No.
1-2, 39–48 (1991)

[Mo92a]   Moore, C.: Fixed point iterations for a class of nonlinear mappings. Math.
Japon., **37**, No. 5, 955–960 (1992)

[Mo92b]   Moore, C.: Iterative approximation of the solution to certain nonlinear
problems arising in mathematical physics. J. Nig. Ass. Math. Phys., **1**,
15–22 (1992)

[Moo94]   Moore, C.: Correction to the paper: "On fixed point iterations for a class
of nonlinear mappings". Math. Japon., **39**, No. 1, 199 (1994)

[Moo99]   Moore, C.: The solution by iteration of nonlinear equations involving psi-
strongly accretive operators. Nonlinear Anal. TMA, **37**, No. 1, 125–138
(1999)

[Mo00a]   Moore, C.: Picard iterations for solution of nonlinear equations in certain
Banach spaces. J. Math. Anal. Appl., **245**, No. 2, 317–325 (2000)

[Mo00b]   Moore, C.: Iterative solution of nonlinear equations involving K-accretive
operator equations. Sci. Math., **3**, No. 3, 309–318 (2000)

[Mo02a]   Moore, C.: The solution by iteration of nonlinear equations of Hammerstein type. Nonlinear Anal. TMA, **49**, No. 5, 631–642 (2002)

[Mo02b]   Moore, C.: A double-sequence iteration process for fixed points of continuous pseudocontractions. Comput. Math. Appl., **43**, No. 12, 1585–1589
(2002)

[MN01a]   Moore, C., Nnoli, B.V.C.: Strong convergence of averaged approximants
for Lipschitz pseudocontractive maps. J. Math. Anal. Appl., **260**, No. 1,
269–278 (2001)

[MN01b]   Moore, C., Nnoli, B.V.C.: Local iterations for fixed points of uniformly
hemicontractive maps in arbitrary normed linear spaces. Soochow J.
Math., **27**, No. 1, 59–72 (2001)

[MN01c]   Moore, C., Nnoli, B.V.C.: Iterative solution of nonlinear equations involving set-valued uniformly accretive operators. Comput. Math. Appl., **42**,
No. 1-2, 131–140 (2001)

[MoN05]   Moore, C., Nnoli, B.V.C.: Iterative sequence for asymptotically demicontractive maps in Banach spaces. J. Math. Anal. Appl., **302**, No. 2, 557–562
(2005)

[MNN01]   Moore, C., Nnoli, B.V.C., Ntatin, B.: Local iterations for nonlinear systems involving uniformly accretive operators in arbitrary normed linear
spaces. Bol. Soc. Mat. Mexicana III Ser., **7**, No. 2, 223–233 (2001)

[Mrl85]   Morales, C.: Zeros for accretive operators satisfying certain boundary conditions. J. Math. Anal. Appl., **105**, 167–175 (1985)

[Mrl90]   Morales, C.: Strong convergence theorems for pseudo-contractive mappings in Banach space. Houston J. Math., **16**, No. 4, 549–557 (1990)

[Mrl97]   Morales, C.: Approximation of fixed points for locally nonexpansive mappings. Ann. Univ. Mariae Curie-Sklodowska, Sect. A, **51**, No. 2, 203–212
(1997)

[MoJ00]   Morales, C., Jung, J.S.: Convergence of paths for pseudo-contractive mappings in Banach spaces. Proc. Amer. Math. Soc., **128**, No. 11, 3411–3419
(2000)

[MoM95] Morales, C., Mutangadura, S.A.: On the approximation of fixed points for locally pseudo-contractive mappings. Proc. Amer. Math. Soc., **123**, No. 2, 417–423 (1995)

[Mor78] Moreau, J.: Un cas de convergence des iterees d'une espace Hilbertien. C. R. Acad. Sci. Paris, **286**, 143-144 (1978)

[Mou00] Moudafi, A.: Viscosity approximation methods for fixed-points problems. J. Math. Anal. Appl., **241**, 46-55 (2000)

[Muk86] Mukherjee, R.N.: Construction of fixed points of strictly pseudocontractive mappings in generalized Hilbert spaces and related applications. Indian J. Pure Appl. Math., **15**, 276–284 (1986)

[MuS86] Mukherjee, R.N., Som, T.: Approximating common fixed points of nonexpansive mappings on a generalized Hilbert space. J. Sci. Res., **8**, No. 1, 27–28 (1986)

[MuS89] Mukherjee, R.N., Som, T., Verma, V.: On weak convergence to the fixed point of a generalized asymptotically nonexpansive map. Publ. Inst. Math., Nouv. Ser., **45(59)**, 179–183 (1989)

[MuR77] Muller, G., Reinermann, J.: Fixed point theorems for pseudocontractive mappings and a counter-example for compact maps. Comment. Math. Univ. Carolin., **18**, No. 2, 281–298 (1977)

[MuR80] Muller, G., Reinermann, J.: A theorem on strong convergence in Banach spaces with applications to fixed points of nonexpansive and pseudocontractive mappings. In: Constructive function theory, Proc. Int. Conf., Blagoevgrad 1977 (1980)

[MuA87] Muresan, A.S.: Some remarks on the comparison functions. Prepr., "Babes-Bolyai" Univ., Fac. Math., Res. Semin. 9, 99–108 (1987)

[MuA88] Muresan, A.S.: Some fixed point theorems of Maia type. Prepr., "Babes-Bolyai" Univ., Fac. Math. Phys., Res. Semin. 1988, No. 3, 35–42 (1988)

[MuA96] Muresan, A.S.: Mappings of Picard, Bessaga and Janos type. Bul. Stiint. Univ. Baia Mare, Ser. B, **12**, No. 1, 85–89 (1996)

[MuV00] Muresan, V.: Some applications of the fiber contraction theorem. Stud. Univ. Babes-Bolyai, Math., **45**, No. 4, 87–96 (2000)

[MV03a] Muresan, V.: Functional-integral equations. Editura Mediamira, Cluj-Napoca (2003)

[MvV03b] Muresan, V.: On a class of differential equations with linear modification of the argument. PU.M.A., Pure Math. Appl., **13**, No. 1-2, 253–258 (2003)

[Nad68] Nadler, S.B.: Sequences of contractions and fixed points. Pacific J. Math., **27**, No. 3, 579–585 (1968)

[Nad69] Nadler, S.B.: Multi-valued contraction mappings. Pacific J. Math., **30**, 282–291 (1969)

[Nad73] Nadler, S.B.: Some problems concerning stability of fixed points. Colloq. Math., **27**, 263–268, 332 (1973)

[NaP86] Naidu, S.V.R., Prasad, J.R.: Ishikawa iterates for a pair of maps. Indian J. Pure Appl. Math., **17**, 193–200 (1986)

[NaS82] Naimpally, S.A.; Singh, K.L.: Sequence of iterates in locally convex spaces. In: Nonlinear phenomena in mathematical sciences, Proc. Int. Conf., Arlington/Tex. 1980, 725–736 (1982)

[NaS83] Naimpally, S.A.; Singh, K.L.: Extensions of some fixed point theorems of Rhoades. J. Math. Anal. Appl., **96**, 437–446 (1983)

[NSW83] Naimpally, S.A., Singh, K.L., Whitfield, J.H.M.: Fixed points and sequences of iterates in locally convex spaces. Contemp. Math., **21**, 159–166 (1983)

[NST03] Nakajo, K., Shimoji, K., Takahashi, W.: Weak and strong convergence theorems by Mann's type iteration and the hybrid method in Hilbert spaces. J. Nonlinear Convex Anal., **4**, No. 3, 463–478 (2003)

[NaT01] Nakajo, K., Takahashi, W.: A nonlinear strong ergodic theorem for families of asymptotically nonexpansive mappings with compact domains. Sc. Math. Japon., **54**, No. 2, 301–310 (2001)

[NaT03] Nakajo, K., Takahashi, W.: Strong convergence theorems for nonexpansive mappings and nonexpansive semigroups. J. Math. Anal. Appl., **279**, 372–379 (2003)

[Nem36] Nemytzki, V.V.: The fixed point method in analysis (Russian). Usp. Mat. Nauk, **1**, 141–174 (1936)

[Ne82a] Neumaier, A.: A better estimate for fixed points of contractions. Freib. Intervall-Ber., **82**/5, 13–16 (1982)

[Ne82b] Neumaier, A.: A better estimate for fixed points of contractions. Z. Angew. Math. Mech., **62**, 627 (1982)

[Nev79] Nevanlinna, O.: Global iteration schemes for monotone operators. Nonlinear Anal., **3** , No. 4, 505–514 (1979)

[NeR79] Nevanlinna, O., Reich, S.: Strong convergence theorems of contraction semigroup and of iterative methods for accretive operators in Banach spaces. Israel J. Math., **32**, 44–58 (1979)

[NiR00] Ni, R.X.: A necessary and sufficient condition for strong convergence of iterative methods for accretive operators and contraction semigroups (Chinese). Appl. Math. J. Chinese Univ. Ser. A, **15**, No. 4, 433–439 (2000)

[Ni01a] Ni, R.X.: Ishikawa iteration procedures with errors for certain nonlinear operators and their stability (Chinese). Appl. Math. J. Chinese Univ. Ser. A, **16**, No. 3, 309–316 (2001)

[Ni01b] Ni, R.X.: Ishikawa iteration procedures with errors for certain generalized Lipschitzian nonlinear operators (Chinese). Acta Math. Sinica, **44**, No. 4, 701–712 (2001)

[Ni01c] Ni, R.X.: Ishikawa-type iterative procedures with errors of nonlinear equations involving m-accretive operators in Banach spaces. J. Shaoxing Coll. Arts Sci., Nat. Sci., **21**, No. 1, 1–8 (2001)

[Ni02a] Ni, R.X.: Convergence of Ishikawa iteration procedures with errors and application (Chinese). J. Ningxia Univ. Nat. Sci. Ed., **23**, No. 1, 1–5 (2002)

[Ni02b] Ni, R.X.: Some convergence theorems for iterative sequence of certain generalized Lipschitzian nonlinear operators in Banach spaces. Numer. Math., Nanjing, **24**, No. 1, 87–96 (2002)

[NiR05] Ni, R.X.: A characteristic condition for convergence of generalized steepest descent method for quasi-accretive operator equations. (Chinese). Acta Math. Sinica, **48**, No. 1, 115–124 (2005)

[NiY01] Ni, R.X., Ye, X.: Ishikawa iteration procedures with errors for certain nonlinear operator without Lipschitz assumption (Chinese). Numer. Math., Nanjing, **23**, No. 1, 29–37 (2001)

[Nus72] Nussbaum, R.D.: Some asymptotic fixed point theorems. Trans. Amer. Math. Soc., **171**, 349–375 (1972)

[Obl68]   Oblomskaja, L.: Methods of successive approximation for linear equations in Banach spaces. USSR Compt. Math. and Math. Phys., **8**, 239–253 (1968)

[Ofo06]   Ofoedu, E.U.: Strong convergence theorem for uniformly L-Lipschitzian asymptotically pseudocontractive mapping in real Banach space. J. Math. Anal. Appl., **321** 722-728 (2006)

[OPX03]   O'Hara, J.G., Pillay, P., Xu, H.K.: Iterative approaches to finding nearest common fixed points of nonexpansive mappings in Hilbert spaces. Nonlinear Anal., **54**, No. 8, 1417–1426 (2003)

[Op67a]   Opial, Z.: Weak convergence of the sequence of successive approximations for nonexpansive mappings. Bull. Amer. Math. Soc., **73**, 591–597 (1967)

[Op67b]   Opial, Z.: Nonexpansive and monotone mappings in Banach spaces. Lectures Notes 67-1. Center for Dynamical Systems, Division of Applied Mathematics, Brown University (1967)

[Opo76]   Opojzev, V.I.: The converses of contraction theorem (Russian). Usp. Math. Nauk, **21**, 1, 169–198 (1976)

[ORe99]   O'Regan, D.: Existence and approximation of fixed points for multivalued maps. Appl. Math. Letters, **12**, 37–43 (1999)

[ORh70]   Ortega, J.M., Rheinboldt, W.C.: Iterative Solution of Nonlinear Equation in Several Variables. Academic Press, New York (1970)

[Osi92]   Osilike, M.O.: Ishikawa and Mann iteration methods for nonlinear strongly accretive mappings. Bull. Austral. Math. Soc., **46**, No. 3, 413–424 (1992)

[Osi93]   Osilike, M.O.: Iterative method for nonlinear monotone-type operators in uniformly smooth Banach spaces. J. Nigerian Math. Soc., **12**, 73–79 (1993)

[Os95a]   Osilike, M.O.: Fixed point iterations for a certain class of nonlinear mappings. Soochow J. Math., **21**, No. 4, 441–449 (1995)

[Os95b]   Osilike, M.O.: Stability results for the Ishikawa fixed point iteration procedure. Indian J. Pure Appl. Math., **26**, No. 10, 937–945 (1995)

[Os95c]   Osilike, M.O.: Stability results for fixed point iteration procedures. J. Nigerian Math. Soc., **14/15**, 17–29 (1995/96)

[Os96a]   Osilike, M.O.: Iterative construction of fixed points of multi-valued operators of the accretive type. Soochow J. Math., **22**, No. 4, 485–494 (1996)

[Os96b]   Osilike, M.O.: Stable iteration procedures for strong pseudo-contractions and nonlinear operator equations of the accretive type. J. Math. Anal. Appl., **204**, No. 3, 677–692 (1996)

[Os96c]   Osilike, M.O.: Iterative solution of nonlinear equations of the ϕ-strongly accretive type. J. Math. Anal. Appl., **200**, No. 2, 259–271 (1996)

[Os96d]   Osilike, M.O.: A stable iteration procedure for quasi-contractive maps. Indian J. Pure Appl. Math., **27**, No. 1, 25–34 (1996)

[Os97a]   Osilike, M.O.: Stability of the Ishikawa iteration method for quasi-contractive maps. Indian J. Pure Appl. Math., **28**, No. 9, 1251–1265 (1997)

[Os97b]   Osilike, M.O.: Stable iteration procedures for nonlinear pseudocontractive and accretive operators in arbitrary Banach spaces. Indian J. Pure Appl. Math., **28**, No. 8, 1017–1029 (1997)

[Os97c]   Osilike, M.O.: Ishikawa and Mann iteration methods with errors for nonlinear equations of the accretive type. J. Math. Anal. Appl., **213**, No. 1, 91–105 (1997)

[Os97d]   Osilike, M.O.: Approximation methods for nonlinear m-accretive operator equations. J. Math. Anal. Appl., **209**, No. 1, 20–24 (1997)

[Os98a]   Osilike, M.O.: Iterative construction of fixed points of multi-valued operators of the accretive type. II. Soochow J. Math., **24**, No. 2, 141–146 (1998)

[Os98b]   Osilike, M.O.: Iterative approximation of fixed points of asymptotically demicontractive mappings. Indian J. Pure Appl. Math., **24**, No. 12, 1291–1300 (1998)

[Os98c]   Osilike, M.O.: Stability of the Mann and Ishikawa iteration procedures for ϕ-strong pseudocontractions and nonlinear equations of the ϕ-strongly accretive type. J. Math. Anal. Appl., **227**, No. 2, 319–334 (1998)

[Os99a]   Osilike, M.O.: Iterative solutions of nonlinear ϕ-strongly accretive operator equations in arbitrary Banach spaces. Nonlinear Anal., **36**, 1–9 (1999)

[Os99b]   Osilike, M.O.: Short proofs of stability results for fixed point iteration procedures for a class of contractive-type mappings. Indian J. Pure Appl. Math., **30**, No. 12, 1229–1234 (1999)

[Os00a]   Osilike, M.O.: Convergence of the Ishikawa-type iteration procedure for multi-valued operators of the accretive type. Indian J. Pure Appl. Math., **31**, No. 2, 117–127 (2000)

[Os00b]   Osilike, M.O.: Strong and weak convergence of the Ishikawa iteration method for a class of nonlinear equations. Bull. Korean Math. Soc., **37**, No. 1, 153–169 (2000)

[Os00c]   Osilike, M.O.: A note on the stability of iteration procedures for strong pseudocontractions and strongly accretive type equations. J. Math. Anal. Appl., **250**, No. 2, 726–730 (2000)

[Os00d]   Osilike, M.O.: Iterative solutions of nonlinear equations of the accretive type. Nonlinear Anal. TMA, **4**, No. 2, 291–300 (2000)

[Os00e]   Osilike, M.O.: Nonlinear equations of the ϕ-strongly pseudocontractive type in arbitrary Banach spaces. In: Fixed point theory and applications (Chinju, 1998), 227-236. Nova Sci. Publ., Huntington (2000)

[Os04a]   Osilike, M.O.: Implicit iteration process for common fixed points of a finite family of pseudocontractive maps. Panamer. Math. J., **14**, No. 3, 89–98 (2004)

[Os04b]   Osilike, M.O.: Implicit iteration process for common fixed points of a finite family of strictly pseudocontractive maps. J. Math. Anal. Appl., **294**, No. 1, 73–81 (2004)

[OsA04]   Osilike, M.O., Akuchu, B.G.: Common fixed points of a finite family of asymptotically pseudocontractive maps. Fixed Point Theory Appl., No. 2, 81–88 (2004)

[OsA00]   Osilike, M.O., Aniagbosor, S.C.: Weak and strong convergence theorems for fixed points of asymptotically nonexpansive mappings. Math. Comput. Modelling, **32**, No. 10, 1181–1191 (2000)

[OsA01]   Osilike, M.O., Aniagbosor, S.C.: Fixed points of asymptotically demicontractive mappings in certain Banach spaces. Indian J. Pure Appl. Math., **32**, No. 10, 1519–1537 (2001)

[OAA02]   Osilike, M.O., Aniagbosor, S.C., Akuchu, B.G.: Fixed points of asymptotically demicontractive mappings in arbitrary Banach spaces. Panamer. Math. J., **12**, No. 2, 77–88 (2002)

[OsI00]   Osilike, M.O., Igbokwe, D.I.: Weak and strong convergence theorems for fixed points of pseudocontractions and solutions of monotone type operator equations. Comput. Math. Appl., **40**, No. 4-5, 559–567 (2000)

[OsI01]   Osilike, M.O., Igbokwe, D.I.: Approximation of fixed points of asymptotically nonexpansive mappings in certain Banach spaces. In: Fixed Point Theory and Applications. Vol. 2 (Chinju/Masan, 2000), 27–42, Nova Sci. Publ., Huntington, NY (2001)

[OsU99]   Osilike, M.O., Udomene, A.: Short proofs of stability results for fixed point iteration procedures for a class of contractive-type mappings. Indian J. Pure Appl. Math., **30**, No. 12, 1229–1234 (1999)

[OU01a]   Osilike, M.O., Udomene, A.: Demiclosedness principle and convergence theorems for strictly pseudocontractive mappings of Browder-Petryshyn type. J. Math. Anal. Appl., **256**, No. 2, 431–445 (2001)

[OU01b]   Osilike, M.O., Udomene, A.: A note on approximation of solutions of a K-positive definite operator equations. Bull. Korean Math. Soc., **38**, No. 2, 231–236 (2001)

[OUI07]   Osilike, M.O., Udomene, A., Igbokwe, D.I., Akuchu, B.G.: Demiclosedness principle and convergence theorems for k-strictly asymptotically pseudocontractive maps. J. Math. Anal. Appl., **326**, No. 2, 1334–1345 (2007)

[Ost66]   Ostrowski, A.M.: Solution of Equations and Systems of Equations. Academic Press, New York (1966)

[Ost75]   Ostrowski, A.M.: An estimate of the approximation to a fixed point of an operator. Comput. Math. Appl., **1**, 427 (1975)

[Out69]   Outlaw, C.L.: Mean value iteration of nonexpansive mappings in a Banach space. Pacific J. Math., **30**, 747–750 (1969)

[OGr69]   Outlaw, C.L., Groetsch, C.W.: Averaging iteration in a Banach space. Bull. Amer. Math. Soc., **75**, 430–432 (1969)

[OwI01]   Owojori, O.O., Imoru, C.: On a general Ishikawa fixed point iteration process for continuous hemicontractive maps in Hilbert spaces. Adv. Stud. Contemp. Math. (Kyungshang), **4**, No. 1, 1–15 (2001)

[OI02a]   Owojori, O.O., Imoru, C.: A general Ishikawa iteration sequence for nonlinear uniformly continuous pseudocontractive operators in arbitrary Banach spaces. In: Proceed. Jangjeon Mathematical Society, 91–100, 4. Jangjeon Math. Soc., Hapcheon (2002)

[OI02b]   Owojori, O.O., Imoru, C.: Convergence of an Ishikawa type iteration process for quasi-contractive maps in arbitrary Banach spaces. Proc. Jangjeon Math. Soc., **5**, No. 2, 105–113 (2002)

[OI03a]   Owojori, O.O., Imoru, C.: On generalized fixed point iterations for asymptotically nonexpansive operators in Banach spaces. Proc. Jangjeon Math. Soc., **6**, No. 1, 49–58 (2003)

[OI03b]   Owojori, O.O., Imoru, C.: New iteration methods for pseudocontractive and accretive operators in arbitrary Banach spaces. Kragujevac J. Math., **25**, 97–110 (2003)

[Pac89]   Pachpatte, B.G.: Existence and uniqueness of solutions of higher order hyperbolic partial differential equations. Chin. J. Math., **17**, No. 3, 181–189 (1989)

[Pan96]   Pan, Y.H.: Generalized Ishikawa-type iteration for generalized quasicontractions. (Chinese) Sichuan Shifan Daxue Xuebao Ziran Kexue Ban, **19**, No. 3, 38–41 (1996)

[PaW99] Pandhare, D.M., Waghmode, B.B. Common fixed point theorem in Hilbert space with Mann iteration scheme. J. Bihar Math. Soc., **19**, 1–6 (1999)

[PaS80] Panja, C., Samanta, S.K.: On determination of a common fixed point. Indian J. Pure Appl. Math., **11**, 120–127 (1980)

[Pap72] Papp, F.J.: Fixed points and iteration homotopies. J. Austral. Math. Soc., **13**, 17–20 (1972)

[PkJ94] Park, J.A.: Mann iteration process for the fixed of strictly pseudocontractive mappings in some Banach spaces. J. Korean Math. Soc., **31**, 333–337 (1994)

[PkJ95] Park, J.Y.: An iterative process for nonexpansive mappings in Banach spaces. Nonlinear Anal. Forum, **1**, 13–20 (1995)

[PJe94] Park, J.Y., Jeong, J.U.: Weak convergence to a fixed point of the sequence of Mann type iterates. J. Math. Anal. Appl., **184**, No. 1, 75–81 (1994)

[PJe96] Park, J.Y., Jeong, J.U.: Ishikawa and Mann iteration methods for strongly accretive operators. Nonlinear Anal. Forum, **2**, 39–48 (1996)

[PJe98] Park, J.Y., Jeong, J.U.: Ishikawa and Mann iteration methods for strongly accretive operators. Commun. Korean Math. Soc., **13**, No. 4, 765–773 (1998)

[PJe00] Park, J.Y., Jeong, J.U.: Ishikawa-type and Mann-type iterative processes with errors for m-accretive operators. Commun. Korean Math. Soc., **15**, No. 2, 309–323 (2000)

[PJe01] Park, J.Y., Jeong, J.U.: Iteration processes of asymptotically pseudo-contractive mappings in Banach spaces. Bull. Korean Math. Soc., **38**, No. 3, 611–622 (2001)

[PJJ96] Park, J.Y., Jung, J.S., Jeong, J.U.: Strong convergence theorems for nonexpansive mappings in Banach spaces. Commun. Korean Math. Soc., **11**, No. 1, 71–79 (1996)

[PCC00] Park, K.S., Chang, S.S., Cho, Y.J.: Iterative approximations of fixed points for Φ-pseudo-contractive type mappings. In: Fixed Point Theory and Applications (Chinju, 1998), 237–247. Nova Sci. Publ., Huntington, NY (2000)

[PkS79] Park, S.: On f-nonexpansive maps. J. Korean Math. Soc., **16**, No. 1, 29–38 (1979/80)

[PkS80] Park, S.: A general principle of fixed point iterations on compact intervals. J. Korean Math. Soc., **17**, No. 2, 229–234 (1980/81)

[PkS81] Park, S.: On the asymptotic behavior of nonexpansive maps in Banach spaces. Bull. Korean Math. Soc., **18**, No. 1, 1–2 (1981/82)

[PkS82] Park, S.: Remarks on subsequential limit points of a sequence of iterates. J. Korean Math. Soc., **19**, 19–22 (1982)

[PkS83] Park, S.: A note on: "Generalized iteration process". [Tamkang J. Math. 11 (1980), No. 1, 135-140; MR 83m:26006] by T. Hu and K. S. Yang. Bull. Korean Math. Soc., **20**, No. 2, 69–70 (1983)

[PkS91] Park, S.: Best approximations, inward sets, and fixed points. In: Progress in approximation theory. Academic Press, Boston (1991)

[PkS96] Park, S.: Fixed points of approximable maps. Proc. Amer. Math. Soc., **124**, No. 10, 3109–3114 (1996)

[PkS97] Park, S.: Best approximations and fixed points of nonexpansive maps in Hilbert spaces. Numer. Funct. Anal. Optim., **18**, No. 5-6, 649–657 (1997)

[PSW94] Park, S., Singh, S.P., Watson, B.: Remarks on best approximations and fixed points. Indian J. Pure Appl. Math., **25**, No. 5, 459–462 (1994)

[PSb78] Pascali, D., Sburlan, S.: Nonlinear Mappings of Monotone Type. Editura Academiei, Sijhoff & Noordhoff (1978)

[Pas82] Passty, G.B.: Construction of fixed points for asymptotically nonexpansive mappings. Proc. Amer. Math. Soc., **84**, 212–216 (1982)

[PCK98] Pathak, H.K., Cho, Y.J., Kang, S.M.: Common fixed points of biased maps of type (*A*) and applications. Int. J. Math. Math. Sci., **21**, No. 4, 681–693 (1998)

[PKh01] Pathak, H.K., Khan, M.S.: Approximating fixed points of nonexpansive type mappings. Int. J. Math. Math. Sci., **26**, No. 3, 183–188 (2001)

[PKJ01] Pathak, H.K., Khan, M.S., Jung, J.S.: Approximation of fixed points via biased-nonexpansive mappings. Math. Sci. Res. Hot-Line, **5**, No. 8, 39–47 (2001)

[PAc91] Patil, S.T., Achari, J.: Convergence of sequence of iterates for a pair of contractive type mappings. Math. Educ., **25**, No. 2, 123–125 (1991)

[Pat74] Patterson, W.M.: Iterative Methods for the Solution of a Linear Operator Equation in Hilbert Space - A Survey. Lectures Notes in Mathematics, Springer, Berlin, Heidelberg, New York (1974)

[Pav76] Pavaloiu, I.: Introduction to the Theory of Approximating Solutions of Operator Equations (Romanian). Editura Dacia, Cluj-Napoca (1976)

[Pav89] Pavaloiu, I.: Sur l'approximation des racines des equations dans un espace metrique. Seminar on Functional Analysis and Numerical Methods, **1**, 95–104 (1989)

[PSe83] Pavaloiu, I., Serb, I.: Sur des methodes iteratives optimales. Seminar on Functional Analysis and Numerical Methods, **1**, 175–182 (1983)

[PWa95] Pawar, S.K., Waghmode, B.B.: Construction of fixed points of generalized nonexpansive mappings in generalized Hilbert space. Math. Ed. (Siwan), **29**, No. 2, 102–105 (1995)

[Paz77] Pazy, A.: On the asymptotic behavior of iterates of nonexpansive mappings in Hilbert space. Israel J. Math., **26**, 197–204 (1977)

[PeW79] Peitgen, H.-O. (ed.), Walther, H.-O. (ed.): Functional Differential Equations and Approximation of Fixed Points. Proceedings, Bonn, July 1978 Lecture Notes in Mathematics,730. Springer-Verlag, Berlin Heidelberg New York (1979)

[Pl66a] Pelczar, A.: On the method of successive approximations for some operator equations with applications to partial differential hyperbolic equations. Zeszyty nauk. Uniw. Jagiellonski. Prace Mat., **11**, 59–68 (1966)

[Pl66a] Pelczar, A.: On the method of successive approximations for some operator equations with applications to partial differential hyperbolic equations. Zeszyty Nauk. Univ. Jagiello. Prace Mat. Zeszyt, **11**, 59–68 (1966)

[Pel69] Pelczar, A.: On the convergence of successive approximations in some abstract spaces. Bull. Acad. Pol. Sc., **17**, 727–731 (1969)

[Pel76] Pelczar, A.: The method of successive approximations. (Polish). Wiadom. Mat. (2), **20**, No. 1, 80–84 (1976)

[Pet96] Petrusel, A.: *A*-fixed point theorems for locally contractive multivalued operators and applications to fixed point stability. Studia Univ. Babes-Bolyai Math., **41**, No. 1, 79–92 (1996)

[Pe04a] Petrusel, A.: Multivalued weakly Picard operators and applications. Sci. Math. Jpn., **59**, No. 1, 169–202 (2004)

[Pe04b]   Petrusel, A.: Fixed point theory: the Picard operators technique. In: Girela lvarez, Daniel (ed.) et al. Seminar of Mathematical Analysis. Proceedings of the lecture notes of the seminar, Universities of Malaga and Seville, Spain, September 2003-June 2004. Sevilla: Univ. de Sevilla, Secretariado de Publicaciones. Coleccin Abierta 71, 175–193 (2004)

[PeR01]   Petrusel, A., Rus, I.A.: Dynamics on $(P_{cp}(X), H_d)$ generated by a finite family of multi-valued operators on (X, d). Math. Moravica, **5**, 103–110 (2001)

[Pt66a]   Petryshyn, W.V.: Construction of fixed points of demicompact mappings in Hilbert space. J. Math. Anal. Appl., **14**, 276–284 (1966)

[Pt66b]   Petryshyn, W.V.: On nonlinear P-compact operators in Banach space with applications to constructive fixed-point theorems. J. Math. Anal. Appl., **15**, 228–242 (1966)

[Pet67]   Petryshyn, W.V.: Iterative construction of fixed points of contractive type mappings in Banach spaces. Centro Internationale Matematico Estivo, Ispra, Italy, July, 309–340 (1967)

[Pet68]   Petryshyn, W.V.: Fixed-point theorems involving P-compact, semicontractive, and accretive operators not defined on all of a Banach space. J. Math. Anal. Appl., **23**, 336–354 (1968)

[Pet70]   Petryshyn, W.V.: A characterization of strictly convexity of Banach spaces and other uses of duality mappings. J. Funct. Anal., **6**, 282–291 (1970)

[Pet71]   Petryshyn, W.V.: Structure of fixed points sets of k-set-contractions. Archive Rat. Mech. and Anal., **40**, 312–328 (1971)

[PTu69]   Petryshyn, W.V., Tucker, T.S.: On the functional equations involving nonlinear generalized P-compact operators. Trans. Amer. Math. Soc., **135**, 343–373 (1969)

[PWi73]   Petryshyn, W.V., Williamson, T.E.: Strong and weak convergence of the sequence of successive approximations for quasi-nonexpansive mappings. J. Math. Anal. Appl., **43**, 459–497 (1973)

[Pi890]   Picard, E.: Memoire sur la theorie des equations aux derivees partielles et la methode des approximations successives. J. Math. Pures et Appl., **6**, 145–210 (1890)

[PlW05]   Plubtieng, S., Wangkeeree, R.: Fixed point iteration for asymptotically quasi-nonexpansive mappings in Banach spaces. Int. J. Math. Math. Sci. **2005**, No. 11, 1685–1692 (2005)

[Pot85]   Potra, F.: On superadditive rates of convergence. Math. Modelling Num. Anal., **19**, 671–685 (1985)

[Pot89]   Potra, F.: On Q-order and R-order of convergence. J. Optim. Theory Appl., **63**, 415–431 (1989)

[PoP84]   Potra, F.A., Ptak, V.: Nondiscrete Induction and Iterative Processes. Pitman, London (1984)

[Pre97]   Precup, R.: Existence and approximation of positive fixed points of nonexpansive maps. Rev. Anal. Numer. Theor. Approx., **26**, No. 1-2, 203–208 (1997)

[Rad01]   Radovanovic, R.: Approximation of a fixed point of some nonself mappings. Math. Balkanica, New Series, **15**, 213–218 (2001)

[Rad02]   Radu, V.: Teoremele Barone-Hillam si recurente de tip Mann-Krasnoselski. Revista de Matematica din Timisoara, Seria a IV-a, **7**, nr. 1, 5–8 (2002)

[RaG99] Rajput, A., Gupta, A.: Fixed point iteration of generalized nonexpansive mappings. J. Maulana Azad College Tech., **32**, 93–112 (1999)

[RG00a] Rajput, A., Gupta, A.: Fixed point approximation for λ-firmly nonexpansive mappings. Acta Cienc. Indica Math., **26**, No. 3, 195–198 (2000)

[RG00b] Rajput, A., Gupta, A.: Approximating common fixed points of two nonexpansive mappings in Banach spaces. Jnanabha, **30**, 81–90 (2000)

[RaG02] Rajput, A., Gupta, A.: Weak and strong convergence to fixed points of asymptotically nonexpansive mappings. Ganita, **53**, No. 1, 89–95 (2002)

[RaM04] Rajput, A., Malhotra, S.K.: A general principle for Mann iteration. Ultra Sci. Phys. Sci., **16**, No. 2M, 151–158 (2004)

[Rak62] Rakotch, E.: On ε-contracting mappings. Bull. Res. Counc. Israel, **10F**, 53–58 (1962)

[RaC94] Rani, D., Chugh, R.: Some fixed point theorems on contractive type mappings. Pure Appl. Math. Sci., **39**, No. 1-2, 153–158 (1994)

[Ra90a] Rashwan, R.A.: On the convergence of Mann iterates to a common fixed point for a pair of mappings. Demonstratio Math., **23**, No. 3, 709–712 (1990)

[Ra90b] Rashwan, R.A.: On the convergence of sequence of iterates to a common fixed point for two mappings in Banach spaces. J. Inst. Math. Comput. Sci., Math. Ser., **3**, No. 1, 67–73 (1990)

[Ra90c] Rashwan, R.A.: Some fixed point theorems for iterates of quasi-nonexpansive mappings in locally convex spaces. Punjab Univ. J. Math., **23**, 99–107 (1990)

[Ras91] Rashwan, R.A.: On the Ishikawa fixed point iterations for some contractive mappings. Punjab Univ. J. Math., **24**, 85–93 (1991)

[Ras94] Rashwan, R.A.: On the convergence of the Ishikawa iterates to a common fixed point for a pair of mappings. Ganita, **45**, No. 1-2, 121–123 (1994)

[Ras95] Rashwan, R.A.: On the convergence of the Ishikawa iterates to a common fixed point for a pair of mappings. Demonstratio Math., **28**, No. 2, 271–274 (1995)

[Ra96a] Rashwan, R.A.: Common fixed points by Ishikawa iterates in metric linear spaces. Southwest J. Pure Appl. Math., **2**, 29–33 (1996)

[Ra96b] Rashwan, R.A.: Some common fixed point theorems in paranormed spaces. Demonstratio Math., **29**, No. 1, 143–148 (1996)

[Ra98a] Rashwan, R.A.: Common fixed points by Ishikawa iterates in metric linear spaces. Demonstratio Math., **31**, No. 1, 19–23 (1998)

[Ra98b] Rashwan, R.A.: Common fixed points by Ishikawa iterates in metric linear spaces. Math. Balkanica, New Ser., **12**, No. 1-2, 237–242 (1998)

[Ra98c] Rashwan, R.A.: A common fixed point theorem in uniformly convex Banach spaces. Ital. J. Pure Appl. Math., **3**, 117–126 (1998)

[RS94a] Rashwan, R.A., Saddek, A.M.: On the Ishikawa iteration process in Hilbert spaces. Collect. Math., **45**, No. 1, 45–52 (1994)

[RS94b] Rashwan, R.A., Saddek, A.M.: Some fixed point theorems without continuity. Punjab Univ. J. Math., **27**, 85–95 (1994)

[RSa99] Rashwan, R.A., Saddek, A.M.: Approximating common fixed points by iteration processes. J. Qufu Norm. Univ., Nat. Sci., **25**, No. 3, 12–16 (1999)

[Ray79] Ray, B.K.: On an extension of a theorem of Ljubomir Ciric. Indian J. Pure Appl. Math., **10**, 145–146 (1979)

[Re73a]   Reich, S.: Asymptotic behavior of contractions in Banach spaces. J. Math. Anal. Appl., **44**, 57–70 (1973)

[Re73b]   Reich, S.: Fixed points of condensing functions. J. Math. Anal. Appl., **41**, 460–467 (1973)

[Re73c]   Reich, S.: Fixed points via Toeplitz iteration. Bull. Calcutta Math. Soc., **65**, No. 4, 203–207 (1973)

[Re73d]   Reich, S.: Iterative solution of linear operator equations in Banach spaces. Atti Accad. Naz. Lincei Rend. Cl. Sci. Fis. Mat. Natur., **(8)54**, 551–554 (1973)

[Re75a]   Reich, S.: Fixed point iterations of nonexpansive mappings. Pacific J. Math., **60**, No. 2, 195–198 (1975)

[Re75b]   Reich, S.: Approximating zeros of accretive operators. Proc. Amer. Math. Soc., **51**, No. 2, 381–384 (1975)

[Rei77]   Reich, S.: Nonlinear evolution equations and nonlinear ergodic theorems. Nonlinear Anal., **1**, 319–330 (1977)

[Re78a]   Reich, S.: An iterative procedure for constructing zeros of accretive sets in Banach spaces. Nonlinear Anal., **2**, 85–92 (1978)

[Re78b]   Reich, S.: Almost convergence and nonlinear ergodic theorems. J. Approx. Theory, **24**, 269–272 (1978)

[Re78c]   Reich, S.: Iterative methods for accretive sets in linear spaces and approximation. In: Proc. Conf., Math. Res. Inst., Oberwolfach, 1977, pp. 317–326, Internat. Schriftenreihe Numer. Math., 40. Birkhauser, Basel-Boston, Mass. (1978)

[Re78d]   Reich, S.: Constructing zeros of accretive operators. Appl. Anal., **8**, No. 4, 349–352 (1978/79)

[Re78e]   Reich, S.: Iterative methods for accretive sets, In: Nonlinear Equations in Abstract Spaces, pp. 317-326. Academic Press, New York (1978)

[Re79a]   Reich, S.: Weak convergence theorems for nonexpansive mappings in Banach spaces. J. Math. Anal. Appl., **67**, 274–276 (1979)

[Re79b]   Reich, S.: Constructing zeros of accretive operators II. Appl. Anal., **9**, 159–163 (1979)

[Re79c]   Reich, S.: Constructive techniques for accretive and monotone operators. In: V. Lakshmikantham, (Ed.) Applied Nonlinear Analysis. Academic Press, New York (1979)

[Rei80]   Reich, S.: Strong convergence theorems for resolvents of accretive operators in Banach spaces. J. Math. Anal. Appl., **75**, No. 1, 287–292 (1980)

[Rei81]   Reich, S.: On the asymptotic behavior of nonlinear semigroups and the range of accretive operators. J. Math. Anal. Appl., **79**, 113–126 (1981)

[Re83a]   Reich, S.: Some problems and results in fixed point theory. Contemp. Math., **21**, 179–187 (1983)

[Re83b]   Reich, S.: Convergence, resolvent consistency, and the fixed point property for nonexpansive mappings. Contemp. Math., **18**, 167–174 (1983)

[Rei85]   Reich, S.: Averaged mappings in the Hilbert ball. J. Math. Anal. Appl., **109**, 199–206 (1985)

[Rei86]   Reich, S.: Nonlinear semigroups, holomorphic mappings and integral equations. Proc. Sympos. Pure Math., **45**, No. 2, 307–324 (1986)

[Rei92]   Reich, S.: Approximating fixed points of holomorphic mappings. Math. Japon., **37**, No. 3, 457–459 (1992)

[Rei94]   Reich, S.: Approximating fixed points of nonexpansive maps. Panamer. Math. J., **4**, 23–28 (1994)

[RS87a]   Reich, S., Shafrir, I.: The asymptotic behavior of firmly nonexpansive mappings. Proc. Amer. Math. Soc., **101**, No. 2, 246–250 (1987)

[RS87b]   Reich, S., Shafrir, I.: On the method of successive approximations for nonexpansive mappings. In: Nonlinear and Convex Analysis, Proc. Conf. Hon. Ky Fan, Santa Barbara/Calif. 1985, Lect. Notes Pure Appl. Math. 107 (1987)

[RZ00a]   Reich, S., Zaslavski, A.J.: Almost all nonexpansive mappings are contractive. C. R. Math. Rep. Acad. Sci. Canada, **22**, 118–124 (2000)

[RZ00b]   Reich, S., Zaslavski, A.J.: Convergence of Krasnoselskii-Mann iterations of nonexpansive operators. Nonlinear operator theory. Math. Comput. Modelling, **32** (2000), No. 11-13, 1423–1431 (2000)

[RZa01]   Reich, S., Zaslavski, A.J.: The set of noncontractive mappings is σ-porous in the space of all nonexpansive mappings. C. R. Acad. Sci. Paris, Ser. I, **333**, 539–544 (2001)

[RZa04]   Reich, S., Zaslavski, A.J.: Generic convergence of iterates for a class of nonlinear mappings. Fixed Point Theory Appl. 2004, **3**, 211–220 (2004)

[Rr69a]   Reinermann, J.: Uber Toeplitzsche Iterationsverfahren und einige ihrer Anwendung in der konstruktiven Fixpunkttheorie. Studia Math., **32**, 209–227 (1969)

[Rr69b]   Reinermann, J.: Uber Fixpunkte kontrahierender Abbildungen und schwach konvergente Toeplitz-Verfahren. Arch. Math. (Basel), **20**, 59–64 (1969)

[RrS76]   Reinermann, J., Schoneberg, R.: Some results and problems in the fixed point theory for nonexpansive and pseudocontractive mappings in Hilbert-space. In: Fixed Point Theory and its Applications (Proc. Sem., Dalhousie Univ., Halifax, N.S., 1975). Academic Press, New York (1976)

[Ren98]   Ren, X.: On Chidume's open problems and fixed point theorems. Xichuan Daxue Xuebao, **35**, No. 4, 505–508 (1998)

[RHe03]   Ren, W.Y., He, Z.: Convergence of Ishikawa sequences with errors for k-subaccretive operators (Chinese). Acta Anal. Funct. Appl., **5**, No. 4, 343–350 (2003)

[Rih01]   Rihm, R.: Acceleration of iteration methods for interval fixed point problems. Linear Algebra Appl., **324**, No. 1-3, 189–207 (2001)

[Rh74a]   Rhoades, B.E.: Fixed point iterations using infinite matrices. Trans. Amer. Math. Soc., **196**, 161–176 (1974)

[Rh74b]   Rhoades, B.E.: Fixed point iterations using infinite matrices. II. In: Constructive and Computational Methods for Differential and Integral Equations, Vol. 430, pp. 390-395. Springer Verlag, New York Berlin (1974)

[Rho76]   Rhoades, B.E.: Comments on two fixed point iteration methods. J. Math. Anal. Appl., **56**, No. 2, 741–750 (1976)

[Rh77a]   Rhoades, B.E.: Some fixed point theorems in Banach spaces. Math. Semin. Notes, **5**, 69–74 (1977)

[Rh77b]   Rhoades, B.E.: A comparison of various definitions of contractive mappings. Trans. Amer. Math. Soc., **226**, 257–290 (1977)

[Rh77c]   Rhoades, B.E.: Fixed point iterations using infinite matrices. III. In: Fixed points. Algorithms and applications. Academic Press, New York (1977)

[Rho81]   Rhoades, B.E.: A fixed point theorem for asymptotically nonexpansive mappings. Kodai Math. J., **4**, 293–297 (1981)

[Rho83]   Rhoades, B.E.: Contractive definitions revisited. Contemp. Math., **21**, 189–205 (1983)

[Rh88a]   Rhoades, B.E.: Contractive definitions and continuity. Contemp. Mathem., **72**, 233–245 (1988)

[Rh88b]   Rhoades, B.E.: Fixed point iterations of generalized nonexpansive mappings. J. Math. Anal. Appl., **130**, No. 2, 564–576 (1988)

[Rho90]   Rhoades, B.E.: Fixed point theorems and stability results for fixed point iteration procedures. Indian J. Pure Appl. Math., **21**, No. 1, 1–9 (1990)

[Rho91]   Rhoades, B.E.: Some fixed point iteration procedures. Int. J. Math. Math. Sci., **14**, No. 1, 1–16 (1991)

[Rh93a]   Rhoades, B.E.: Fixed point theorems and stability results for fixed point iteration procedures. II. Indian J. Pure Appl. Math., **24**, No. 11, 691–703 (1993)

[Rh93b]   Rhoades, B.E.: Some fixed point iterations. Soochow J. Math., **19**, No. 4, 377–380 (1993)

[Rh94a]   Rhoades, B.E.: Convergence of an Ishikawa-type iteration scheme for a generalized contraction. J. Math. Anal. Appl., **185**, No. 2, 350–355 (1994)

[Rh94b]   Rhoades, B.E.: Fixed point iterations for certain nonlinear mappings., J. Math. Anal. Appl., **183**, 118–120 (1994)

[Rh95a]   Rhoades, B.E.: Some properties of Ishikawa iterates of nonexpansive mappings. Indian J. Pure Appl. Math., **26**, No. 10, 953–957 (1995)

[Rh95b]   Rhoades, B.E.: A general principle for Mann iterates. Indian J. Pure Appl. Math., **26**, No. 8, 751–762 (1995)

[Rh96a]   Rhoades, B.E.: Corrigendum: "Convergence of an Ishikawa-type iteration scheme for a generalized contraction.". J. Math. Anal. Appl., **199**, No. 2, 636 (1996)

[Rh96b]   Rhoades, B.E.: A fixed point theorem in search of an example. Panamer. Math. J., **6**, No. 3, 35–39 (1996)

[Rh97a]   Rhoades, B.E.: A general principle for Ishikawa iterations. Proceed. Int. Workshop in Analysis and Its Applications, Math. Moravica, 21–26 (1997)

[Rh97b]   Rhoades, B.E.: A general principle for Ishikawa iterations for multi-valued mappings. Indian J. Pure Appl. Math., **28**, No. 8, 1091–1098 (1997)

[Rho00]   Rhoades, B.E.: Finding common fixed points of nonexpansive mappings by iteration. Bull. Austral. Math. Soc., **62**, No. 2, 307–310 (2000)

[Rh01a]   Rhoades, B.E.: Iteration to obtain random solutions and fixed points of operators in uniformly convex Banach spaces. Soochow J. Math., **27**, No. 4, 401–404 (2001)

[Rh01b]   Rhoades, B.E.: Corrigendum: "Finding common fixed points of nonexpansive mappings by iteration". Bull. Austral. Math. Soc., **63**, No. 2, 345–346 (2001)

[Rh01c]   Rhoades, B.E.: Some theorems on weakly contractive maps. Nonlinear Anal., **47**, 2683–2693 (2001)

[Rho04]   Rhoades, B.E.: Comments on some iteration processes with errors. Fixed Point Theory, **5**, 121-124 (2004)

[Rho07]   Rhoades, B.E.: A biased discussion on fixed point theory. Carpathian J. Math., **23** (2007) (in press)

[Rh01d]   Rhoades, B.E., Saliga, L.: Some fixed point iteration procedures. II. Nonlinear Anal. Forum, **6**, No. 1, 193–217 (2001)

[RSK87]   Rhoades, B.E., Sessa, S., Khan, M.S., Swaleh, M.: On fixed points of asymptotically regular mappings. J. Austral. Math. Soc. Ser. A, **43**, No. 3, 328–346 (1987)

[RS03a]   Rhoades, B.E., Soltuz, S.: The equivalence of Mann iteration and Ishikawa iteration for non-Lipschitzian operators. Int. J. Math. Math. Sci., No. **42**, 2645–2651 (2003)

[RS03b]   Rhoades, B.E., Soltuz, S.: The equivalence between the convergences of Ishikawa and Mann iterations for an asymptotically pseudocontractive map. J. Math. Anal. Appl., **283**, No. 2, 681–688 (2003)

[RS03c]   Rhoades, B.E., Soltuz, S.: On the equivalence of Mann and Ishikawa iteration methods. Int. J. Math. Math. Sci., No. **7**, 451–459 (2003)

[RS04a]   Rhoades, B.E., Soltuz, S.: The equivalence of Mann iteration and Ishikawa iteration for a Lipschitzian ψ-uniformly pseudocontractive and ψ-uniformly accretive maps. Tamkang J. Math., **35**, No. 3, 235–245 (2004)

[RS04b]   Rhoades, B.E., Soltuz, S.: The equivalence between the convergences of Ishikawa and Mann iterations for an asymptotically nonexpansive in the intermediate sense and strongly successively pseudocontractive maps. J. Math. Anal. Appl., **289**, No. 1, 266–278 (2004)

[RS04c]   Rhoades, B.E., Soltuz, S.: The equivalence of Mann and Ishikawa iteration for Ψ-uniformly pseudocontractive or Ψ-uniformly accretive maps. Int. J. Math. Math. Sci., **46**, 2443–2452 (2004)

[RS04d]   Rhoades, B.E., Soltuz, S.: The equivalence between Mann-Ishikawa iterations and multistep iteration. Nonlinear Anal., **58**, No. 1-2, 219–228 (2004)

[RS04e]   Rhoades, B.E., Soltuz, S.: The equivalence of Mann and Ishikawa iteration dealing with strongly pseudocontractive or strongly accretive maps. Panamer. Math. J., **14**, No. 4, 51–59 (2004)

[Rob86]   Robert, F.: Discrete iterations. Springer, New York (1986)

[RCM75]   Robert, F., Charnay, M., Musy, F.: Iterations chaotiques serie-parallele pour des equations non-linaires de point fixe. Appl. Mat., **20**, 1–38 (1975)

[Rob80]   Robinson, S.M. (ed.): Analysis and Computation of Fixed Points. Proc. Symp. Mathematics Research Center, Univ. of Wisconsin-Madison, May 7-8, 1979; Publication No. 43, M. R. C., Univ. of Wisconsin - Madison. New York, Academic Press (1980)

[Rod82]   Rode, G.: An ergodic theorem for semigroups of nonexpansive mappings in a Hilbert space. J. Math. Anal. Appl., **85**, 172–178 (1982)

[Rou02]   Rouhani, B.D.: Remarks on asymptotically non-expansive mappings in Hilbert space. Nonlinear Anal., **49**, 1099–1104 (2002)

[Rou77]   Roux, D.: Applicazioni quasi non-espansive: approssimazione dei punti fissi. Rend. Mat., **10**, No. 6, 597–605 (1977)

[RZa77]   Roux, D., Zanco, C.: Quasi-nonexpansive mappings: Strong and weak convergence to a fixed point of the sequence of iterates. Sem. Mat. Univ. Catania, **32**, 307–315 (1977)

[Rus72]   Rus, I.A.: On the method of successive approximations. Rev. Roum. Math. Pures Appl., **17**, 1433–1437 (1972)

[Rus75]   Rus, I.A.: Approximation of fixed points of generalized contractions mappings. In: Topics in Numerical Analysis. Academic Press, New York (1975)

[Ru79a]   Rus, I.A.: Approximation of common fixed point in a generalized metric space. Rev. Anal. Numer. Theor. Approx., **8**, 83–87 (1979)

[Ru79b]   Rus, I.A.: Metrical Fixed Point Theorems. Univ. of Cluj-Napoca (1979)

[Ru79c]   Rus, I.A.: Principles and Applications of the Fixed Point Theory (Romanian). Editura Dacia, Cluj-Napoca (1979)

[Rus81]   Rus, I.A.: An iterative method for the solution of the equation $x = f(x, x, ..., x)$. Rev. Anal. Numer. Theor. Approx., **10**, No. 1, 95–100 (1981)

[Rus82]   Rus, I.A.: Surjectivity and iterated mappings. Math. Sem. Notes, **10**, 179–181 (1982)

[Rus83]   Rus, I.A.: Generalized contractions. Seminar on Fixed Point Theory, **3**, 1–130 (1983)

[Rus87]   Rus, I.A.: Picard mappings. Results and problems. Itin. Sem. Funct. Eq. Approx. Conv., Cluj-Napoca, 55–64 (1987)

[Rus88]   Rus, I.A.: Picard mappings I. Studia Univ. Babes-Bolyai, **33**, 2, 70–73 (1988)

[Rus93]   Rus, I.A.: Weakly Picard mappings. Comment. Math. Univ. Carolin., **34**, No. 4, 769–773 (1993)

[Rus96]   Rus, I.A.: Picard operator and applications. Seminar on Fixed Point Theory, Babes-Bolyai Univ., **3**, 1–36 (1996)

[Rus98]   Rus, I.A.: Stability of attractor of a φ-contractions system. Seminar on Fixed Point Theory, **3**, 31–34 (1998)

[Rus01]   Rus, I.A.: Generalized Contractions and Applications. Cluj University Press, Cluj-Napoca (2001)

[Rus02]   Rus, I.A.: Iterates of Stancu operators, via contraction principle. Studia Univ. Babes-Bolyai Math., **47**, No. 4, 101–104 (2002)

[Ru03a]   Rus, I.A.: Picard operators and applications., Sci. Math. Japon., **58**, No. 1, 191–219 (2003)

[Ru03b]   Rus, I.A.: Some applications of weakly Picard operators. Studia Univ. Babes-Bolyai Math., **48**, No. 1, 101–107 (2003)

[Ru04a]   Rus, I.A.: Iterates of Bernstein operators, via contraction principle. J. Math. Anal. Appl., **292**, 259–261 (2004)

[Ru04b]   Rus, I.A.: Sequences of operators and fixed points. Fixed Point Theory, **5**, No. 2, 349–368 (2004)

[RMu98]   Rus, I.A., Muresan, S.: Data dependence of the fixed points set of weakly Picard operators. Studia Univ. Babeş-Bolyai Math., **43**, No. 1, 79–83 (1998)

[RPP02]   Rus, I.A., Petrusel, A., Petrusel, G.: Fixed point theory 1950-2000. Romanian contributions. House of the Book of Science, Cluj-Napoca (2002)

[RPS01]   Rus, I.A., Petrusel, A., Sintamarian, A.: Data dependence of the fixed point set of c-multivalued weakly Picard operators. Studia Univ. Babeş-Bolyai Math., **46**, No. 2, 111–121 (2001)

[RPS03]   Rus, I.A., Petrusel, A., Sintamarian, A.: Data dependence of the fixed point set of some multivalued weakly Picard operators. Nonlinear Analysis, **52**, 1947–1959 (2003)

[Sah99]   Sahu, D.R.: Strong convergence theorems for nonexpansive type and nonself multi-valued mappings. Nonlinear Anal., **37**, 401–407 (1999)

[Sah03]   Sahu, D.R.: On generalized Ishikawa iteration process and nonexpansive mappings in Banach spaces. Demonstratio Math., **36**, No. 3, 721–734 (2003)

[SaD03]   Sahu, D.R., Dashputre, S.: On Ishikawa iteration process with errors. Nanjing Daxue Xuebao Shuxue Bannian Kan, **20**, No. 2, 131–138 (2003)

[SaJ03]   Sahu, D.R., Jung, J.S.: Fixed-point iteration processes for non-Lipschitzian mappings of asymptotically quasi-nonexpansive type. Int. J. Math. Math. Sci., No. **33**, 2075–2081 (2003)

[SJC04]   Sahu, D.R., Jung, J.S., Cho, Y.J.: Strong convergence of approximants to fixed points of asymptotically nonexpansive mappings in Banach spaces without uniform convexity. Demonstratio Math., **37**, No. 2, 419–428 (2004)

[SJV04]   Sahu, D.R., Jung, J.S., Verma, R.K.: Strong convergence of weighted averaged approximants of asymptotically nonexpansive mappings in Banach spaces without uniform convexity. Bull. Malays. Math. Sci. Soc., **(2)27**, No. 2, 225–235 (2004)

[Sam81]   Samanta, S.K.: Fixed point theorems in a Banach space satisfying Opial's condition. J. Indian Math. Soc., **45**, 251–258 (1981)

[SaB99]   Sastry, K.P.R., Babu, G.V.R.: Convergence of Ishikawa and Mann iteration schemes for a sequence of selfmaps in a Hilbert space. Proc. Natl. Acad. Sci. India, Sect. A, **69**, No. 4, 447–458 (1999)

[SaB00]   Sastry, K.P.R., Babu, G.V.R.: Approximation of fixed points of strictly pseudo-contractive mappings on arbitrary closed convex sets in a Banach space. Proc. Amer. Math. Soc., **128**, No. 10, 2907–2909 (2000)

[SBS01]   Sastry, K.P.R., Babu, G.V.R., Srinivasa Rao, C.: Convergence of an Ishikawa iteration scheme for nonlinear quasi-contractive mappings in convex metric spaces. Tamkang J. Math., **32**, No. 2, 117–126 (2001)

[SBS02]   Sastry, K.P.R., Babu, G.V.R., Srinivasa Rao, C.: Convergence of an Ishikawa iteration scheme for a nonlinear quasi-contractive pair of selfmaps of convex metric spaces. Indian J. Pure Appl. Math., **33**, No. 2, 203–214 (2002)

[SBa01]   Sayyed, F., Badshah, V.H.: Generalized contraction and common fixed point theorem in Hilbert space. J. Indian Acad. Math., **23**, No. 2, 267–275 (2001)

[Sca67]   Scarf, H.: The approximation of fixed points of a continuous mapping. SIAM J. Appl. Math., **15**, 1328–1343 (1967)

[SHa73]   Scarf, H.E., Hansen, T.: The Computation of Economic Equilibria. Yale Univ. Press (1973)

[Sch57]   Schaefer, H.: Uber die Methode sukzessiver Approximationen. Jahresber. Deutsch. Math. Verein., **59**, 131–140 (1957)

[Sch86]   Schilling, K.: Simpliziale Algorithmen zur Brechnung von Fixpunkten mengen-wertiger Operatoren. WVT Wissenschaftlicher Verlag Trier, Trier (1986)

[Sch77]   Schoneberg, R.: On the structure of fixed point sets of pseudocontractive mappings. II. Comment. Math. Univ. Carolin., **18**, No. 2, 299–310 (1977)

[Sch79]   Schoneberg, R.: Matrix-Limitierungen von Picardfolgen nichtexpansiver Abblidungen im Hilbertraum. Math. Nachr., **91**, 263–267 (1979)

[Sch99]   Schroder, B.: Algorithms for the fixed point property. Theoret. Comput. Sci., **217**, 301–358 (1999)

[Sch89]   Schu, J.: Approximating fixed points of Lipschitz pseudocontractive mappings. Preprint No. 17, RWTH Aachen, Lehrstuhl C fur Mathematik (1989)

[Sc90a]   Schu, J.: Iterative approximation of fixed points of nonexpansive mappings with starshaped domain. Comment. Math. Univ. Carolin., **31**, No. 2, 277–282 (1990)

[Sc90b]   Schu, J.: Weak convergence to fixed point of asymptotically nonexpansive mappings in uniformly convex Banach spaces with a Frechet differentiable norm. Lehrenstuhl C für Mathematik, Preprint No. 21 (1990)

[Sc91a]   Schu, J.: Weak and strong convergence to fixed points of asymptotically nonexpansive mappings. Bull. Austral. Math. Soc., **43**, No. 1, 153–159 (1991)

[Sc91b]   Schu, J.: Iterative construction of fixed points of strictly pseudocontractive mappings. Appl. Anal., **40**, No. 2/3, 67–72 (1991)

[Sc91c]   Schu, J.: Approximation for fixed points of asymptotically nonexpansive mappings., Proc. Amer. Math. Soc., **112**, No. 1, 143–151 (1991)

[Sc91d]   Schu, J.: Iterative construction of fixed points of asymptotically nonexpansive mappings. J. Math. Anal. Appl., **158**, No. 2, 407–413 (1991)

[Sc91e]   Schu, J.: A fixed point theorem for non-expansive mappings on star-shaped domains. Z. Anal. Anwend., **10**, No. 4, 417–431 (1991)

[Sc91f]   Schu, J.: On a theorem of C. E. Chidume concerning the iterative approximation of fixed points. Math. Nachr., **153**, 313–319 (1991)

[Sch93]   Schu, J.: Approximating fixed points of Lipschitzian pseudocontractive mappings. Houston J. Math., **19**, No. 1, 107–115 (1993)

[SeD74]   Senter, H.F., Dotson, W.G.: Approximating fixed points of nonexpansive mappings. Proc. Amer. Math. Soc., **44**, 375–380 (1974)

[SSa00]   Sharma, B.K., Sahu, D.R.: Existence and approximation results for asymptotically pseudocontractive mappings. Indian J. Pure Appl. Math., **31**, No. 2, 185–196 (2000)

[SSB01]   Sharma, B.K., Sahu, D.R., Bounias, M.: Weak almost-convergence theorem without Opial's condition. J. Math. Anal. Appl., **254**, No. 2, 636–644 (2001)

[STh95]   Sharma, B.K., Thakur, B.S.: Iterative approximation of fixed points in Banach space of type $(U, \lambda, m + 1, M)$ for local strictly hemi-contractive maps. Bull. Calcutta Math. Soc., **87**, No. 6, 557–562 (1995)

[SD02a]   Sharma, S., Deshpande, B.: Approximation of fixed points and convergence of generalized Ishikawa iteration. Indian J. Pure Appl. Math., **33**, No. 2, 185–191 (2002)

[SD02b]   Sharma, S., Deshpande, B.: Iterative approximation of fixed points for strongly pseudo-contractive mappings. Bull. Korean Math. Soc., **39**, No. 1, 43–51 (2002)

[ShS96]   Sharma, S., Sahu, D.R.: Fixed point approximation for λ-firmly nonexpansive mappings. Bull. Calcutta Math. Soc., **88**, No. 4, 285–290 (1996)

[ShS00]   Sharma, S., Sahu, D.R.: Existence and approximation results for asymptotically pseudocontractive mappings. Indian J. Pure Appl. Math., **31**, No. 2, 185–196 (2000)

[ShS02]   Shellman, S., Sikorski, K.: A two-dimensional bisection envelope algorithm for fixed points. J. Complexity, **18**, 641–659 (2002)

[ShS03]   Shellman, S., Sikorski, K.: A recursive algorithm for the infinity-norm fixed point problem. J. Complexity, **19**, 799–834 (2003)

[SLi01]   Shen, P., Liu, Q.: Fixed point iterations for generalized asymptotically hemicontractive mapping (Chinese). J. Henan Norm. Univ., Nat. Sci., **29**, No. 4, 6–12 (2001)

[SXu01]   Sheng, S., Xu, H.F., Picard iteration for nonsmooth equations. J. Comput. Math., **19**, No. 6, 583–590 (2001)

[Shi78]   Shimi, T.N.: Approximation of fixed points of certain nonlinear mappings. J. Math. Anal. Appl., **65**, No. 3, 565–571 (1978)

[Shi97]   Shimizu, T.: A strong convergence theorem for an iteration of nonexpansive mappings. Nihonkai Math. J., **8**, No. 1, 85–89 (1997)

[ShT96]   Shimizu, T., Takahashi, W.: Strong convergence theorems for asymptoti-
          cally nonexpansive mappings in Banach spaces. Nonlinear Anal., **26**, 265–
          272 (1996)
[ShT97]   Shimizu, T., Takahashi, W.: Strong convergence to common fixed points
          of families of nonexpansive mappings. J. Math. Anal. Appl., **211**, No. 1,
          71–83 (1997)
[ST01a]   Shimoji, K., Takahashi, W.: Strong convergence to common fixed points
          of infinite nonexpansive mappings and applications. Taiwanese J. Math.,
          5, No. 2, 387–404 (2001)
[ST01b]   Shimoji, K., Takahashi, W.: Approximating fixed points of infinite non-
          expansive mappings. Commun. Appl. Nonlinear Anal., **8**, No. 4, 47–61
          (2001)
[Shj97]   Shioji, N.: Strong convergence theorems for nonexpansive mappings and
          nonexpansive semigroups. In: Proceedings of Workshop on Fixed Point
          Theory (Kazimierz Dolny, 1997). Ann. Univ. Mariae Curie-Skłodowska
          Sect. A, **51**, No. 2, 261–276 (1997)
[ST97a]   Shioji, N., Takahashi, W.: Strong convergence theorems of approximated
          sequences for nonexpansive mappings in Banach spaces. Proc. Amer.
          Math. Soc., **125**, No. 12, 3641–3645 (1997)
[ST97b]   Shioji, N., Takahashi, W.: Convergence of approximated sequences for
          nonexpansive mappings. Proceedings of the Second World Congress of
          Nonlinear Analysts, Part 7 (Athens, 1996). Nonlinear Anal., **30**, No. 7,
          4497–4507 (1997)
[ST97c]   Shioji, N., Takahashi, W.: Convergence of approximated sequences for
          nonexpansive mappings. Investigations on nonlinear analysis and convex
          analysis (Japanese) (Kyoto, 1996). Sūrika isekikenkyūsho Kōkyūroku No.
          985, (1997)
[SjT98]   Shioji, N., Takahashi, W.: Strong convergence theorems for asymptot-
          ically nonexpansive semigroups in Hilbert spaces. Nonlinear Anal., **34**,
          No. 1, 87–99 (1998)
[ST99a]   Shioji, N., Takahashi, W.: Strong convergence of averaged approximants
          for asymptotically nonexpansive mappings in Banach spaces. J. Approx.
          Theory, **97**, 53–64 (1999)
[ST99b]   Shioji, N., Takahashi, W.: Strong convergence theorems for continuous
          semigroups in Banach spaces. Math. Japon., **50**, No. 1, 57–66 (1999)
[ST99c]   Shioji, N., Takahashi, W.: A strong convergence theorem for asymptoti-
          cally nonexpansive mappings in Banach spaces. Arch. Math. (Basel), **72**,
          No. 5, 354–359 (1999)
[SjT00]   Shioji, N., Takahashi, W.: Strong convergence theorems for asymptoti-
          cally nonexpansive semigroups in Banach spaces. J. Nonlinear Convex
          Anal., **1**, No. 1, 73–87 (2000)
[SAg93]   Shridharan, R., Agarwal, R.P.: Stationary and nonstationary iterative
          methods for nonlinear boundary value problems. Math. Comput. Mod-
          elling, **18**, No. 2, 43–62 (1993)
[SLi03]   Shu, X.B., Li, Y.J.: Ishikawa iterative process for constructing solutions
          of k-subaccretive operator equations. Far East J. Math. Sci., **11**, No. 2,
          215–228 (2003)
[Sik89]   Sikorski, K.: Fast algorithm for the computation of fixed points. In: Mi-
          lanese, M. et al. (eds) Robustness in Identification and control. Plenum,
          New York (1989)

[STW93] Sikorski, K., Tsay, C.V., Wozniakowski, H.: An ellipsoid algorithm for the computation of fixed points. J. Complexity, **9**, 181–200 (1993)

[SWo87] Sikorski, K., Wozniakowski, H.: Complexity of fixed points. J. Complexity, **3**, 388–405 (1987)

[SgS90] Singh, A.K., Singh, S.B.: Sequence of iterates for nonexpansive type mapping in Banach space. Proc. Math. Soc., **6** (1990), 105–106 (1991)

[SgK78] Singh, K.L.: Generalized contractions and the sequence of iterates. In: Lakshmikantham, V. (ed.) Nonlinear Equations in Abstract Spaces. Academic Press, New York (1978)

[SgK79] Singh, K.L.: Fixed point iterations using infinite matrices. In: Lakshmikantham, V. (ed.) Applied Nonlinear Analysis. Academic Press, New York (1979)

[SgS76] Singh, K.L., Srivastava, S.: Construction of fixed points for quasi-nonexpansive mappings. II. Bull. Math. Soc. Sci. Math. Math. R. S. Roumanie (N.S.), **18(66)** (1974), No. 3–4, 367–378 (1976)

[SSL74] Singh, S.L.: On the convergence of sequence of iterates. Atti Accad. Naz. Lincei, VIII. Ser., Rend., Cl. Sci. Fis. Mat. Natur., **57**, 502–505 (1974)

[SL77a] Singh, S.L.: A note on the convergence of sequence of iterates. II. J. Nat. Sci. Math., **17**, No. 2, 15–17 (1977)

[SL77b] Singh, S.L.: A note on the convergence of sequence of iterates. J. Nat. Sci. Math., **17**, 67–71 (1977)

[SSL82] Singh, S.L.: A note on the convergence of sequence of iterates. III. Punjab Univ. J. Math., **14**, 123–128 (1982)

[SSL88] Singh, S.L.: Approximating fixed points of multivalued maps. J. Nat. Phys. Sci., **2**, No. 1/2, 51–61 (1988)

[SGM90] Singh, S.L., Gairola, U.C., Mishra, S.N.: Convergence of sequences of multivalued operators (Hindi). J. Nat. Phys. Sci., **4**, No. 1-2, 187–198 (1990)

[SGM96] Singh, S.L., Gairola, U.C., Mishra, S.N.: Convergence of sequences of iterates of multivalued operators. J. Nat. Phys. Sci., **9-10**, 13–24 (1996)

[SMa85] Singh, S.L., Mall, R.: Ishikawa iteration process for a pair of nonlinear maps. J. Uttar Pradesh Gov. Colleges Acad. Soc., **2**, No. 2, 136–138 (1985)

[SMi81] Singh, S.L., Mishra, S.N.: Common fixed points and convergence theorems in uniform spaces. Mat. Vesnik **5(18)(33)**, No. 4, 403–410 (1981)

[SMi02] Singh, S.L., Mishra, S.N.: On a Ljubomir Ciric fixed point theorem for nonexpansive type maps with applications. Indian J. Pure Appl. Math., **33**, No. 4, 531–542 (2002)

[SMR83] Singh, S.P., Massa, S., Roux, D.: Approximation technique in fixed point theory. Rend. Semin. Mat. Fis. Milano, **53**, 165–172 (1983)

[SWa88] Singh, S.P., Watson, B.: On approximating fixed points. In: Nonlinear Functional Analysis and its Applications, Proc. Sympos. Pure Math. 45, part 2, Amer. Math. Soc., Providence (1988)

[SWa93] Singh, S.P., Watson, B.: On convergence results in fixed point theory. Rend. Semin. Mat., Torino, **51**, No. 2, 73–91 (1993)

[SYa73] Singh, S.P., Yadav, R.K.: On the convergence of the sequence of iterates, Ann. Soc. Sci. Bruxelles, Ser. I, **87**, 279–284 (1973)

[Sint01] Sintamarian, A.: Data dependence of the fixed points of some Picard operators. Semin. Fixed Point Theory Cluj-Napoca, **2**, 81–86 (2001)

[Sint02] Sintamarian, A.: Picard pairs and weakly Picard pairs of operators. Stud. Univ. Babes-Bolyai, Math., **47**, No. 1, 89–103 (2002)

[Sma74] Smart, D.R.: Fixed Point Theorems. Cambridge Tracts in Mathematics, 66. Cambridge University Press (1974)

[Sma80] Smart, D.R.: When does $T_x^{n+1} - T_x^n \to 0$ imply convergence ?. Amer. Math. Monthly, **87**, 748–749 (1980)

[Sod84] Soderlind, G.: An error bound for fixed point iterations. BIT, **24**, 391-393 (1984)

[So00a] Soltuz, S.: A sequence given by a inequality. Octogon Math. Magazine, **8**, No. 1, 171–172 (2000)

[So00b] Soltuz, S.: Some sequences supplied by inequalities and their applications. Rev. Anal. Numer. Theor. Approx., **29**, No. 2, 207–212 (2000)

[So00c] Soltuz, S.: An example for the convergence of Mann iteration. Lect. Mat., **21**, No. 2, 113–118 (2000)

[So01a] Soltuz, S.: Mann iteration for generalized pseudocontractive maps in Hilbert spaces. Math. Commun., **6**, No. 1, 97–100 (2001)

[So01b] Soltuz, S.: Mean value iteration. Octogon Math. Magazine, **9**, No. 1A, 457–459 (2001)

[So01c] Soltuz, S.: Three proofs for the convergence of a sequence. Octogon Math. Magazine, **9**, No. 1A, 503–505 (2001)

[So01d] Soltuz, S.: Mann iteration for direct pseudocontractive maps. Bul. Stiint. Univ. Baia Mare, **17**, No. 1-2, 141–144 (2001)

[So01e] Soltuz, S.: Data dependence for Mann iteration. Octogon Math. Magazine, **9**, No. 2, 825–828 (2001)

[So01f] Soltuz, S.: The converses of two fixed point theorems. Octogon Math. Magazine, **9**, No. 2, 832–836 (2001)

[So01g] Soltuz, S.: A mean value iteration for a Holder map. Lect. Matem., **21**, No. 2, 121–125 (2001)

[So01h] Soltuz, S.: Two Mann iteration types for generalized pseudocontractive maps. In: Fixed Point Theory and Applications. Vol. 2 (Chinju/Masan, 2000), 105–110. Nova Sci. Publ., Huntington, NY (2001)

[So02a] Soltuz, S.: Sequences supplied by inequalities and an application to the convergence of Mann iteration with delay. Octogon Math. Magazine, **10**, No. 1, 103–105 (2002)

[So02b] Soltuz, S.: Mann iteration for weakly quasicontractive maps in real Banach spaces. In: Fixed Point Theory and Applications. Vol. 3, 205–208. Nova Sci. Publ., Huntington, NY (2002)

[So02c] Soltuz, S.: The convergence of Mann iteration for an asymptotic hemicontractive map. Bul. Stiint. Univ. Baia Mare, **18**, No. 1, 115–118 (2002)

[So02d] Soltuz, S.: A correction for a result on convergence of Ishikawa iteration for strongly pseudocontractive maps. Math. Commun., **7**, No. 1, 61–65 (2002)

[So03a] Soltuz, S.: An equivalence between the convergences of Ishikawa, Mann and Picard iterations. Math. Commun., **8**, No. 1, 15–22 (2003)

[So03b] Soltuz, S.: Mann-Ishikawa iterations and Mann-Ishikawa iterations with errors are equivalent models. Math. Commun., **8**, No. 2, 139–149 (2003)

[So04a] Soltuz, S.: A remark concerning the paper "An equivalence between the convergences of Ishikawa, Mann and Picard iterations", Rev. Anal. Numer. Theor. Approx., **33**, No. 1, 95–96 (2004)

[So04b] Soltuz, S.: Contributions to the theory of Mann and Ishikawa iterations. PhD Thesis, "Babes-Bolyai" University, Cluj-Napoca (2004)

[SVi98]   Sommariva, A., Vianello, M.: Approximating fixed-points of decreasing operators in spaces of continuous functions. Numer. Funct. Anal. Optimization, **19**, No. 5-6, 635–646 (1998)

[SVi00]   Sommariva, A., Vianello, M.: Computing positive fixed-points of decreasing Hammerstein operators by relaxed iterations. J. Integral Equations Appl., **12**, No. 1, 95–112 (2000)

[Son06]   Song, Y., Chen, R.D.: Viscosity approximation methods for nonexpansive nonself-mappings. J. Math. Anal. Appl., **321**, 316–326 (2006)

[Ste74]   Steinlein, H.: An approximation method in asymptotic fixed point theory. Math. Ann., **211**, 199–218 (1974)

[St03a]   Stevic, S.: Stability results for ϕ-strongly pseudocontractive mappings. Yokohama Math. J., **50**, No. 1-2, 71–85 (2003)

[St03b]   Stevic, S.: Stability of a new iteration method for strongly pseudocontractive mappings. Demonstratio Math., **36**, No. 2, 405–412 (2003)

[Ste05]   Stevic, S.: Approximating fixed points of strongly pseudocontractive mappings by a new iteration method. Appl. Anal., **84**, No. 1, 89–102 (2005)

[SuK01]   Su, K.: Convergence theorems of Ishikawa iteration for nonexpansive mappings in a uniformly convex Banach space (Chinese). Huaihua Shizhuan Xuebao, **20**, No. 5, 6–10 (2001)

[SHe03]   Su, K., He, Z.: Convergence of Ishikawa iterative sequences for k-subaccretive operators (Chinese). Acta Anal. Funct. Appl., **5**, No. 3, 255–264 (2003)

[Sub80]   Subrahmanyan, P.V.: On the convergence of iterates. Nonlinear Anal., **4**, 1203–1211 (1980)

[SLi92]   Sun, J., Liu, L.S.: Iterative method for coupled quasi-solutions of mixed monotone operator equations. Appl. Math. Comput., **52**, No. 2-3, 301–308 (1992)

[SnZ03]   Sun, Z.H.: Strong convergence of an implicit iteration process for a finite family of asymptotically quasi-nonexpansive mappings. J. Math. Anal. Appl., **286**, 351–358 (2003)

[SnZ04]   Sun, Z.H.: Convergence theorems of iteration sequences for asymptotically quasi-nonexpansive mappings. Nonlinear Funct. Anal. Appl., **9**, No. 2, 245–250 (2004)

[SnH04]   Sun, Z.H., He, C.: Iterative approximation of fixed points for asymptotically nonexpansive type mappings with error term (Chinese). Acta Math. Sinica, **47**, No. 4, 811–818 (2004)

[SHN03]   Sun, Z.H., He, C., Ni, Y.Q.: Strong convergence of an implicit iteration process for nonexpansive mappings in Banach spaces. Nonlinear Funct. Anal. Appl., **8**, No. 4, 595–602 (2003)

[SN04a]   Sun, Z.H., Ni, Y.Q., He, C.: An implicit iteration process for nonexpansive mappings with errors in Banach spaces. Nonlinear Funct. Anal. Appl., **9**, No. 4, 619–626 (2004)

[SN04b]   Sun, Z.H., Ni, Y.Q., He, C.: Convergent problem of iterative sequences for nonlinear mappings with error members in Banach spaces. Appl. Math., Ser. B (Engl. Ed.), **19**, No. 1, 81–89 (2004)

[Sz02a]   Suzuki, T.: Strong convergence theorem to common fixed points of two nonexpansive mappings in general Banach spaces. J. Nonlinear Convex Anal., **3** (2002), No. 3, 381–391 (2002)

[Sz02b]   Suzuki, T.: Convergence theorems for common fixed points of non-expansive mapping families in general Banach spaces (Japanese). Nonlinear analysis and convex analysis (Japanese) (Kyoto, 2001). Sūrikaisekikenkyūsho Kōkyūroku No. **1246**, 195–199 (2002)

[Suz03]   Suzuki, T.: On strong convergence to common fixed points of nonexpansive semigroups in Hilbert spaces. Proc. Amer. Math. Soc., **131**, No. 7, 2133–2136 (2003)

[Suz04]   Suzuki, T.: Common fixed points of two nonexpansive mappings in Banach spaces. Bull. Austral. Math. Soc., **69**, No. 1, 1–18 (2004)

[Sz05a]   Suzuki, T.: Strong convergence of Krasnoselskii and Mann's type sequences for one-parameter nonexpansive semigroups without Bochner integrals. J. Math. Anal. Appl., **305**, No. 1, 227–239 (2005)

[Suz05b]   Suzuki, T.: Strong convergence theorems of Browder's type sequences for infinite families of nonexpansive mappings in Hilbert spaces. Bull. Kyushu Inst. Technol., **52**, 21–28 (2005)

[Suz06]   Suzuki, T.: Browder's type strong convergence theorems for infinite families of nonexpansive mappings in Banach spaces. Fixed Point Theory Appl., **2006**, Art. ID 59692, 16 pp. (2006)

[Sz07a]   Suzuki, T.: A sufficient and necessary condition for Halpern-type strong convergence to fixed points of nonexpansive mappings. Proc. Amer. Math. Soc., **135**, 99–106 (2007)

[Sz07b]   Suzuki, T.: Moudafi's viscosity approximations with MeirKeeler contractions. J. Math. Anal. Appl., **325**, No. 1, 342–352 (2007)

[SzT98]   Suzuki, T., Takahashi, W.: On weak convergence to fixed points of nonexpansive mappings in Banach spaces. RIMS Kokyuroku, **1031**, 149–156 (1998)

[SzT01]   Suzuki, T., Takahashi, W.: Weak and strong convergence theorems for nonexpansive mappings in Banach spaces. Nonlinear Anal., **47**, 2805–2815 (2001)

[Tk84a]   Takahashi, W.: On Reich's strong convergence theorems for resolvents of accretive operators. J. Math. Anal. Appl., **104**, 546–553 (1984)

[Tk84b]   Takahashi, W.: Fixed point theorems for families of nonexpansive mappings on unbounded sets. J. Math. Soc. Japan, **36**, 543–553 (1984)

[Tk97a]   Takahashi, W.: Fixed point theorems and nonlinear ergodic theorems for nonlinear semigroups and their applications. Nonlinear Anal., **30**, 1283–1293 (1997)

[Tk97b]   Takahashi, W.: Weak and strong convergence theorems for families of nonexpansive mappings and their applications. Ann. Univ. Marie Curie-Sklodowska Sect. A, **51**, 277–292 (1997)

[Tak99]   Takahashi, W.: Fixed point theorems, convergence theorems and their applications. Nonlinear analysis and convex analysis (Niigata, 1998), 87–94. World Sci. Publishing, River Edge (1999)

[Tk00a]   Takahashi, W.: Convex Analysis and Approximation of Fixed Points (Japanese). Suri-kaiseki Shirizu. 2. Yokohama Publishers, Yokohama, (2000)

[Tk00b]   Takahashi, W.: Iterative methods for approximation of fixed points and feasibility problems (Japanese. RIMS Kokyuroku **1136**, 60–75 (2000)

[Tk00c]   Takahashi, W.: Nonlinear Functional Analysis. Fixed Point Theory and its Applications. Yokohama Publ., Yokohama (2000)

[Tak01] Takahashi, W.: Weak and strong convergence of approximating fixed points and applications. Nonlinear Anal. **47**, 4981–4993 (2001)

[Tak03] Takahashi, W.: Weak and strong convergence theorems for nonlinear operators of accretive and monotone type and applications. In: Nonlinear analysis and applications: to V. Lakshmikantham on his 80th birthday. Vol. 1, 2, 891–912, Kluwer Acad. Publ., Dordrecht (2003)

[TJe94] Takahashi, W., Jeong, D.: Fixed point theorem for nonexpansive semi-groups on Banach spaces. Proc. Amer. Math. Soc., **122**, 1175–1179 (1994)

[TK98a] Takahashi, W., Kim, G.E.: Approximating fixed points of nonexpansive mappings in Banach spaces. Math. Japon., **48**, No. 1, 1–9 (1998)

[TK98b] Takahashi, W., Kim, G.E.: Strong convergence of approximants to fixed points of nonexpansive nonself-mappings in Banach spaces. Nonlinear Anal., **32**, No. 3, 447–454 (1998)

[TSh00] Takahashi, W., Shimoji, K.: Convergence theorems for nonexpansive mappings and feasibility problems., Math. Comput. Modelling, **32**, 1463–1471 (2000)

[TTa98] Takahashi, W., Tamura, T.: Convergence theorems for a pair of nonexpansive mappings. J. Convex Anal., **5**, 45–56 (1998)

[TTT02] Takahashi, W., Tamura, T., Toyoda, M.: Approximation of common fixed points of a family of finite nonexpansive mappings in Banach spaces. Sc. Math. Japon., **56**, No. 3, 475–480 (2002)

[TTo03] Takahashi, W., Toyoda, M.: Weak convergence theorems for nonexpansive mappings and monotone mappings. J. Optim. Theory Appl., **118**, No. 2, 417-428 (2003)

[TTs00] Takahashi, W., Tsukiyama, N.: Approximating fixed points of nonexpansive mappings with compact domains. Commun. Appl. Nonlinear Anal., **7**, No. 4, 39–47 (2000)

[TUe84] Takahashi, W., Ueda, Y.: On Reich's strong convergence theorems for resolvents of accretive operators. J. Math. Anal. Appl., **100**, 546–553 (1984)

[Tam98] Tamura, T.: Strong convergence theorems of iterations for a pair of nonexpansive mappings in Banach spaces. Nihonkai Math. J., **9**, No. 1, 1–16 (1998)

[Tam00] Tamura, T.: Convergence theorems of iterations for a pair of nonexpansive mappings in a Banach space. In: Hudzik, H. (ed.) et al. Function Spaces. Proceed. 5th Int. Conf., Poznan, Poland, August 28-September 3, 1998, Lect. Notes Pure Appl. Math. 213. Marcel Dekker, New York (2000)

[TXu91] Tan, K.K., Xu, H.K.: Inequalities in Banach spaces with applications. Nonlinear Anal. **16**, No. 2, 1127–1138 (1991)

[TX92a] Tan, K.K., Xu, H.K.: A nonlinear ergodic theorem for asymptotically nonexpansive mappings. Bull. Austral. Math. Soc., **45**, No. 1, 25–36 (1992)

[TX92b] Tan, K.K., Xu, H.K.: The nonlinear ergodic theorem for asymptotically nonexpansive mappings in Banach spaces. Proc. Amer. Math. Soc., **114**, 399–404 (1992)

[TX93a] Tan, K.K., Xu, H.K.: Approximating fixed points of nonexpansive mappings by the Ishikawa iteration process. J. Math. Anal. Appl., **178**, 301–308 (1993)

[TX93b] Tan, K.K., Xu, H.K.: Asymptotic behavior of almost-orbits of nonlinear semigroups of non-Lipschitzian mappings in Hilbert spaces. Proc. Amer. Math. Soc., **117**, No. 2, 385–393 (1993)

[TX93c]   Tan, K.K., Xu, H.K.: Iterative solutions to nonlinear equations of strongly accretive operators in Banach spaces. J. Math. Anal. Appl., **178**, 9–21 (1993)

[TXu94]   Tan, K.K., Xu, H.K.: Fixed point iteration processes for asymptotically nonexpansive mappings. Proc. Amer. Math. Soc., **122**, No. 3, 733–739 (1994)

[TDe98]   Tang, C., Deng, L.: Approximation of fixed points of strict hemi-contraction mappings. Xinan Shifan Daxue Xuebao Ziran Kexue Ban, **23**, No. 5, 501–504 (1998)

[Tas86]   Taskovic, M.: Osnove teorije fiksne tacke (Fundamental Elements of Fixed Point Theory). Matematicka biblioteka 50, Beograd (1986)

[TBa98]   Thera, M.A., Baillon, J.-B. (eds): Fixed Point Theory and Applications. Proceed. Intern. Conf., held at CIRM, Marseille-Luminy, France, June 5-9, 1989, Pitman Research Notes in Mathematics Series, 252. Harlow: Longman Scientific & Technical. John Wiley & Sons, New York (1991)

[Tho85]   Thorlund-Petersen, L.: Fixed point iterations and global stability in economics. Math. Oper. Res., **10**, 642–649 (1985)

[Tia93]   Tian, Y.X.: A sequence of quasi-contractive mappings and generalized Ishikawa's iteration. J. Sichuan Univ., Nat. Sci. Ed., **30**, No. 3, 331–334 (1993)

[Ti00a]   Tian, Y.X.: Generalized Ishikawa-type iteration for more generalized quasi contraction mappings (Chinese). J. Sichuan Univ., Nat. Sci. Ed., **37**, No. 5, 688–691 (2000)

[Ti00b]   Tian, Y.X.: Quasi-contractive mapping and Ishikawa iteration process with errors. (Chinese). J. Sichuan Univ., Nat. Sci. Ed., **37**, No. 6, 839–843 (2000)

[Ti00c]   Tian, Y.X.: An iterative algorithm of common fixed point for a generalized quasi-contractive mappings sequence (Chinese). Xinan Shifan Daxue Xuebao Ziran Kexue Ban, **25**, No. 6, 640–644 (2000)

[Ti01a]   Tian, Y.X.: On the stability problem of an iterative process for accretive and pseudo-contractive mappings (Chinese). J. Sichuan Univ., Nat. Sci. Ed., **38**, No. 5, 653–657 (2001)

[Ti01b]   Tian, Y.X.: Stability for the Ishikawa iteration procedure of contractive functions in convex metric spaces (Chinese). J. Sichuan Univ., Nat. Sci. Ed., **38**, No. 6, 827–830 (2001)

[Ti01c]   Tian, Y.X.: Stability of fixed point iteration procedures of the generalized quasi-contractive mapping (Chinese). J. Sichuan Univ., Nat. Sci. Ed., **38**, No. 4, 495–498 (2001)

[Ti01d]   Tian, Y.X.: Ishikawa type iteration with errors and generalized quasi-contractive mappings (Chinese). Xinan Shifan Daxue Xuebao Ziran Kexue Ban, **26**, No. 6, 635–639 (2001)

[Ti02a]   Tian, Y.X.: Convergence and stability of the Ishikawa iteration procedures for asymptotically nonexpansive mappings (Chinese). Sichuan Daxue Xuebao, **39**, No. 6, 1019–1022 (2002)

[Ti02b]   Tian, Y.X.: Ishikawa iterative process with errors for a sequence of more generalized quasicontractive mappings (Chinese). Sichuan Shifan Daxue Xuebao Ziran Kexue Ban, **25**, No. 5, 472–475 (2002)

[Tia03]   Tian, Y.X.: Convergence and stability of the Mann iteration for asymptotically nonexpansive mappings (Chinese). Sichuan Shifan Daxue Xuebao Ziran Kexue Ban, **26**, No. 4, 348–351 (2003)

[TiX03] Tian, Y.X., Xu, W.-J.: Convergence of Ishikawa iterative sequences for asymptotically quasi-nonexpansive mappings in convex metric spaces. Sichuan Daxue Xuebao, **40**, No. 6, 1027–1031 (2003)

[TZ02a] Tian, Y.X., Zhang, S.S.: Convergence of Ishikawa-type iterative sequence with errors for quasi-contractive mappings in convex metric spaces. Appl. Math. Mech., Engl. Ed., **23**, No. 9, 1001–1008 (2002)

[TZ02b] Tian, Y.X., Zhang, S.S.: Convergence of Ishikawa type iterative sequence with errors for quasi-contractive mappings in convex metric spaces. Appl. Math. Mech. (English Ed.), **23**, No. 9, 1001–1008 (2002). Translated from Appl. Math. Mech. (Chinese), **23**, No. 9, 889–895 (2002)

[TiZ03] Tian, Y.X., Zhang, S.S.: Convergence of iterative sequences for asymptotically nonexpansive mappings (Chinese). J. Northwest Univ., **33**, No. 6, 641–644 (2003)

[TDe95] Tiwary, K., Debnath, S.C.: On Ishikawa iterations. Indian J. Pure Appl. Math., **26**, No. 8, 743–750 (1995)

[Tod76] Todd, M.: The computation of fixed points and applications. Lecture Notes in Economics and Mathematical Systems. 124. Springer Verlag, Berlin Heidelberg New York (1976)

[Ton03] Tong, H.: Convergence problems of Ishikawa iterative processes with error for generalized Φ-pseudo-contractive type mappings (Chinese). J. Hebei Univ. Nat. Sci., **23**, No. 3, 244–248 (2003)

[Trc16] Tricomi, F.: Una teorema sulla convergenza delle successioni formate delle successive iterate di una funzione di una variabile reale. Giorn. Mat. Bataglini, **54**, 1–9 (1916)

[Trf87] Trif, D.: The approximation of fixed points of C^1-mappings. Seminar on Fixed Point Theory, **3**, 31–38 (1987)

[Tru87] Trubnikov, Y.V.: Hanner inequality and convergence of iteration processes. Soviet Math. (Izvestiya), **31**, 74–83 (1987)

[Tur78] Turinici, M.: Sequentially iterative processes and applications to Volterra functional equations. Ann. Univ. "Mariae Curie-Sklodowska" Sect. A, **32**, 127–134 (1978)

[Tur81] Turinici, M.: Multiple iterative processes based on simple fixed points and applications. Mathematica, **23(46)**, 141–148 (1981)

[Tyc35] Tychonoff, A.: Ein Fixpunktsatz. Math. Ann., **111**, 767–776 (1935)

[Udo01] Udomene, A.: Construction of zeros of accretive mappings. J. Math. Anal. Appl., **262**, No. 2, 623–632 (2001)

[Ume96] Ume J.: Convergence of the Ishikawa iteration process for two mappings. Nonlinear Anal. Forum, **2**, 1–9 (1996)

[Ume98] Ume, J.: Convergence theorems for two mappings in Banach spaces. Far East J. Math. Sci., **6**, No. 3, 415–423 (1998)

[UKK97] Ume, J., Kim, K.W., Kim, T.H.: Common fixed point theorems for a generalized contraction. Math. Japon., **46**, No. 3, 387–392 (1997)

[UKT01] Ume, J., Kim, T.H.: Common fixed point theorems for weak compatible mappings. Indian J. Pure Appl. Math., **32**, No. 4, 565–571 (2001)

[UKY99] Ume, J., Kim, Young-Ho: Ishikawa's iteration method to construct fixed point of nonlinear mappings in convex metric spaces. Far East J. Math. Sci., **1**, No. 6, 873–887 (1999)

[Vaj60] Vajnberg, M.M.: On the convergence of the method of steepest descent for nonlinear equations. Sov. Math., Dokl., **1**, 1–4 (1960). Translation from Dokl. Akad. Nauk SSSR **130**, 9–12 (1960)

294   References

[Vaj61]   Vajnberg, M.M.: On the convergence of the process of steepest descent
for nonlinear equations. Sibirsk Math. Zh., **2**, 201–220 (1961)
[vCr72]   van de Craats, J. On the region of convergence of Picard's iteration. Z.
Angew. Math. Mech., **52**, No. 9, 487–491 (1972)
[vDu82]   van Dulst, D.: Equivalent norms and the fixed point property for nonexpansive mappings. J. London Math. Soc., **25**, 139–144 (1982)
[Vas96]   Vasilyev, N.S., The search for a fixed point of a consistently monotone
mapping. Contemp. Math. & Math. Physics, **36**, No. 12, 1671–1677 (1996)
[Vas92]   Vasin, V.V. Ill-posed problems and iterative approximation of fixed points
of pseudo-contractive mappings. Ill-posed problems in natural sciences
(Moscow, 1991), 214–223, VSP, Utrecht (1992)
[VeP82]   Veeramani, P., Pai, D.V.: On a fixed point theorem on uniformly convex
Banach spaces. Indian J. Pure Appl. Math., **13**, 647–650 (1982)
[VSJ03]   Verma, R.K., Sahu, D.R., Jung, J.S., Dubey, R.P. Mann and Ishikawa
iterative sequence with errors for m-accretive operator equations. Acta
Cienc. Indica Math., **29**, No. 4, 693–698 (2003)
[Ver93]   Verma, R.U.: Iterative algorithms for the approximation of fixed points of
strongly monotone operators (Spanish). Bol. Acad. Cienc. Fis. Mat. Nat.,
53, No. 173-174, 72–76 (1993)
[Ver96]   Verma, R.U.: An iterative procedure for approximating fixed points of
relaxed monotone operators. Numer. Funct. Anal. Optim., **17**, No. 9-10,
1045–1051 (1996)
[Ve97a]   Verma, R.U.: A fixed-point theorem involving Lipschitzian generalised
pseudo contractions. Proc. Roy. Irish Acad. Sect. A, **97**, No. 1, 83–86
(1997)
[Ve97b]   Verma, R.U.: An approximation procedure for fixed points of strongly
Lipschitz operators. Portugal. Math., **54**, No. 4, 461–465 (1997)
[Ve97c]   Verma, R.U.: On fixed points of Lipschitzian strongly Lipschitz operators.
Math. Sci. Res. Hot-Line, **1**, No. 7, 20–26 (1997)
[Ve97d]   Verma, R.U.: An iterative algorithm on fixed points of relaxed Lipschitz
operators. J. Appl. Math. Stochastic Anal., **10**, No. 2, 187–189 (1997)
[Ve97e]   Verma, R.U.: An approximation procedure for fixed points of strongly
Lipschitz operators. Portugal. Math., **54**, No. 4, 461–465 (1997)
[Ver98]   Verma, R.U.: Mann type algorithms for the fixed points of Lipschitzian
strongly Lipschitz operators. Math. Sci. Res. Hot-Line, **2**, No. 10, 7–12
(1998)
[Vij95]   Vijayaraju, P.: Fixed points and their approximations for asymptotically
nonexpansive mappings in locally convex spaces. Int. J. Math. Math. Sci.,
18, No. 2, 293–298 (1995)
[Vij97]   Vijayaraju, P.: Iterative construction of fixed points of asymptotic 1-set
contractions in Banach spaces. Taiwanese J. Math., **1**, No. 3, 315–325
(1997)
[Wal81]   Walter, W.: Remarks on a paper by F. Browder about contractions. Nonlinear Anal. TMA, **5**, 21–25 (1981)
[WaL06]   Wang, L.: Strong and weak convergence theorems for common fixed points
of nonself asymptotically nonexpansive mappings. J. Math. Anal. Appl.,
323, No. 1, 550–557 (2006)
[Wa04a]   Wang, S.R.: Some new strong convergence theorems for Ishikawa iterative sequences with errors for asymptotically nonexpansive mappings
(Chinese). Acta Anal. Funct. Appl., **6**, No. 2, 187–192 (2004)

[Wa04b]   Wang, S.R.: Ishikawa iterative sequences with errors for asymptotically nonexpansive mappings (Chinese). Sichuan Daxue Xuebao, **41**, No. 4, 881–883 (2004)

[Wa04c]   Wang, S.R.: Ishikawa iterative approximation of fixed points with errors for asymptotically quasi-nonexpansive type mappings in Banach spaces (Chinese). Sichuan Shifan Daxue Xuebao Ziran Kexue Ban, **27**, No. 3, 255–258 (2004)

[Wa04d]   Wang, S.R.: The Ishikawa iterative approximation of fixed points with errors for asymptotically quasi-nonexpansive type mappings in Banach spaces (Chinese). Sichuan Daxue Xuebao, **41**, No. 2, 231–235 (2004)

[Wa04e]   Wang, S.R., Xiong, M.: Iterative approximation problems for the fixed points of asymptotically quasi-nonexpansive type mappings (Chinese). Pure Appl. Math. (Xi'an), **20**, No. 1, 18–23 (2004)

[WaC05]   Wang, S.R., Chen, Ying: Implicit iterative approximation for a finite family of asymptotically nonexpansive mappings in Banach spaces. (Chinese). Appl. Math., Ser. A (Chin. Ed.), **20**, No. 1, 63–69 (2005)

[Wa04f]   Wang, S.R., Zuo, G.C.: Iterative approximation for the fixed points of asymptotically quasi-nonexpansive mappings (Chinese). J. Yunnan Univ. Nat. Sci., **26**, No. 4, 279–283 (2004)

[Wa89a]   Wang, T.: On fixed point theorems and fixed point stability for multivalued mappings on metric spaces. J. Nanjing Univ., Math. Biq., **6** (1989), No. 1, 16–23 (1989)

[Wa89b]   Wang, T.: Fixed-point theorems and fixed-point stability for multivalued mappings on metric spaces. Nanjing Daxue Xuebao Shuxue Bannian Kan, **6**, No. 1, 16–23 (1989)

[WHe04]   Wang, X., He, Z.: Fixed-point iteration for $(L\text{-}\alpha)$ uniform Lipschitz asymptotically nonexpansive mapping of uniform convex Banach space. J. Hebei Univ. Nat. Sci., **24**, No. 2, 126–129 (2004)

[WaY03]   Wang, Y.: Ishikawa iterative sequences for asymptotically quasi-nonexpansive mappings with errors (Chinese). Xinan Shifan Daxue Xuebao Ziran Kexue Ban, **28**, No. 1, 52–54 (2003)

[We91a]   Weng, X.: Fixed point iteration for local strictly pseudo-contractive mapping. Proc. Amer. Math. Soc., **113**, No. 3, 727–731 (1991)

[We91b]   Weng, X.: The iterative solution of the equation $f \in x + Tx$ for a accretive operator T in uniformly smooth Banach spaces. J. Shanghai Univ. Sci. Technol., **14**, No. 4, 23–27 (1991)

[We92a]   Weng, X.: Iterative solution of nonlinear equations of the accretive and dissipative type in certain Banach spaces. Bull. Calcutta Math. Soc., **84**, No. 2, 103–108 (1992)

[We92b]   Weng, X.: The iterative solution of nonlinear equations in certain Banach spaces. Special issue in honour of Professor Chike Obi. J. Nigerian Math. Soc., **11**, No. 1, 1–7 (1992)

[We92c]   Weng, X.: Iterative construction of fixed points of a dissipative type operator, Tamkang J. Math., **23**, No. 3, 205–212 (1992)

[We92d]   Weng, X.: Iterative solutions of a class of nonlinear equations in reflexive Banach spaces. J. Inst. Math. Comput. Sci., Math. Ser., **5**, No. 3, 325–328 (1992)

[We92e]   Weng, X.: Approximating fixed points of quasi-asymptotically nonexpansive mappings. J. Inst. Math. Comput. Sci., Math. Ser., **5**, No. 2, 201–206 (1992)

[Wit90]   Wittmann, R.: Mean ergodic theorems for nonlinear operators. Proc. Amer. Math. Soc., **108**, No. 3, 781–788 (1990)

[Wit92]   Wittmann, R.: Approximation of fixed points of nonexpansive mappings. Arch. Math., **58**, 486–491 (1992)

[Wol79]   Wolf, R.: Approximation of fixed points of condensing mappings. Appl. Anal. **9**, 125-136 (1979)

[Won76]   Wong, C.S.: Approximation to fixed points of generalized nonexpansive mappings. Proc. Amer. Math. Soc., **54**, 93–97 (1976)

[WuW04]   Wu, C.X., Wu, Y.: A common fixed point problem for mean nonexpansive mappings (Chinese). J. Yantai Univ. Nat. Sci. Eng., **17**, No. 3, 161–163, 175 (2004)

[XiD02]   Xia, X., Deng, L.: Ishikawa iterative process with errors for asymptotically quasi-nonexpansive mappings in Banach spaces. Math. Appl. (Wuhan), **15**, suppl., 181–185 (2002)

[XuA02]   Xu, B., Aslam Noor, M.: Fixed-point iterations for asymptotically nonexpansive mappings in Banach spaces. J. Math. Anal. Appl., **267**, No. 2, 444–453 (2002)

[XuC02]   Xu, C.Z.: Mann iteration process with errors for the solution of the nonlinear equation $x + Tx = f$. Xinan Shifan Daxue Xuebao Ziran Kexue Ban, **27**, No. 5, 652–657 (2002)

[XuC03]   Xu, C.Z.: The Ishikawa iterative solution of the equation $x + Tx = f$ for a k-subaccretive operator T (Chinese). Acta Anal. Funct. Appl., **5**, No. 3, 249–254 (2003)

[XuC04]   Xu, C.Z.: The Ishikawa iterative process with errors for the solution of the equation $x + Tx = f$ for a k-subaccretive operator T (Chinese). Sichuan Shifan Daxue Xuebao Ziran Kexue Ban, **27**, No. 2, 160–164 (2004)

[XH91a]   Xu, H.K.: Inequalities in Banach spaces with applications. Nonlinear Anal. TMA, **16**, No. 2, 1127–1138 (1991)

[XH91b]   Xu, H.K.: Existence and convergence for fixed points of mappings of asymptotically nonexpansive type. Nonlinear Anal. TMA, **16**, No. 12, 1139–1146 (1991)

[XuH92]   Xu, H.K.: A note on the Ishikawa iteration scheme. J. Math. Anal. Appl., **167**, 582–587 (1992)

[XuH97]   Xu, H.K.: Approximating curves of nonexpansive nonself mappings in Banach spaces. C.R. Acad. Sci. Paris Ser. I Math., **325**, 179–184 (1997)

[XuH98]   Xu, H.K.: Approximations to fixed points of contraction semigroups in Hilbert spaces. Numer. Funct. Anal. Optim., **19**, No. 1-2, 157–163 (1998)

[XuH00]   Xu, H.K.: Convergence of an iteration process for nonexpansive mappings. Nonlinear Funct. Anal. Appl., **5**, No. 2, 107–111 (2000)

[XuH02]   Xu, H.K.: Another control condition in an iterative method for nonexpansive mappings. Bull. Austral. Math. Soc., **65**, No. 1, 109–113 (2002)

[XuH03]   Xu, H.K.: Remarks on an iterative method for nonexpansive mappings. Commun. Appl. Nonlinear Anal., **10**, No. 1, 67–75 (2003)

[XuH04]   Xu, H.K.: Viscosity approximation methods for nonexpansive mappings. J. Math. Anal. Appl., **298**, No. 1, 279–291 (2004)

[XuH06]   Xu, H.K.: Strong convergence of an iterative method for nonexpansive and accretive operators. J. Math. Anal. Appl., **314**, No. 2, 631–643 (2006)

[XuO01]   Xu, H.K., Ori, R.G.: An implicit iteration process for nonexpansive mappings. Numer. Funct. Anal. Optim., **22**, 767–773 (2001)

[XYi95]   Xu, H.K., Yin, X.M.: Strong convergence theorems for nonexpansive nonself-mappings. Nonlinear Anal., **24**, 223–228 (1995)

[XuJ04]   Xu, S.Y., Jia, B.: Fixed-point theorems of Φ concave-$(-\Psi)$ convex mixed monotone operators and applications. J. Math. Anal. Appl., **295**, 645–657 (2004)

[XuY98]   Xu, Y.G.: Ishikawa and Mann iterative processes with errors for nonlinear strongly accretive operator equations. J. Math. Anal. Appl., **224**, 91–101 (1998)

[XuY04]   Xu, Y.G.: Iterative processes with errors for fixed points of nonlinear Φ-pseudocontractive mappings. (Chinese). Acta Math. Sci., Ser. A, Chin. Ed., **24**, No. 6, 730–736 (2004)

[XuL03]   Xu, Y.G., Liu, Z.: On estimation and control of errors of the Mann iteration process. J. Math. Anal. Appl., **286**, No. 2, 804–806 (2003)

[XXi04]   Xu, Y.G., Xie, F.: Stability of Mann iterative process with random errors for the fixed point of strongly-pseudocontractive mapping in arbitrary Banach spaces. Rostock. Math. Kolloq., No. **58**, 93–100 (2004)

[XuR91]   Xu, Z.B., Roach, G.F.: Characteristic inequalities for uniformly and uniformly smooth Banach spaces. J. Math. Anal. Appl., **157**, 189–210 (1991)

[XuR92]   Xu, Z.B., Roach, G.F.: A necessary and sufficient condition for the convergence of steepest descent approximation to accretive operator equations. J. Math. Anal. Appl., **167**, 340–354 (1992)

[Xue97]   Xue, Z.Q.: Ishikawa iterative methods with errors for solving Lipschitzian strongly accretive operator equations (Chinese). Qufu Shifan Daxue Xuebao Ziran Kexue Ban, **23**, No. 4, 26–31 (1997)

[Xue98]   Xue, Z.Q.: Iterative solution of operator equations of the Φ-strongly accretive type. J. Qufu Norm. Univ., Nat. Sci., **24**, No. 3, 31–34 (1998)

[Xue99]   Xue, Z.Q.: Ishikawa iterative process with errors for Φ-strongly accretive mappings in arbitrary Banach spaces. Math. Sci. Res. Hot-Line, **3**, No. 8, 55–64 (1999)

[XLL04]   Xue, Z.Q., Liu, G., Li, X.H.: Mann iteration process for certain nonlinear mappings in q-uniformly smooth Banach spaces (Chinese). J. Hebei Norm. Univ. Nat. Sci. Ed., **28**, No. 2, 116–119 (2004)

[XTi02]   Xue, Z.Q., Tian, H.: Remark on stability of Ishikawa iterative procedures. Appl. Math. Mech. (English Ed.), **23**, No. 12, 1472–1476 (2002)

[XWL03]   Xue, Z.Q., Wang, Z.J., Li, X.H.: Stability of Ishikawa iteration procedures for non-Lipschitz Φ-strongly pseudocontractive operators (Chinese). J. Hebei Norm. Univ. Nat. Sci. Ed., **27**, No. 6, 563–566 (2003)

[XZh97]   Xue, Z.Q., Zhou, H.: Ishikawa iterative methods with errors for a solution of nonlinear Lipschitzian strongly accretive operators equations (Chinese). J. Qufu Norm. Univ., Nat. Sci., **23**, No. 4, 26–31 (1997)

[XZh98]   Xue, Z.Q., Zhou, H.: Iterative solution of operator equations of the ϕ-strongly accretive type. J. Qufu Norm. Univ., Nat. Sci., **24**, No. 3, 31–34 (1998)

[XZh99]   Xue, Z.Q., Zhou, H.: Iterative approximation with errors of fixed point for a class of nonlinear operators with a bounded range. Appl. Math. Mech., English Ed., **20**, No. 1, 99–104 (1999)

[XZh00]   Xue, Z.Q., Zhou, H.: A remark on the stability of Mann and Ishikawa iteration procedures. Math. Sci. Res. Hot-Line, **4**, No. 7, 47–54 (2000)

[YaJ80]   Yadav, B.S., Jaggi, D.S.: Weak convergence of the sequence of succes-
          sive approximations for para-nonexpansive mappings. Publ. Inst. Math.
          (Beograd) (N.S.), **28(42)**, 89–93 (1980)
[YOY98]   Yamada, I., Ogura, N., Yamashita, Y., Sakaniwa, K.: Quadratic approxi-
          mations of fixed points of nonexpansive mappings in Hilbert spaces. Nu-
          mer. Funct. Anal. Optim., **19**, No. 1, 165–190 (1998)
[Yan99]   Yan, J.: A Mann iteration process with errors for nonexpansive mappings
          (Chinese). Xinan Shifan Daxue Xuebao Ziran Kexue Ban, **24**, No. 3, 267–
          269 (1999)
[Yan00]   Yan, J.: The theorem for approximation couple fixed point sequences of
          semicompact nonexpansive mappings approximating couple fixed points.
          J. Sichuan Univ., Nat. Sci. Ed., **37**, No. 4, 512–515 (2000)
[YLH00]   Yan, M., Li, S.M., He, Z.: Convergence of Ishikawa iterative sequences
          with errors for M-accretive operators and ϕ-pseudo-contractive mappings.
          (Chinese) Acta Anal. Funct. Appl., **2**, No. 4, 359–370 (2000)
[YKo84]   Yang, B., Kojima, M.: Improving the computational efficiency of fixed
          point algorithms. J. Oper. Res. Soc. Jap., **27**, 59–77 (1984)
[Yg98a]   Yang, X.: Approximating zeros of accretive operators by Ishikawa itera-
          tion with errors (Chinese). J. Hebei Norm. Univ., Nat. Sci. Ed., **22**, No.
          2, 151–153 (1998)
[Yg98b]   Yang, Y.Q.: The Ishikawa iteration process for strongly pseudocontractive
          operators in Banach spaces (Chinese). Xinan Shifan Daxue Xuebao Ziran
          Kexue Ban, **23**, No. 6, 642–646 (1998)
[Yng00]   Yang, Y.Q.: The Ishikawa iteration process with errors for nonexpansive
          mappings (Chinese). Xinan Shifan Daxue Xuebao Ziran Kexue Ban, **25**,
          No. 6, 637–639 (2000)
[Yng02]   Yang, Y.Q.: Stability of the Ishikawa iteration procedure with errors for
          the solution of the nonlinear equation $x + Tx = f$ (Chinese). Xinan Shifan
          Daxue Xuebao Ziran Kexue Ban, **27**, No. 4, 486–489 (2002)
[Yng99]   Yang, Z.: Computing Equilibria and Fixed Points. Kluwer Academic Pub-
          lishers, Boston Dordrecht London (1999)
[YCL91]   Yang, Z., Chen, K., Liang, Z.: A new variable dimension algorithm for
          computing fixed points (Chinese). Appl. Math., J. Chin. Univ., **6**, No. 3,
          382–391 (1991)
[YJi02]   Yao, L., Jin, M.: Stability of the Ishikawa iteration procedure with errors
          for Lipschitz strictly hemicontractive mappings (Chinese). Xinan Shifan
          Daxue Xuebao Ziran Kexue Ban, **27**, No. 2, 133–137 (2002)
[YaC04]   Yao, Y.H., Chen, R.D.: Some strong convergence theorems for Ishikawa
          iterative schemes for asymptotically nonexpansive mappings in uniformly
          convex Banach spaces. (Chinese). Acta Anal. Funct. Appl., **6**, No. 3, 262–
          266 (2004)
[YaA07]   Yao, Y.H., Aslam Noor, M.: On viscosity iterative methods for variational
          inequalities. J. Math. Anal. Appl., **325**, No. 2, 776–787 (2007)
[YeN01]   Ye, X., Ni, R.X.: On convergence of Ishikawa iteration procedures with
          errors. Numer. Math. J. Chinese Univ. (English Ser.), **10**, No. 1, 105–120
          (2001)
[YLL00]   Yin, Q., Liu, Z., Lee, B.S.: Iterative solutions of nonlinear equations with
          Φ-strongly accretive operators, Nonlinear Anal. Forum, **5**, 87–99 (2000)
[Yos02]   Yoshimoto, T.: Nonlinear ergodic theorems of Dirichlet's type in Hilbert
          space. Nonlinear Anal., **48**, 551–565 (2002)

[Yos04]   Yoshimoto, T.: Strong nonlinear ergodic theorems for asymptotically non-expansive semigroups in Banach spaces. J. Nonlinear Convex Anal., 5, No. 3, 307-319 (2004)

[YHe02]   You, C.L., He, Z.: Convergence of Ishikawa iteration sequences with errors for nonexpansive mappings and strongly pseudo-contractive mappings (Chinese). Acta Anal. Funct. Appl., 4, No. 4, 361–370 (2002)

[YXu90]   You, Z.Y., Xu, H.K.: An ergodic convergence theorem for mappings of asymptotically non-expansive type (Chinese). Chin. Ann. Math., Ser. A, 11, No. 4, 519–523 (1990)

[YXu83]   You, Z.Y., Xu, Z.B.: Pseudo-monotonic sequence and convergence of convex combined iteration with errors (Chinese). Numer. Math. J. Chinese Univ., 5, 335–341 (1983)

[YXu84]   You, Z.Y., Xu, Z.B.: Resolvent iteration processes for constructing a solution of a nonlinear equation in Banach space (Chinese). Math. Numer. Sin., 6, 407–413 (1984)

[YXu85]   You, Z.Y., Xu, Z.B.: Ergodic convergence for fixed points and variational inequality of a class of nonlinear mappings (Chinese). J. Xi'an Jiaotong Univ., 19, No. 2, 1–10 (1985)

[YXu82]   You, Z.H., Xu, Z.B.: The resolvent iteration processes to construct a solution of a nonlinear equation in Banach space (Chinese). J. Xi'an Jiaotong Univ., 16, No. 6, 109–110 (1982)

[YuG03]   Yu, L., Guo, Y.: The convergence of Mann iteration sequences to the unique solution of a class of nonlinear operator equations and its applications. (Chinese) Gongcheng Shuxue Xuebao, 20, No. 1, 49–54 (2003)

[Yua96]   Yuan, D.: Convergence rate and acceleration on Ishikawa's iterations of real Lipschitz mappings. Math. Appl., 9, No. 3, 311–314 (1996)

[Zmn85]   Zaman, S.I.: On fixed points of operators on a Banach space. Ganita, 5, No. 1-2, 75–79 (1985)

[Zam72]   Zamfirescu, T.: Fix point theorems in metric spaces. Arch. Math. (Basel), 23, 292–298 (1972)

[Zam84]   Zamfirescu, T.: Convergence to fixed points on normed linear spaces. Math. Japon., 29, 63–67 (1984)

[Zar60]   Zarantonello, E.: Solving functional equations by constructive averaging. Tech. Report 160, US Army research Centre, Madison (1960)

[ZPr02]   Zegeye, H., Prempeh, E.: Strong convergence of approximants to fixed points of Lipschitzian pseudocontractive maps. Comput. Math. Appl., 44, No. 3-4, 339–346 (2002)

[Zei93]   Zeidler, E.: Nonlinear Functional Analysis and its Applications. Volume I: Fixed-point theorems. Springer-Verlag., New York (1993)

[Zen95]   Zeng, L.: Iterative construction of solutions to nonlinear equations of Lipschitzian and local strongly accretive operators. Appl. Math. Mech., English Edition, 16, No. 6, 583–592 (1995)

[Ze97a]   Zeng, L.: An iterative process for finding approximate solutions to nonlinear equations of strongly accretive operators. Numer. Math. J. Chin. Univ., 6, No. 2, 132–141 (1997)

[Ze97b]   Zeng, L.: Error bounds for approximation solutions to nonlinear equations of strongly accretive operators in uniformly smooth Banach spaces. J. Math. Anal. Appl., 209, No. 1, 67–80 (1997)

300   References

[Ze98a]   Zeng, L.: Iterative construction of solutions to nonlinear equations of strongly accretive operators in Banach spaces. J. Math. Res. Expo., **18**, No. 3, 329–334 (1998)

[Ze98b]   Zeng, L.: A note on approximating fixed points of nonexpansive mappings by Ishikawa iteration process. J. Math. Anal. Appl., **226**, No. 1, 245–250 (1998)

[Ze98c]   Zeng, L.: Iterative approximation of solutions to nonlinear equations of strongly accretive operators in Banach spaces. Nonlinear Anal. TMA, **31**, No. 5-6, 589–598 (1998)

[Zen99]   Zeng, L.: Iterative approximation of fixed points for multivalued operators of the monotone type in uniformly smooth Banach spaces., Numer. Math., J. Chin. Univ., **8**, No. 1, 59–66 (1999)

[Ze01a]   Zeng, L.: Ishikawa type iterative sequences with errors for Lipschitzian strongly pseudocontractive mappings in Banach spaces (Chinese). Chin. Ann. Math., Ser. A, **22**, No. 5, 639–644 (2001)

[Ze01b]   Zeng, L.: Iterative approximation of fixed points of (asymptotically) nonexpansive mappings. Appl. Math. J. Chinese Univ. Ser. B, **16**, No. 4, 402–408 (2001)

[Ze02a]   Zeng, L.: Ishikawa iterative process for solutions of m-accretive operator equations. Appl. Math. Mech. (English Ed.), **23**, 686–693 (2002)

[Ze02b]   Zeng, L.: Approximating fixed points of strictly pseudocontractive mappings by Ishikawa iterative procedure. Math. Appl. (Wuhan), **15**, No. 1, 7–10 (2002)

[Ze02c]   Zeng, L.: Convergence rate estimate of Ishikawa iterative sequence for strictly pseudocontractive mappings. Appl. Math. J. Chinese Univ. Ser. B, **17**, No. 2, 189–192 (2002)

[Ze02d]   Zeng, L.: Ishikawa iteration process with errors for approximate solutions to equations of Lipschitz strongly accretive operators (Chinese). Acta Anal. Funct. Appl., **4**, No. 3, 274–279 (2002)

[Ze02e]   Zeng, L.: Ishikawa type iterative sequences with errors for Lipschitzian ϕ-strongly accretive operator equations in arbitrary Banach spaces. Numer. Math. J. Chinese Univ. (English Ser.), **11**, No. 1, 25–33 (2002)

[Ze02f]   Zeng, L.: Iterative approximation of solutions to nonlinear equations involving m-accretive operators in Banach spaces. J. Math. Anal. Appl., **270**, 319–331 (2002)

[Ze03a]   Zeng, L.: Ishikawa iteration process for approximation of fixed points of nonexpansive mappings. J. Math. Res. Expo., **23**, No. 1, 33–39 (2003)

[Ze03b]   Zeng, L.: Ishikawa iterative procedure for approximating fixed points of strictly pseudocontractive mappings. Appl. Math. J. Chinese Univ. Ser. B, **18**, No. 3, 283–286 (2003)

[Ze03c]   Zeng, L.: Iterative approximation of fixed points of non-Lipschitzian asymptotically pseudocontractive mappings. Numer. Math. J. Chinese Univ. (English Ser.), **12**, No. 1, 66–70 (2003)

[Ze03d]   Zeng, L.: Convergence rate estimate of Ishikawa iterative approximations for strictly pseudocontractive mappings (Chinese). Gongcheng Shuxue Xuebao, **20**, No. 1, 123–126 (2003)

[Ze03e]   Zeng, L.: On the characteristics of the convergence of Ishikawa type iterative sequences for strong pseudocontractions and strongly accretive operators. J. Math. Res. Exposition, **23**, No. 3, 403–409 (2003)

[Ze03f] Zeng, L.: Convergence rate estimate of Ishikawa iteration method for equations involving accretive operators (Chinese). Numer. Math. J. Chinese Univ., **25**, No. 1, 74–80 (2003)

[Ze03g] Zeng, L.: Modified Ishikawa iteration process for asymptotically nonexpansive mappings. Math. Appl. (Wuhan), **16**, No. 2, 28–31 (2003)

[Ze03h] Zeng, L.: Iterative approximation of fixed points for almost asymptotically nonexpansive type mappings in Banach spaces. Appl. Math. Mech. (English Ed.), **24**, No. 12, 1421–1430 (2003). Translated from Appl. Math. Mech. (Chinese), **24**, No. 12, 1258–1266 (2003)

[Ze04a] Zeng, L.: Modified Ishikawa iteration process with errors in Banach spaces (Chinese). Acta Math. Sinica, **47**, No. 2, 219–228 (2004)

[Ze04b] Zeng, L.: Iterative construction of fixed points of asymptotically pseudocontractive type mappings (Chinese). J. Systems Sci. Math. Sci., **24**, No. 2, 261–270 (2004)

[Ze04c] Zeng, L.: Modified Ishikawa iteration process with errors in Banach spaces (Chinese). Acta Math. Sinica, **47**, No. 2, 219–228 (2004)

[Ze04d] Zeng, L.: Ishikawa iteration process with errors for solutions to equations involving accretive operators. (Chinese). Acta Math. Sci., Ser. A, Chin. Ed., **24**, No. 6, 654–660 (2004)

[Ze04e] Zeng, L.: Ishikawa iterative approximation of solutions to equations of Lipschitz strongly accretive operators. (Chinese). J. Math., Wuhan Univ., **24**, No. 5, 524–530 (2004)

[ZeY99] Zeng, L., Yang, Y.L.: Iterative approximation of Lipschitz strictly pseudocontractive mappings in Banach spaces (Chinese). Chin. Ann. Math., Ser. A, **20**, No. 3, 389–398 (1999)

[ZSK04] Zhang, C.H., Shi, F., Kim, Y.S., Kang, S.M.: Iterative approximations of fixed points for asymptotically nonexpansive mappings in Banach spaces. In: Fixed Point Theory and Applications. Vol. 5, 183–189. Nova Sci. Publ., Hauppauge, NY (2004)

[ZhF03] Zhang, F.X.: The Ishikawa iterative solution of a nonlinear k-subaccretive operator equation (Chinese). Xinan Shifan Daxue Xuebao Ziran Kexue Ban, **28**, No. 2, 177–180 (2003)

[Zh03a] Zhang, G.W.: Convergence theorems of Mann and Ishikawa iterative processes with errors for multivalued Φ-strongly accretive mapping. Northeast. Math. J., **19**, No. 2, 174–180 (2003)

[Zh03b] Zhang, G.W.: Stability of Mann and Ishikawa iterative processes for a class of nonlinear equations (Chinese). Acta Anal. Funct. Appl., **5**, No. 2, 183–188 (2003)

[ZSo00] Zhang, G.W., Song, S.: A note on Ishikawa and Mann iterative processes with errors for strongly accretive operators. Math. Appl. (Wuhan), **13**, No. 3, 63–66 (2000)

[ZhS99] Zhang, S.S.: Mann and Ishikawa iterative approximation of solutions for m-accretive operator equations. Appl. Math. Mech. (English Ed.), **20**, 1310–1318 (1999)

[ZhS00] Zhang, S.S.: On the convergence problems of Ishikawa and Mann iterative processes with error for Φ-pseudo contractive type mappings. Appl. Math. Mech., Engl. Ed., **21**, No. 1, 1–12 (2000)

[ZS01a] Zhang, S.S.: Iterative approximation problem of fixed point for asymptotically nonexpansive mappings in Banach spaces. Acta Math. Appl. Sinica, **24**, 236–241 (2001)

[ZS01b]   Zhang, S.S.: On the iterative approximation problem of fixed points for as-
           ymptotically nonexpansive type mappings in Banach spaces. Appl. Math.
           Mech., Engl. Ed., **22**, No. 1, 25–34 (2001)
[ZS01c]   Zhang, S.S.: On the iterative approximation problem of fixed points for as-
           ymptotically nonexpansive type mappings in Banach spaces. Appl. Math.
           Mech. (English Ed.), **22**, No. 1, 25–34 (2001)
[ZGu02]   Zhang, S.S., Gu, F: Ishikawa iterative approximations of fixed points and
           solutions for multivalued Φ-strongly accretive and multivalued Φ-strongly
           pseudo-contractive mappings (Chinese). J. Math. Res. Exposition, **22**,
           No. 3, 447–454 (2002)
[ZGZ00]   Zhang, S.S., Gu, F., Zhang, X.L.: Convergence problem of iterative ap-
           proximation for pseudo-contractive type mappings (Chinese). J. Sichuan
           Univ., Nat. Sci. Ed., **37**, No. 6, 795–802 (2000)
[ZXH03]   Zhang, S.S., Xu, Y.G., He, C.: Some convergence theorems for asymp-
           totically nonexpansive mappings in Banach spaces (Chinese). Acta Math.
           Sinica, **46**, No. 4, 665–672 (2003)
[Zha85]   Zhao, H.: Successive approximations of fixed points for some nonlinear
           mappings (Chinese). Math. Numer. Sin., **7**, 131–137 (1985)
[ZSu94]   Zhao, X., Sun, X.: The construction and convergence of a type of Mann it-
           erative sequence for the operator with boundary condition. Numer. Math.,
           Nanjing, **16**, No. 4, 297–303 (1994)
[ZH97a]   Zhou, H.: Some convergence theorems for the Ishikawa iterative sequences
           of certain nonlinear operations in uniformly smooth Banach spaces (Chi-
           nese. Acta Math. Sin., **40**, No. 5, 751–758 (1997)
[ZH97b]   Zhou, H.: A remark on Ishikawa iteration. Chin. Sci. Bull., **42**, No. 8,
           631–633 (1997)
[ZH97c]   Zhou, H.: Iterative solution of nonlinear equations involving strongly ac-
           cretive operators without the Lipschitz assumption. J. Math. Anal. Appl.,
           213, 296–307 (1997)
[ZH97d]   Zhou, H.: Remarks on Ishikawa iteration. Chinese Sci. Bull., **42**, 126–128
           (1997)
[ZH97e]   Zhou, H.: Some convergence theorems for the Ishikawa iterative sequences
           of certain nonlinear operators in uniformly smooth Banach spaces. Acta
           Math. Sinica, **40**, 751–758 (1997)
[ZH98a]   Zhou, H.: Approximating zeros of φ-strongly accretive operators by the
           Ishikawa iteration procedures with errors. Acta Math. Sinica, **41**, 1091–
           1100 (1998)
[ZH98b]   Zhou, H.: A note on a theorem of Xu and Roach. J. Math. Anal. Appl.,
           227, 300–304 (1998)
[ZH99a]   Zhou, H.: Stable iteration procedures for strong pseudocontractions and
           nonlinear equations involving accretive operators without Lipschitz as-
           sumption J. Math. Anal. Appl., **230**, 1–10 (1999)
[ZH99b]   Zhou, H.: Ishikawa iteration process with errors for Lipschitzian and ϕ-
           hemicontractive mappings in normed linear spaces. Panamer. Math. J.,
           9, No. 3, 65–77 (1999)
[ZH99c]   Zhou, H.: Iterative approximation of fixed points for uniformly continuous
           and strongly pseudocontractive mappings in smooth Banach spaces. Chin.
           Q. J. Math., **14**, No. 2, 42–46 (1999)

[ZH99d]   Zhou, H.: Iterative approximation of fixed points of Lipschitz Φ-hemicontractive mappings (Chinese). Chin. Ann. Math., Ser. A, **20**, No. 3, 399–402 (1999)

[ZH00a]   Zhou, H.: Iterative approximation of fixed points for Φ-hemicontractions in uniformly smooth Banach spaces (Chinese). Numer. Math., Nanjing, **22**, No. 1, 23–27 (2000)

[ZH00b]   Zhou, H.: Ishikawa iterative process with errors for Lipschitzian and ϕ-hemicontractive mappings in Banach spaces. J. Math. Res. Expo., **20**, No. 2, 159–165 (2000)

[ZH01a]   Zhou, H.: Iterative approximation of fixed points of φ-hemicontractive maps in Banach spaces (Chinese). J. Math. Res. Expo., **21**, No. 2, 237–240 (2001)

[ZH01b]   Zhou, H.: A new inequality in Banach space with applications. Preprint

[ZAC02]   Zhou, H., Agarwal, R.P., Cho, Y.J., Kim, Y.S.: Nonexpansive mappings and iterative methods in uniformly convex Banach spaces. Georgian Math. J., **9**, No. 3, 591–600 (2002)

[ZCC01]   Zhou, H., Chang, S.S., Cho, Y.J.: Weak stability of the Ishikawa iteration procedures for ϕ-hemicontractions and accretive operators. Appl. Math. Lett., **14**, No. 8, 949–954 (2001)

[ZCA02]   Zhou, H., Chang, S.S., Agarwal, R.P., Cho, Y.J.: Stability results for the Ishikawa iteration procedures. Dyn. Contin. Discrete Impuls. Syst. Ser. A Math. Anal., **9**, No. 4, 477–486 (2002)

[ZC98a]   Zhou, H., Chen, D.Q.: Iterative processes for certain nonlinear mappings in uniformly smooth Banach spaces. (Chinese). Math. Appl., **11**, No. 4, 70–73 (1998)

[ZC98b]   Zhou, H., Chen, D.Q.: Iterative approximation of fixed points for nonlinear mappings of Φ-hemicontractive type in normed linear spaces. Math. Appl., **11**, No. 3, 118–121 (1998)

[ZCh99]   Zhou, H., Chen, D.Q.: Iterative solution of nonlinear involving φ-quasi-accretive operators without Lipschitz assumption. Math. Sci. Res. Hot-Line, **3**, 15–26 (1999)

[ZCX00]   Zhou, H., Chen, D.Q., Xue, Z.Q.: A necessary and sufficient condition for convergence of the Ishikawa iteration for ϕ-strongly accretive operators in Banach spaces. In: Fixed Point Th. Appl. (Chinju, 1998), 255-262. Nova Sci. Publ., Huntington (2000)

[ZhC99]   Zhou, H., Cho, Y.J.: Ishikawa and Mann iterative processes with errors for nonlinear Φ-strongly quasi-accretive mappings in normed linear spaces. J. Korean Math. Soc., **36**, No. 6, 1061–1073 (1999)

[ZCC01]   Zhou, H., Cho, Y.J., Chang, S.S.: Approximating the fixed points of ϕ-hemicontractions by the Ishikawa iterative process with mixed errors in normed linear spaces. Nonlinear Anal., **47**, No. 7, 4819–4826 (2001)

[ZCG00]   Zhou, H., Cho, Y.J., Guo, J.T.: Approximation of fixed point and solution for Φ-hemicontraction and Φ-strongly quasi-accretive operator without Lipschitz assumption. Math. Sci. Res. Hot-Line, **4**, No. 3, 45–51 (2000)

[ZK01a]   Zhou, H., Cho, Y.J., Kang, S.M.: Approximating the zeros of accretive operators by the Ishikawa iterative scheme with mixed errors. Commun. Appl. Nonlinear Anal., **8**, No. 3, 27–35 (2001)

[ZK01b]   Zhou, H., Cho, Y.J., Kang, S.M.: Iterative approximations for solutions of nonlinear equations involving non-self-mappings. J. Inequal. Appl., **6**, No. 6, 577–597 (2001)

[ZG03a]   Zhou, H., Gao, G.L., Chen, D.Q.: Some new strong convergence theorems for iterative schemes for asymptotically nonexpansive mappings in uniformly convex Banach spaces. Acta Anal. Funct. Appl., **5**, No. 3, 234–239 (2003)

[ZG03b]   Zhou, H., Gao, G.L., Guo, J.T., Cho, Y.J.: Some general convergence principles with applications. Bull. Korean Math. Soc., **40**, 351–363 (2003)

[ZGK04]   Zhou, H., Gao, G.L., Kang, J.I.: On the iteration methods for asymptotically nonexpansive mappings in uniformly convex Banach spaces. In: Fixed Point Th. Appl. Vol. 5, 213–221. Nova Sci. Publ., Hauppauge, NY (2004)

[ZGH04]   Zhou, H., Guo, G.T., Hwang, H.J., Cho, Y.J.: On the iterative methods for nonlinear operator equations in Banach spaces. Panamer. Math. J., **14**, No. 4, 61–68 (2004)

[ZJ96a]   Zhou, H., Jia, Y.: Approximating the zeros of accretive operators by the Ishikawa iteration process. Abstr. Appl. Anal., **1**, 19–33 (1996)

[ZJ96b]   Zhou, H., Jia, Y.: Approximating the zeros of accretive operators by the Ishikawa iteration process. Abstr. Appl. Anal., **1**, No. 2, 153–167 (1996)

[ZJ96c]   Zhou, H., Jia, Y.: On the Mann and Ishikawa iteration processes. Abstr. Appl. Anal., **1**, No. 4, 341–349 (1996)

[ZJi97]   Zhou, H., Jia, Y.: Approximation of fixed points of strongly pseudocontractive maps without Lipschitz assumption. Proc. Amer. Math. Soc., **125**, No. 6, 1705–1709 (1997)

[ZKK04]   Zhou, H., Kang, J.I., Kang, S.M., Cho, Y.J.: Convergence theorems for uniformly quasi-Lipschitzian mappings. Int. J. Math. Math. Sci. 2004, No. 13-16, 763–775 (2004)

[ZLu99]   Zhou, H., Luo, S.: Remarks on the iterative process with errors of nonlinear equations involving m-accretive operators. J. Math. Res. Expo., **19**, No. 2, 471–474 (1999)

[ZZG03]   Zhou, H., Zhao, L.J., Guo, J.T.: Approximation of fixed points and solution for φ-hemicontractive and φ-strongly quasi accretive operators without Lipschitz assumption. J. Math. Res. Exp., **23**, No. 1, 40–46 (2003)

[ZZh04]   Zhou, L., Zhang, S.Q.: Convergence of Ishikawa iterative sequences in uniformly convex Banach spaces (Chinese). J. Huazhong Univ. Sci. Technol. Nat. Sci., **32**, No. 6, 47–48 (2004)

[ZhC02]   Zhou, Y.Y., Chang, S.S.: Convergence of implicit iteration process for a finite family of asymptotically nonexpansive mappings in Banach spaces. Numer. Funct. Anal. Optimization, **23**, No. 7-8, 911–921 (2002)

[ZHe04]   Zhou, Y.Y., He, Z.: Fixed-point iteration for asymptotically pseudocontractive mappings with error member. J. Hebei Univ. Nat. Sci., **24**, No. 1, 23–27 (2004)

[Zhu94]   Zhu, L.: Iterative solution of nonlinear equations involving m-accretive operators in Banach spaces. J. Math. Anal. Appl., **188**, No. 2, 410–416 (1994)

Note. We used the short names *Nonlinear Anal. TMA* for Nonlinear Analysis, Ser. A: Theory Methods Appl. and, respectively, *Nonlinear Anal.* for Nonlinear Analysis, Ser. B: Real World Appl.

List of Symbols

$\mathbf{N} = \{0, 1, ..., n, ...\}$;

$\mathbf{Z} = \{... - n, ..., -2, -1, 0, 1, 2, ..., n, ...\}$

$\mathbf{N}^* = \{1, 2, ..., n, ...\}$;

$\mathbf{R} =$ the set of all real numbers

$[a, b]$ - the closed interval , $a, b \in \mathbf{R}$

(a, b) - the open interval , $a, b \in \mathbf{R}$

∂D - the boundary of the domain D

$|x|$ - the absolute value of x, $x \in \mathbf{R}$

$\emptyset$ - the empty set

For $T : X \to X$ a mapping,

$\quad D(T)$ is the domain of T

$\quad R(T)$ - the range of T

$\quad F_T = \{x \in X : \ Tx = x\}$ or $Fix\,(T)$ - the set of fixed points of T

$\quad I = I_X$ - the identity map

$\quad T^0 = 1_X, \ T^1 = T, ..., T^n = T \circ T^{n-1}, ...$ - the iterates of T

$\quad 0_T(x, n) = \{x, Tx, ..., T^n x\}$;

For (X, d) a metric space,

$\quad B(a, R) = \{x \in X : d(x, a) < R\}$, $R > 0$ is the open ball

$\quad \overline{B}(a, R) = \{x \in X : d(x, a) \leq R\}$, $R > 0$ - the closed ball

$\quad \delta\,(A) = \sup\{d(a, b) : a, b \in A\}$ - the diameter of $A \subset X$

For $(E, \|\cdot\|)$ a normed space,

$\quad E^*$ is the dual of E

$\quad E^{**}$ - the bidual of E

$\quad Jx\,(jx)$ - the normalized (single valued) duality mapping

$\quad \rho_E$ - the modulus of smoothness of E

$\quad \delta_E$ - the modulus of convexity of E

$\quad co\,K$ - the convex hull of K

$\quad diam(K)$ - the diameter of the set K

$\quad x_n \rightharpoonup x$ means that x_n converges weakly to x

For $(E, \|\cdot\|)$ a normed space and $T : X \to X$ a mapping,

$K(x_0, \lambda, T)$ is the Krasnoselskij iteration associated to the operator T, the initial guess x_0 and parameter λ

$M(x_0, A, T)$ - the (general) Mann iteration associated to the operator T, the initial guess x_0 and matrix A

$M(x_0, \alpha_n, T)$ - the (normal) Mann iteration associated to the operator T, the initial guess x_0 and parameter sequence $\{\alpha_n\}$

$I(x_0, \alpha_n, \beta_n, T)$ - the Ishikawa iteration associated to the operator T, the initial guess x_0 and parameter sequences $\{\alpha_n\}$, $\{\beta_n\}$

Author Index

Subject Index

Lecture Notes in Mathematics

For information about earlier volumes
please contact your bookseller or Springer
LNM Online archive: springerlink.com

Vol. 1868: J. Jorgenson, S. Lang, Pos$_n$(R) and Eisenstein Series. (2005)
Vol. 1869: A. Dembo, T. Funaki, Lectures on Probability Theory and Statistics. Ecole d'Eté de Probabilités de Saint-Flour XXXIII-2003. Editor: J. Picard (2005)
Vol. 1870: V.I. Gurariy, W. Lusky, Geometry of Müntz Spaces and Related Questions. (2005)
Vol. 1871: P. Constantin, G. Gallavotti, A.V. Kazhikhov, Y. Meyer, S. Ukai, Mathematical Foundation of Turbulent Viscous Flows, Martina Franca, Italy, 2003. Editors: M. Cannone, T. Miyakawa (2006)
Vol. 1872: A. Friedman (Ed.), Tutorials in Mathematical Biosciences III. Cell Cycle, Proliferation, and Cancer (2006)
Vol. 1873: R. Mansuy, M. Yor, Random Times and Enlargements of Filtrations in a Brownian Setting (2006)
Vol. 1874: M. Yor, M. Émery (Eds.), In Memoriam Paul-André Meyer - Séminaire de probabilités XXXIX (2006)
Vol. 1875: J. Pitman, Combinatorial Stochastic Processes. Ecole d'Eté de Probabilités de Saint-Flour XXXII-2002. Editor: J. Picard (2006)
Vol. 1876: H. Herrlich, Axiom of Choice (2006)
Vol. 1877: J. Steuding, Value Distributions of L-Functions (2007)
Vol. 1878: R. Cerf, The Wulff Crystal in Ising and Percolation Models, Ecole d'Eté de Probabilités de Saint-Flour XXXIV-2004. Editor: Jean Picard (2006)
Vol. 1879: G. Slade, The Lace Expansion and its Applications, Ecole d'Eté de Probabilités de Saint-Flour XXXIV-2004. Editor: Jean Picard (2006)
Vol. 1880: S. Attal, A. Joye, C.-A. Pillet, Open Quantum Systems I, The Hamiltonian Approach (2006)
Vol. 1881: S. Attal, A. Joye, C.-A. Pillet, Open Quantum Systems II, The Markovian Approach (2006)
Vol. 1882: S. Attal, A. Joye, C.-A. Pillet, Open Quantum Systems III, Recent Developments (2006)
Vol. 1883: W. Van Assche, F. Marcellàn (Eds.), Orthogonal Polynomials and Special Functions, Computation and Application (2006)
Vol. 1884: N. Hayashi, E.I. Kaikina, P.I. Naumkin, I.A. Shishmarev, Asymptotics for Dissipative Nonlinear Equations (2006)
Vol. 1885: A. Telcs, The Art of Random Walks (2006)
Vol. 1886: S. Takamura, Splitting Deformations of Degenerations of Complex Curves (2006)
Vol. 1887: K. Habermann, L. Habermann, Introduction to Symplectic Dirac Operators (2006)
Vol. 1888: J. van der Hoeven, Transseries and Real Differential Algebra (2006)
Vol. 1889: G. Osipenko, Dynamical Systems, Graphs, and Algorithms (2006)
Vol. 1890: M. Bunge, J. Funk, Singular Coverings of Toposes (2006)
Vol. 1891: J.B. Friedlander, D.R. Heath-Brown, H. Iwaniec, J. Kaczorowski, Analytic Number Theory, Cetraro, Italy, 2002. Editors: A. Perelli, C. Viola (2006)
Vol. 1892: A. Baddeley, I. Bárány, R. Schneider, W. Weil, Stochastic Geometry, Martina Franca, Italy, 2004. Editor: W. Weil (2007)
Vol. 1893: H. Hanßmann, Local and Semi-Local Bifurcations in Hamiltonian Dynamical Systems, Results and Examples (2007)
Vol. 1894: C.W. Groetsch, Stable Approximate Evaluation of Unbounded Operators (2007)
Vol. 1895: L. Molnár, Selected Preserver Problems on Algebraic Structures of Linear Operators and on Function Spaces (2007)

Vol. 1896: P. Massart, Concentration Inequalities and Model Selection, Ecole d'Eté de Probabilités de Saint-Flour XXXIII-2003. Editor: J. Picard (2007)
Vol. 1897: R. Doney, Fluctuation Theory for Lévy Processes, Ecole d'Eté de Probabilités de Saint-Flour XXXV-2005. Editor: J. Picard (2007)
Vol. 1898: H.R. Beyer, Beyond Partial Differential Equations, On linear and Quasi-Linear Abstract Hyperbolic Evolution Equations (2007)
Vol. 1899: Séminaire de Probabilités XL. Editors: C. Donati-Martin, M. Émery, A. Rouault, C. Stricker (2007)
Vol. 1900: E. Bolthausen, A. Bovier (Eds.), Spin Glasses (2007)
Vol. 1901: O. Wittenberg, Intersections de deux quadriques et pinceaux de courbes de genre 1, Intersections of Two Quadrics and Pencils of Curves of Genus 1 (2007)
Vol. 1902: A. Isaev, Lectures on the Automorphism Groups of Kobayashi-Hyperbolic Manifolds (2007)
Vol. 1903: G. Kresin, V. Maz'ya, Sharp Real-Part Theorems (2007)
Vol. 1904: P. Giesl, Construction of Global Lyapunov Functions Using Radial Basis Functions (2007)
Vol. 1905: C. Prévôt, M. Röckner, A Concise Course on Stochastic Partial Differential Equations (2007)
Vol. 1906: T. Schuster, The Method of Approximate Inverse: Theory and Applications (2007)
Vol. 1907: M. Rasmussen, Attractivity and Bifurcation for Nonautonomous Dynamical Systems (2007)
Vol. 1908: T.J. Lyons, M. Caruana, T. Lévy, Differential Equations Driven by Rough Paths, Ecole d'Eté de Probabilités de Saint-Flour XXXIV-2004. (2007)
Vol. 1909: H. Akiyoshi, M. Sakuma, M. Wada, Y. Yamashita, Punctured Torus Groups and 2-Bridge Knot Groups (I) (2007)
Vol. 1910: V.D. Milman, G. Schechtman (Eds.), Geometric Aspects of Functional Analysis. Israel Seminar 2004-2005 (2007)
Vol. 1911: A. Bressan, D. Serre, M. Williams, K. Zumbrun, Hyperbolic Systems of Balance Laws. Lectures given at the C.I.M.E. Summer School held in Cetraro, Italy, July 14-21, 2003. Editor: P. Marcati (2007)
Vol. 1912: V. Berinde, Iterative Approximation of Fixed Points (2007)

Recent Reprints and New Editions

Vol. 1618: G. Pisier, Similarity Problems and Completely Bounded Maps. 1995 – 2nd exp. edition (2001)
Vol. 1629: J.D. Moore, Lectures on Seiberg-Witten Invariants. 1997 – 2nd edition (2001)
Vol. 1638: P. Vanhaecke, Integrable Systems in the realm of Algebraic Geometry. 1996 – 2nd edition (2001)
Vol. 1702: J. Ma, J. Yong, Forward-Backward Stochastic Differential Equations and their Applications. 1999 – Corr. 3rd printing (2007)
Vol. 830: J.A. Green, Polynomial Representations of GL_n, with an Appendix on Schensted Correspondence and Littelmann Paths by K. Erdmann, J.A. Green and M. Schocker 1980 – 2nd corr. and augmented edition (2007)